MAN, ENERGY, SOCIETY

MAN, ENERGY, SOCIETY

Earl Cook

TEXAS A&M UNIVERSITY,
College Station

W. H. Freeman and Company
San Francisco

Library of Congress Cataloging in Publication Data

Cook, Earl Ferguson, 1920–
  Man, energy, society.

  Bibliography: p.
  Includes index.
  1. Power resources. 2. Energy consumption.
3. Technology and civilization. I. Title.
HD9502.A2C657   333.7   75-33774
ISBN 0-7167-0725-X
ISBN 0-7167-0724-1 pbk.

9 8 7 6 5 4 3 2 1

To Helen Louise Royer

CONTENTS

PREFACE

It seems likely that the world will be plagued by energy problems for a long time to come. Such problems—especially if we count food as an energy resource—are nothing new. However, there are some new and disturbing aspects to the contemporary world energy scene. First, there is the dependence of a considerable part of the human population on a food subsidy provided by the fossil fuels, which are resources that can be neither renewed nor recycled. Second, there is a highly uneven geographical distribution of known fossil-energy resources, a distribution that is incongruent with the map of consumption and is becoming more so. Third, there is a momentum of population growth that threatens to overwhelm the productiveness of energy and food technology. And finally, an increasing rate of accumulation, in the environment, of waste materials and unused heat derived directly and indirectly from the eager use of fossil and nuclear fuels. The physical, biological, and social hazards inherent in this picture are subjects of concern.

In this book I have tried to give a geographic view of energy that will provide the background necessary for an understanding of contemporary problems. This geographic view is explanatory as well as descriptive. In addition to locating and classifying people and resources, it outlines the history of man's use of energy and the consequent effects on social fabric and life-style.

At present there is no worldwide energy shortage. The capacity for producing crude oil exceeds the consumption. Much natural gas is being wasted—it is emitted into the atmosphere because there is no use for it where it is being produced. The known world reserves of coal would last for several hundred years at the present rates of extraction. We have only started to consume the global supplies of nuclear fuels.

Why then has there been so much talk of an energy *crisis?*

When we look into this question, we find that an *impaired availability* of energy to specific countries—not a worldwide shortage of energy—is the cause of what is commonly called the energy crisis. Except for the Soviet Union and Canada, the industrialized nations do not produce nearly as much energy as they consume (Norway may soon join the select group of industrialized countries with a surplus of energy); their energy budgets are balanced by imports of oil. Consequently, when the cartel of oil-exporting countries (OPEC) caused the world price of crude oil to quadruple within less than two years, energy became less available to the energy-importing nations because of the sharp rise in price. However, the price rise was perhaps most hurtful to those oil-poor developing countries that were counting on imported oil to subsidize both food production and industrialization.

Not all the energy price rise was due to the cartel's action. In the United States, for example, reserves of oil and natural gas are diminishing and the costs of development and production are increasing. The implementation in recent years of strong air-pollution abatement regulations has created such unprecedented and costly oddities as large power plants that are on or near coal fields but are burning coal hauled about fifteen hundred miles because the nearby coal cannot be burned without transgressing the new standards. In addition to the economic constriction of international energy flows, Holland and the United States have felt the pinch of a political interdiction of oil imports from the Arab exporting countries. The embargo by the Arabs brought into sharp focus the hazard of economic strangulation that faces any industrialized nation dependent on foreign sources for much of its energy supply and it evoked the specter of retaliatory military intervention as well as the hope of becoming "independent" of energy imports.

Shortage of gasoline, fuel oil, and clean coal for power plants may exasperate but they hardly constitute a crisis, especially when they occur in a country that has ample room—as does the United States—for *decreasing* its energy consumption through improved efficiencies of use. The same may be said, but with somewhat less assurance, of the danger of an interdiction of fuel imports; there, the "crisis"—for the United States—is not yet one of survival, but rather one of decision about the strategies to pursue to limit dependence on unreliable external sources of energy at a cost that will not greatly reduce the nation's level of living.

A true energy crisis remains partially veiled by the energy problems of the moment and by a persistent faith in technological salvation. The problem is simple. Only the answer is difficult. How may mankind, short of catastrophe, make its use of energy compatible with survival? Survival may have several meanings—the maintenance of subsistence-level living standards, the survival of political and economic institutions, the survival of national cultures, or the survival of the human species. This crisis does exist and is growing because of the rapid depletion of nonrenewable resources, including those of energy. It may be delayed by the discovery and installation of new energy-delivery systems and by more frugal use of fossil and nuclear fuels, but mankind must fall back eventually on renewable forms of energy, and there is no assurance that energy from renewable sources can be made available at rates and costs compatible with survival in any of its meanings. In any event, it seems unlikely that the growth state, with all its economic mechanisms and philosophic rationalizations, will survive the decline of the nonrenewable fuels. Whether the so-called steady state is a real probability for industrial society is somewhat beside the point. As a theoretical ideal, it is a necessary concept as we try to think ahead, to change our goals and strategies, and to avoid the worst consequences of the heedless depletion of our energy and other resources of the earth.

Most of the good ideas in this book are borrowed. My debt to two persons far exceeds the credit inherent in my citations of their work. My interest in energy resources was provoked, and has been sharpened for more than a dozen years, by discussions with M. King Hubbert, a teacher of great talent. W. Fred Cottrell, whom I have not met, gave me, through his book *Energy and Society,* invaluable insights into the ways that the availability of energy may affect life style, government, and the family.

H. W. Menard gave useful criticism and encouragement at a time when both were needed. John Griffiths, State Climatologist of Texas, provided the temperature data for figure 11.10. A fine office staff helped greatly to make this book possible, notably Sue Mellor, Joyce Reichert, Candi Miller, and Ginger Franks.

August 1975   *Earl Cook*

MAN, ENERGY, SOCIETY

Chapter 1

ENERGY, ENVIRONMENT, AND EVOLUTION

Since energy is an essential ingredient in all terrestrial activity, organic and inorganic, it follows that the history of the evolution of human culture must also be a history of man's increasing ability to control and manipulate energy.

—M. King Hubbert, 1962

THE IMPORTANCE OF ENERGY

Life exists on earth by the grace of solar radiation. It provides the ambient temperature range within which organisms can survive, drives the winds, and powers the hydrologic cycle. It provides the energy for plant growth and the light without which there would be perpetual night. Without plants there would be no animals, no humans. Over millions of years a tiny part of the incoming solar radiation has been trapped in the form of plant and animal detritus, preserved, and turned into deposits of petroleum, coal, and natural gas—the fossil fuels upon which modern industrial man is utterly dependent.

This is a book about energy and mankind. Man's cultural and social development, his distinction and alienation from the other animals, have been made possible by his exploiting energy sources outside his own body and beyond the wild food and the ambient heat of his environment. This is not to say that energy has made man. Man made himself. But, without the progressive acquisition of control over external energy sources—and probably without the unusual climatic variations of the past few million years—man would not be what he is today.

When an animal is not able to make effective use of the muscular energy of, for instance, sharp claws and strong teeth, it may use external objects. These we call tools. A weapon is a tool, although there is often a tacit psychological distinction between tools and weapons. Some other animals do make limited use of natural objects as tools; man uses tools extensively. Perhaps two habits or abilities further distinguish man even from his close relatives: the ability to make and control fire and the ability to design tools and machines for uses not obvious and imminent. In other words, man uses inanimate sources of energy and is capable of abstract reasoning. This potent combination has made him, at least temporarily, the supreme species of the natural world.

Human social systems, mores, and moral codes appear to be related to the availability and the uses of energy within societies, but the causes and effects, beyond a few that seem fundamental, are apt to be obscured by accidents of environment, consequences of cultural interaction, and defects of the historical record. There are today greater differences than ever before among the communities of man in their access to energy, in their material standard of living, and in their potential for survival. It is a paradox that, despite our marvelous ability to find and use energy, to measure it, and to predict its state, we are very far from being able to forecast accurately the social impacts, or to foresee possibly irreversible environmental consequences of energy use, or even to chart the future availability of energy resources. We understand energy scientifically, we control it technologically, but we have not mastered it as the major factor of social well-being.

In order to understand the present and have some chance of influencing the future of man as it will be influenced by the use of energy, we need to study the past and to review the science and technology of energy. These tasks will take us quite a way into this book, for they need to be done carefully.

SETTING THE STAGE

Seventy million years ago (see figure 1.1), the Cenozoic era opened on a world from which the giant reptiles that had dominated the land areas during the preceding era had disappeared, leaving an ecological gap. One of the many mammalian genera of the early Cenozoic era was a small tree shrew that hunted with its eyes rather than its nose and held objects in its claws. As the era wore on, the eyes of the shrew worked more and more around to the front of its head, giving improved depth perception; the ability to distinguish colors was obtained; and the claws evolved toward fingers with flat nails. This tree shrew was either the earliest primate, the order of animals that includes man, or, as George Gaylord Simpson remarks, "the next thing to it."

About 40 million years ago (during the Oligocene epoch) the first apelike (pithecoid) creature shows up in the fossil record and, 10 million years later, the oldest known primate with thirty-two teeth (a characteristic of man, apes, and Old World monkeys) appears, along with the oldest known pithecoid with five-cusped molars, a feature shared by apes and men. All these creatures were still true quadrupeds. About 22 million years ago appeared the dryopithecines. They lived in

Energy, Environment, and Evolution

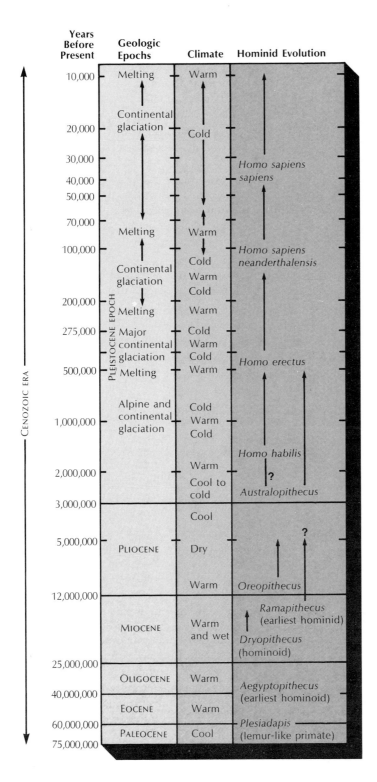

FIGURE 1.1
Geologic evidence indicates that climatic oscillations became more frequent and severe as the Cenozoic era wore on. The Pleistocene epoch has been characterized by ice ages, during which large portions of the northern continents were covered by glaciers, and the emergence of man. The chain of human evolution is not yet complete. Not only is there controversy over Australopithecus *as a human ancestor, but also the paucity of fossils from the Pliocene epoch leaves us uncertain about the position in the family tree of Miocene forms such as the great arboreal ape,* Dryopithecus, *and the earliest known animal with human features,* Ramapithecus.

the humid forest of the Miocene epoch and had flat faces, five-cusped, Y-shaped molars, free-swinging arms, and a semi-erect stance; all of these features are hominoid, characteristic of apes and men.

The oldest known manlike (hominid) primate roamed India and Africa about 14 million years ago, still in the Miocene epoch. He is probably a human ancestor, for he had curving rows of molars and an arched hard palate. In addition, all his teeth were about the same size; in apes, the canines and incisors are relatively longer than the other teeth.

The continental climate became suddenly drier as the Pliocene epoch succeeded the Miocene. In low latitudes, lush forests gave way to savanna, while grasslands changed to deserts. The Pliocene epoch represents a puzzling gap in the fossil record of primate evolution. We can only guess that one or more of the great arboreal, herbivoral apes of the Miocene epoch successfully adapted to the changed environment by descending to the savanna and becoming a meat-eating hunter. Then, by the end of the Pliocene epoch, we are certain that the primates had developed three specialties that were critically important in the ascent to man: binocular (stereoscopic) color vision, long, flexible forelimbs and grasping hands with thumbs and fingernails, and a semi-erect posture facilitated by enlarged buttocks.

MAN'S EPOCH: THE PLEISTOCENE

The cultural development of man and the Pleistocene epoch are coincident. One of the striking features of scientific discovery in the past few decades is that man and the Pleistocene epoch have been growing older together. In the 1940s, the Pleistocene epoch was regarded by geologists as being about 1 million years old and man, by anthropologists, as being perhaps 500,000 years old. Now, on the evidence of glaciation and radioactive dating, the Pleistocene epoch is believed to be somewhere between 2 and 3 million years old, while early man, on entirely separate evidence, has recently achieved the age of about 2.5 million years.

As the Pleistocene epoch dawned in Africa, the australopithecines were already there. They were erect, less than five feet tall, with flat feet and balanced-buttocks. One type was a meat-eater. The australopithecines made bone, tooth, and horn tools, some of which they used as weapons. The position of the australopithecines in man's evolutionary tree is unclear; we are no longer certain that they represent a true missing link. They were more manlike than apelike and they did make and use tools; but Richard Leakey has found, in sediments that are about 2.5 million years old, remains of a more probable human ancestor.

Energy, Environment, and Evolution

Whether Leakey's find or the carnivorous *Australopithecus africanus* turns out to be in man's family tree, it is interesting that both appear in the fossil record at or near the beginning of the Pleistocene epoch and that man is virtually coincident with the Pleistocene epoch (or the Quaternary period, if one prefers to sustain the mild conceit that calls for a separate geologic epoch, the Holocene or Recent, more or less coincident with *Homo sapiens sapiens*).

The tempo of geomorphic activity quickened markedly as the Pliocene epoch gave way to the Pleistocene. The continents became higher and the world's land area greater than the average continental elevation and extent during most of geologic history. Crustal unrest was marked by an explosive ring of volcanoes around the Pacific Basin, by the uplifting of the Alps and the Himalayas, and by much local folding and faulting. The unusual height of the land areas of the world was a major factor in changing world climates drastically. The Pleistocene has been an epoch of repeated continental glaciation in the Northern Hemisphere and of alpine glaciation in mountains far from the poles. During the periods of glaciation climates were severe near the fronts of the ice sheets, cool and humid in the rest of the world.

Four times during the Pleistocene epoch glaciers invaded what is now the United States. In Europe and Asia the continental glaciers covered all of Scandinavia and the northern half of what is now the Soviet Union. Throughout the proglacial world climates changed from hot and dry to cool and wet in the early millennia of the Pleistocene epoch. Rock that had weathered under the arid Pliocene sun provided voluminous comminuted material for Pleistocene torrents to carry down the hillsides into the valleys and out to the seas.

About 32 percent of the present land area of the world has been covered by glacial ice once or more during the Pleistocene epoch. At present, about 10 percent is covered. During the three interglacial stages, glaciers may have disappeared entirely. There is little reason to assume that we are now in a postglacial epoch. It seems more likely that we are in the transition between the fourth glacial stage (called Wisconsin in North America and Würm in Europe) and a fourth interglacial stage.

At the height of each glacial stage, the pluvial continental climate produced rivers (in the Sahara, for instance) and lakes (in many closed basins of the western United States) where none exists now. The atmosphere was laden with water and all the streams were swollen. Much water was turned into glacial ice. As a result of the great transfer of water from the oceans to the glaciers, streams, and lakes of the continents and to the atmosphere over the continents, the oceans were diminished, sea level declined by at least three hundred feet, and the land area of the world was increased substantially. Evidence on which to reconstruct the climates of the three interglacial stages, which together represent more time than the four glacial stages, is meager, but what there is indicates aridity and warmth—in other words, repeated returns to the climatic conditions of the Pliocene epoch.

Thus the Pleistocene epoch was a time of great environmental stress, of rapidly (in the scale of geological and evolutionary time) changing climate. Those plants and animals that could adapt to a constantly changing environment survived; those that could not—and they included some of the largest and strongest mammals yet to appear in the long reel of evolution—became extinct. Man led the contingent of survivors. He not only survived, but also thrived on the challenges of the Pleistocene epoch. He developed culture as a survival mechanism, a mechanism that did not depend on chance mutations and on thousands of generations for evolutionary adaptation. The main survival elements of his culture were weapons and tools, social organization, speech and language, control of fire, and clothing and shelter. These saw him through four glacial and three interglacial stages, in other words, through 2 to 3 million years of climatic change; then he invented agriculture, an innovation that led to an enormous increase in his numbers, to astonishing advances in his control over nature, and to a position of overwhelming dominance in the global ecosystem.

THE START OF POWER TECHNOLOGY

Man is distinguished from other animals mainly by what is called culture—his language, his social organization, his use of tools, his invention of machines, and his creation of structures. Man's cultural evolution is the history of his increasing control of power technology and energy resources. By discovering ways to produce energy and to apply power, man has transformed himself from a puny being at the mercy of his environment into a creature with more power than he knows how to use well, whose environment now depends upon his grace, and whose longevity as a species depends upon his own wisdom.

Power technology began with the lever, the club, and the thrown rock. Sometime at least 2 million years ago one of our remote ancestors picked up a stone and threw it with a purpose defined by reasoning through analogy from memory and on a trajectory determined by his stereoscopic color vision, his semi-erect stance, and the excellent physiological coordination of his eye, brain, and muscle. This was an historic event, for it started man on a different evolutionary path from that of his fellow primates. It led to the refinement of abstract reasoning in increasingly sophisticated hunting and gathering societies; to the development of bola stones, slings, spears, bows and arrows, other weapons, and tools; to the flowering of language; and to the storage and communication from one generation to the next of useful information about the environment.

The development of power technology may have impelled all the physiological and cultural adaptations that came to distinguish man from the rest of the animal world: bipedal locomotion, enlarged brain, erect stance, omnivorousness, well-developed speech, abstract thought, and specialized social organization.

THE CONQUEST OF FIRE

For a long time man's only source of energy was the wild food he hunted, gathered, and ate. By finding the principle of the lever long before Archimedes "discovered" or defined it, man greatly extended the *power* but not the *energy* of his own body and of the bodies of his slaves and other work animals. With *machines*—devices whereby energy or force applied at one point is transmitted, in more useful form, to another point or in another direction—the pry, the pulley, the inclined plane, the wedge, and the wheel, man built the Pyramids and the Great Wall of China; with the bow, bola stones, the hurled spear, and the poisoned dart, man hunted game and waged war. But these devices did not increase the total energy at his command.

Only when he brought fire under his control was he able to break the bonds of the physical limitations of an animal. With fire he was able to live in many diverse habitats, not just the warm ones. With fire, he expanded his food base by cooking vegetables previously inedible and by drying and smoking meat to preserve it. With fire, he had formidable protection against other animals and he could use fire against them to drive them over cliffs or past ambushes and thus increase his hunting productivity. With fire he was free to talk at night and paint in caves. Shelter and clothing probably were invented as extensions of the heated cave. With fire, shelter, and clothing he was able to spread out of the warm lands to which he had been confined, right up to the fronts of great continental glaciers of the Pleistocene epoch.

Fire may have quickened the development of language, although language was necessary to the efficient working of a hunting society before the advent of fire. Before fire, however, the nights must have been periods of self-enforced silence, because man was frail compared to some of the nocturnal predators and must have relied on hearing to warn him of their advance. Once fire was acquired, other animals would not approach, silence was no longer required, and some of the hours spent around the fire were probably devoted to telling tales, to teaching, and to planning the next day's hunt. For the creative who wearied of language as a developing art form, there was cave painting or the elaboration of artifacts. The acquisition of fire thus may have led directly to the development of the arts.

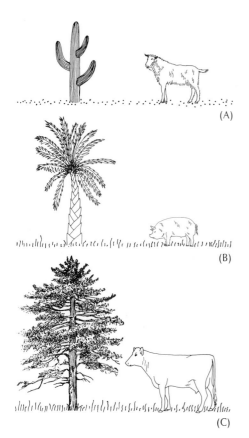

FIGURE 1.2
Although the dog is often called "man's best friend," a good argument can be made for other animals upon which man depends far more than he does the dog. In semiarid regions, such as Iran, the goat increases the conversion efficiency of the human food chain, as does the pig in more humid regions, such as coastal China and the Philippines; in temperate prairie regions, such as the American Midwest, first bison and now cattle play the same role.

AGRICULTURE

The origins of agriculture are still shrouded in the mists of unwritten history. We have good evidence of primitive agriculture about ten thousand years ago in the hills bordering the Fertile Crescent of southwest Asia. There is evidence of agriculture in Mexico seven thousand years ago, and in southeast Asia about six thousand years ago. Whether these were independent developments or represent cultural diffusion from a single source in Asia is not known.

Agriculture allowed man to supplement his own physical power by the power of domesticated animals and to supplement his food supply by the flesh of animals and the produce of plants he raised. Sustained levels of food consumption could be maintained, because surplus grain could be stored and because food animals could supply milk and meat when hunting was poor. The agricultural revolution was of enormous importance to man. It reduced the amount of energy he had to spend to obtain an energy unit of food. He was tapping the solar energy cycle in a more efficient way than by hunting and gathering wild food. It allowed him to increase his population density, regardless of the density of available game, to the point at which towns became possible.

It is not known whether man domesticated plants or animals first. If animals, either the dog or the goat may have been man's first tame companion. In the record, the goat has the edge, being known to have been used in southwest Asia at least eight thousand years ago. The dog was a warner and occasionally a source of food. The goat was first and last a supplementary food source. The goat is less efficient than other animals in turning plant food into edible food for man, but can do it on a terrain where other animals would perish. The goat, since cheese could be made from its milk, allowed man another way to store energy. In warm humid climates, where vegetation was lush and there was less need to store food over the winter, the pig took the place of the goat as scavenger and provider of protein.

WATER AND WIND POWER

The invention of agriculture and its spread from the hills of Asia Minor down into the semiarid floodplains of the Tigris and Euphrates quickly led to irrigation and to the development of the world's first cities, some five or six thousand years ago, when for the first time man harnessed the power of flowing water by the noria, a waterwheel designed to lift water rather than to power a machine. As man's agricultural communities reached the seacoast, the power of the wind was used to propel sailing craft on voyages of discovery and trade.

On land, however, water and wind power were not well developed until the late Middle Ages, when the dim outline of our present technological civilization began to appear in northwest Europe. The great power sources at that time were the horse, the watermill, and the windmill. Teams of horses pulling heavy plows powered an agricultural revolution in the northern plains. Water power was applied to all sorts of industrial processes from crushing ore to making mash for beer. The windmill helped the Dutch reclaim their country from the sea.

FOSSIL FUELS AND THE INDUSTRIAL REVOLUTION

By the thirteenth century, the forests of England and France had been so depleted by the demands for charcoal for the metal-working industry, for timber for ship building and other construction, and for firewood that people in London and Paris turned to burning soft coal, despite its noxious smoke. For some four hundred years, however, coal, although its use increased steadily, remained just a substitute for fuel wood in space and process heating. It provided neither power nor work, which still came from men and horses, falling water, and the wind. Watermills proliferated along the streams of Europe and sailing ships grew in size, complexity, maneuverability, and efficiency. A medieval invention that greatly improved the use of ships was the compass, which works on magnetic energy.

The increasing use of coal for heat indirectly brought about the great technological breakthrough to the power bonanza of the fossil fuels. The coal mines in northeast England were getting deeper as the demand for coal, especially in London, grew. As the mines deepened, it became more difficult to keep water out of them; windlasses, windmills, and hand pumps were used but the battle was being lost.

In 1698 Thomas Savery invented a steam pump designed to raise water by a combination of vacuum lift and high-pressure steam expulsion. Because the Savery pump was dangerous, it was little used. Then, in 1705, the first practical steam pump was designed by Thomas Newcomen and was eagerly adopted by the coal miners. The Newcomen pump was a fine example of provident, or at least serendipitous, technology, coming as it did just four years before Abraham Darby perfected a means of making coke from coal. The result was an enormous increase in the demand for coal, a demand that could not have been met without an effective pump to keep the mines from being flooded.

Coke, coal, and the steam engine powered the Industrial Revolution, which gave man power to change the face of the earth and to devour in a few centuries resources that had taken hundreds of millions of years to accumulate. Although steam engines were used to drive trains and ships, to dig canals and mines, and to power a great variety of industrial machines, rapid growth of power production in terms of present rates began less than one hundred years ago. That growth has been made possible by the invention of electrical means to generate and distribute power from central power stations, the first of which was opened in 1881, and by the development and proliferation of internal-combustion engines for motor vehicles, ships, locomotives, and airplanes.

More recently, man has succeeded in tapping the energy of the atomic nucleus. The first large nuclear power plant went into operation in 1957; by the year 2025, about half the power produced at central generating plants in the United States may be nuclear.

MAN, ENERGY, AND THE ECOSYSTEM

Man is part of a dynamic system, often called the ecosystem, that includes the physical and chemical, living and inert components of his environment. This system is constantly adjusting and changing. By far the greatest amount of its energy enters the system as solar radiation (figure 1.3). The rest comes as heat flow from within the earth and as tidal energy. Part of the solar radiation is captured by green plants, which through photosynthesis convert it to chemical or food energy. Some powers the hydrologic cycle, the main feature of which is a constant transfer, without which terrestrial plants and animals would not exist, of desalted water from sea to land. Some keeps the earth's surface and lower atmosphere warm enough for life to exist. The chemical energy of plants is available to organisms that feed on them. It is also available to be converted to heat by burning, or to be buried and stored as fossil fuel. Food energy may pass through animals that eat plants, and through the herbivores to meat-eating animals, along a series known as a food chain. Food chains are channels of energy flow and dissipation.

The ecosystem acts as a storehouse of energy, held in plants and in animal bodies, as well as in their remains when suitably preserved and it slows down the degradation of energy. Because energy cannot be transferred into work without some being lost in the form of heat, all useful energy eventually is lost to the system. At each step in a food chain, there is a tremendous loss of energy. Only a small part of the living tissue produced in one step is eaten in the next step. It has been calculated, for the food chain that begins in the sea, that for a man or woman to gain a pound, the sea must produce one thousand pounds of living matter.

Energy, Environment, and Evolution

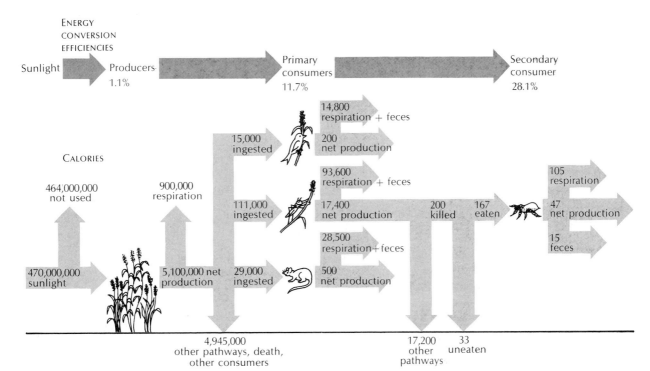

Energy conversion efficiencies

Sunlight → Producers 1.1% → Primary consumers 11.7% → Secondary consumer 28.1%

Calories

464,000,000 not used

900,000 respiration

470,000,000 sunlight

5,100,000 net production

15,000 ingested

14,800 respiration + feces

200 net production

29,000 ingested

111,000 ingested

93,600 respiration + feces

17,400 net production

28,500 respiration + feces

500 net production

200 killed

167 eaten

105 respiration

47 net production

15 feces

4,945,000 other pathways, death, other consumers

17,200 other pathways

33 uneaten

FIGURE 1.3
Energy flow through a grassland ecosystem. All figures represent kilocalories/square meter of surface area/year. Only a little more than 1 percent of the radiant energy of the sunlight received by the plants is used for carbon fixation, in other words, for production of plant material. Although the consumers of plants and animals have much higher energy conversion efficiencies than do plants, only a tiny fraction of the energy of the plants remains in the food chain at the level of the secondary consumer. The invention of agriculture greatly improved the energy conversion efficiency of the human food chain.
(Adapted from Darnell, 1973, p. 68.)

THE PRINCIPLES OF ENERGY

Energy is multiform and variable. To the twentieth century inhabitants of North America, it means electricity, gasoline, and natural gas, used to power air conditioners and electric motors, automobiles and aircraft, to heat buildings, boil water, and melt iron ore. In nineteenth century England, it meant coal, and coke, and steam, and steel, railroads, factories, and steamships. In the Renaissance, it would have meant falling water, blowing winds, fire, and draft animals. To a Roman of the Empire, it would have meant galley slaves and beasts of burden. Yet the most basic, most vital sources of energy are none of these; they are food and solar radiation, the one dependent on the other.

The word *energy* did not exist until 1807; the scientific principles that govern energy were not well established until 1850; and even these principles had to be modified when it was discovered that *mass* was a form of energy. Because some of the scientific principles of energy are important to an understanding of later parts of this book, we shall outline them now.

Most of the energy we use comes from the sun, either directly or indirectly. We live in an atmosphere warmed by solar energy, we eat food whose existence is due to photosynthetic conversion of solar energy. We benefit from a system of winds, rains, and rivers that is driven by solar energy, and we derive more than 95 percent of our inanimate energy from the fossil fuels, which represent stored solar energy. Although most of the energy in the universe is gravitational, man is most concerned with solar radiation and the chemical energy of foods and fuels.

Power, being the rate at which energy is expended or work is done, is not measured in the same units as are energy and work. Two of the most commonly used units of power are the *kilowatt* (kW) and the horsepower (hp). A *kilowatt* is a thousand watts; a watt is a rate of one joule per second. One *horsepower* is a rate of 550 foot pounds per second.

Work also can be expressed in units of power and time (see figure 1.4). Two in common use are the kilowatt-hour (kWh) and the horsepower-hour (hph).

Some conversion factors much used in calculations of energy, power, and work are.

$$1.000 \text{ kcal} = 3.968 \text{ Btu}$$
$$1.000 \text{ hp} = 0.746 \text{ kW}$$
$$1.000 \text{ kWh} = 3,413 \text{ Btu}$$
$$1.000 \text{ kWh} = 1.341 \text{ hph}$$

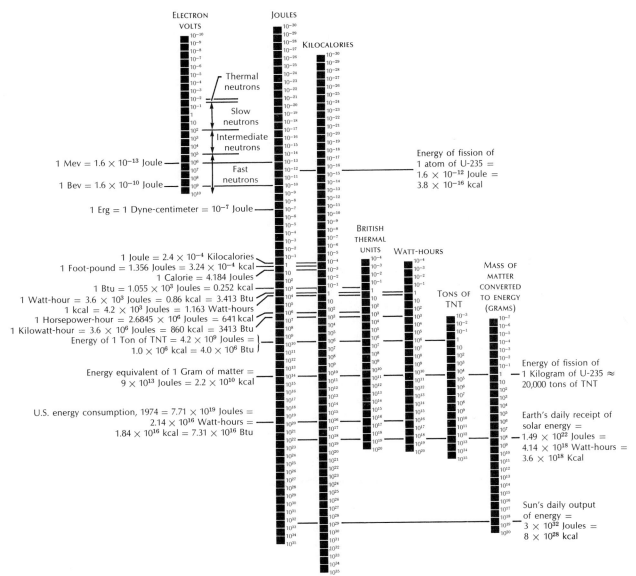

ELECTRON VOLTS

JOULES

KILOCALORIES

Thermal neutrons

Slow neutrons

Intermediate neutrons

Fast neutrons

1 Mev = 1.6 × 10⁻¹³ Joule

1 Bev = 1.6 × 10⁻¹⁰ Joule

1 Erg = 1 Dyne-centimeter = 10⁻⁷ Joule

Energy of fission of
1 atom of U-235 =
1.6 × 10⁻¹² Joule =
3.8 × 10⁻¹⁶ kcal

BRITISH THERMAL UNITS

WATT-HOURS

1 Joule = 2.4 × 10⁻⁴ Kilocalories
1 Foot-pound = 1.356 Joules = 3.24 × 10⁻⁴ kcal
1 Calorie = 4.184 Joules
1 Btu = 1.055 × 10³ Joules = 0.252 kcal
1 Watt-hour = 3.6 × 10³ Joules = 0.86 kcal = 3.413 Btu
1 kcal = 4.2 × 10³ Joules = 1.163 Watt-hours
1 Horsepower-hour = 2.6845 × 10⁶ Joules = 641 kcal
1 Kilowatt-hour = 3.6 × 10⁶ Joules = 860 kcal = 3413 Btu
Energy of 1 Ton of TNT = 4.2 × 10⁹ Joules =
1.0 × 10⁶ kcal = 4.0 × 10⁶ Btu

Energy equivalent of 1 Gram of matter =
9 × 10¹³ Joules = 2.2 × 10¹⁰ kcal

U.S. energy consumption, 1974 = 7.71 × 10¹⁹ Joules =
2.14 × 10¹⁶ Watt-hours =
1.84 × 10¹⁶ kcal = 7.31 × 10¹⁶ Btu

TONS OF TNT

MASS OF MATTER CONVERTED TO ENERGY (GRAMS)

Energy of fission of
1 Kilogram of U-235 ≈
20,000 tons of TNT

Earth's daily receipt of
solar energy =
1.49 × 10²² Joules =
4.14 × 10¹⁸ Watt-hours =
3.6 × 10¹⁸ Kcal

Sun's daily output
of energy =
3 × 10³² Joules =
8 × 10²⁸ kcal

FIGURE 1.4

Several units are used to measure energy. Physicists prefer electron volts and joules. Biologists and nutritionists use the kilocalorie (which they often miscall calorie). Engineers like British thermal units and watt-hours or kilowatt-hours. In this book, the kilocalorie is used except where other standard units take precedence, as in the ranking of coals. Note that this chart represents energy on a logarithmic scale and that energy increases downward in each column.

(The illustration is based on one that appears in The New College Physics: A Spiral Approach *by Albert V. Baez. W. H. Freeman and Company. Copyright © 1967.)*

In this book the kilocalorie will be used as the basis of quantity and efficiency comparisons, unless another unit is the accepted standard, as the British Thermal Unit is for coal.

Heat is a form of energy. One of the greatest discoveries of science was that *heat and work are equivalent.* Another great discovery was *the law of the conservation of energy,* which says that the total energy in the universe is constant and that, when energy is transmitted from one body to another or converted from one form to another, the total quantity of energy (and its mass equivalent) in the system after the transmittal or conversion will be the same as it was before. This first law of thermodynamics—and the equivalence of heat and work—should be the solace and support of any beginner in the mathematics of energy, for they mean that all the standard units for power, energy, and work have fixed mathematical relations to one another.

It is important to understand how heat differs from other forms of energy, because in that difference lies the key to energy conservation as well as to the fate of the universe. Its singularity lies in the fact that heat cannot be converted into work without a substantial loss in available or useful energy. Other forms of energy can be converted to work or heat with little or no loss. The kinetic energy of water falling through a turbine can be converted to mechanical work through a drive shaft with a loss of less than 5 percent. Electrical energy flowing through a wire can be converted to heat (and its equivalent in work) without any loss in quantity. The chemical energy in coal or fuel oil is convertible to heat without loss. The kinetic energy of a cueball is transmitted to a struck ball on the pool table with negligible loss, and the work done in moving the struck ball almost equals the energy of the cueball before impact (a very small amount is converted to heat).

But heat cannot be converted into work without loss, usually a considerable loss, and the loss is always in the form of heat at a temperature nearer to that of the surroundings than that of the heat from which work has been obtained. Once heat has descended to the ambient temperature, it is no longer available to do work. We could raise its temperature, but to no avail, since we could never get more work out of it than we put into it.

Heat is often spoken of as low-grade energy, because heat is not fully convertible into any other form and becomes of less use to man as its temperature falls. When another form is converted to heat, or when high-temperature heat declines in temperature, we speak of the degradation of energy, referring to the permanent loss of its usefulness. In the end, all energy goes down the thermal hill and diffuses into space. The degradation of energy involves an increase in *entropy*, of energy not available to do work. All conversions of heat to work, and all flows of heat from a higher to a lower temperature are accompanied by increases in entropy. The general increase of entropy in the universe is inexorable, but man has hastened the flight of what Eddington called "time's arrow" in a number of ways.

ENTROPY

In his *Principles of Physics* (1945),* Francis W. Sears, who taught physics at Massachusetts Institute of Technology for almost forty years, wrote, "There is no concept in the whole field of physics which

---

*Sources cited in the text are listed in the back of the book.

Energy, Environment, and Evolution

is more difficult to understand than is the concept of entropy, nor is there one which is more fundamental" (p. 447). For his students he pointed out that "there is *no principle of conservation of entropy.* In fact, the reverse is true. Entropy *can* be created at will and there is an increase in entropy in every natural process." Professor Sears then illustrated entropy with a simple example, that of mixing a liter of water at 100° C with a liter of water at 0° C, which results in two liters at 50° C. He showed how the change in entropy could be calculated for engineering purposes, and that it was a decrease. The significance of the example, however, was set forth in these words:

> We *might* have used the hot and cold water as the high and low temperature reservoirs of a heat engine, and in the course of giving heat to the cold water we could have obtained some mechanical work. But once the hot and cold water have been mixed and have come to a uniform temperature, this opportunity of converting heat to mechanical work is lost, and, moreover, it is lost irretrievably. . . . There is no decrease in energy when the hot and cold water are mixed, but there has been a decrease in the availability, or an *increase in the unavailability* of the energy, in the sense that a certain amount of energy is no longer available for conversion to mechanical work. Hence when entropy increases, energy becomes more unavailable, and we say that the Universe has "run down" to that extent. [p. 451–52]

THE FORMS, GUISES, AND SETTINGS OF ENERGY

Energy comes in several forms, among which the more common are chemical, gravitational, mechanical, electrical, radiant, nuclear, magnetic, and thermal or heat energy. Radiant energy includes solar and cosmic radiation, as well as X rays and all electromagnetic-wave phenomena (radio, radar, microwave, light, and so on). In addition, each form of energy is found in several guises and settings. Chemical energy, for example, is found in food, wood, and the fossil fuels; each of these guises has one or two preferred settings or environments, food coming from fields and oceans, wood from forests, and the fossil fuels from mines and wells. The forms, guises, and settings of natural energy stores and flows have a great deal to do with the access to energy enjoyed by a society and with the cost of energy to that society.

Natural stocks of energy are not distributed evenly throughout the world. The great coal beds of the world are mainly in North America and Europe. The productive arable lands are concentrated in the temperate zones. The natural resource of falling water is abundant in Sweden and Switzerland, but scarce in much of Africa and Asia. Petroleum reserves, more widely distributed than those of coal, still favor some regions, such as that of the Persian Gulf, over others.

Uranium deposits appear to be concentrated in perhaps five nations of the world. Even the natural distribution of solar energy favors low-latitude, inland areas over high-latitude, maritime areas.

WORK DONE BY NATURAL ENERGY SYSTEMS

A great deal of work is done in the world by natural energy systems. Continents are raised and mountains are formed by *tectonic* energy systems, which are probably powered by great thermal convection cells within the earth's mantle. Mountains are created and sometimes destroyed by *volcanic* energy systems, which are driven by differences in the earth's heat within the crust. The land masses of the world are sculpted into valleys, hills, and plains by *erosional* energy systems, which are powered by the diurnal variation in solar incidence (insolation) and by the hydrologic cycle, itself driven by solar energy. Some 3 billion tons of rock each year are disintegrated and carried off the continents and into the oceans by this global erosional system. Fortunately for man, the energy of falling water is not all used up in cutting valleys and in transporting terrestrial debris to the sea; much remains for his use to drive mills and generators. A great natural *storm* energy system couples the atmosphere with the continents and oceans; the earth is struck by lightning about one hundred times each second (Viemeister, 1972); winds and waves are unceasingly at work moving rock particles, scouring and abrading desert mountains and rocky coastlines on the one hand while piling up sand dunes and creating beaches on the other. The key factor in the hydrologic cycle is the evaporation into the atmosphere of one hundred thousand cubic miles of sea and lake water each year. This water is carried laterally by winds and much of it is dropped on land, where it helps sustain plants and animals, provides a habitat for fish, creates groundwater reservoirs, and carries away weathered rock particles. The fact that evaporation from the sea produces fresh, not salty, water is of prime importance to life on land. The magnitude of the hydrologic system is illustrated by the fact that the Mediterranean loses more water by evaporation than it receives from streams around its margins. The deficiency is made up by a strong inflow of water from the Atlantic through the Strait of Gibraltar. In the fairly recent geological past, when sea levels were lowered by Pleistocene ice advances, the world sea level fell below the Gibraltar threshold, and the Mediterranean dried up. Later it was revived as a series of coalescing freshwater lakes reflecting a pluvial climate induced by continental glaciation, reappearing as a salt sea only when the retreat of the continental glaciers returned sufficient water to the oceans to allow the flow through the strait to be resumed. Should a fifth Pleistocene glacial stage take place, this sequence probably would recur.

Energy, Environment, and Evolution

INCOME AND FUND SOURCES OF ENERGY

To put them into perspective one might find it useful to divide the sources of energy in our planet into categories of income (renewable) and fund (nonrenewable). Income sources are those of steady direct supply or those that represent readily replenishable reservoirs. Fund sources are those that represent energy traps or reservoirs not replenishable on any practical human time scale.

Income (renewable) energy sources are:
Solar radiation
Water power
Wind power
Tidal energy
Photochemical energy stored in plants and animals (food, wood, vegetable refuse)
Animal (including human) power
Geothermal energy (heat flow)

Fund (nonrenewable) energy sources are:
Fossil hydrocarbons
Coal, lignite, peat
Petroleum
Natural gas and natural gas liquids
Oil shale and tar sands

Frozen nuclear transformations (processes such as fusion and fission)
Fissionable isotopes
Uranium 235 (fissile)
Uranium 238 and thorium 232 (fertile)
Fusion
With deuterium (^2H or hydrogen 2) only
With lithium 6
Geothermal energy (in heat traps)

Man's social systems for more than 1 million years were based on income (renewable) energy resources; only during the past 150 years have systems emerged that have been based largely, or almost entirely, on the use of fund (nonrenewable) energy sources. For about 90 percent of his existence man's only source of energy was the wild food he captured, gathered, and ate (figure 1.5). For only about 10 percent of his brief history on earth has he had fire. For only 1 percent of his history has he planted and harvested food crops, milked, butchered, driven, and ridden domesticated animals. For only 0.1 percent of his terrestrial tenure has man used wind and water power to any great extent. And only in the most recent 0.01 percent of man's history has he switched his energy base from such income resources to the funds of fossil fuel and fissionable uranium.

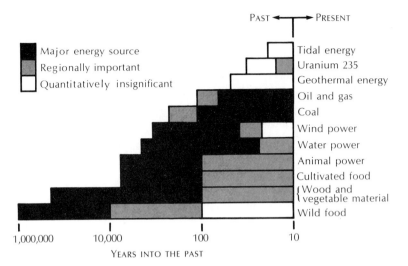

■ Major energy source
▨ Regionally important
□ Quantitatively insignificant

Tidal energy
Uranium 235
Geothermal energy
Oil and gas
Coal
Wind power
Water power
Animal power
Cultivated food
{ Wood and
{ vegetable material
Wild food

1,000,000   10,000   100   10

YEARS INTO THE PAST

FIGURE 1.5
Man's energy sources during the past million years are represented on this chart, which has a logarithmic scale. Only in the past 150 years have nonrenewable resources become more important than the renewable resources on which man depended for so long.

POPULATION GROWTH AND ENERGY USE

Man's population is a direct function of his rate of production of useful energy and power. The population of the world has grown in surges that have been related to major acquisitions of control over energy. Plateaus in the population curve represent periods when population growth was limited by food production and available habitat.

A technological revolution had begun in western Europe in the Middle Ages and was well developed, mainly on a base of water power and fuel wood, when the twin inventions of the steam engine and coke ignited that phase of history known as the Industrial Revolution. The Industrial Revolution clearly divided the world into haves and have-nots. The haves were those who had control of inanimate energy and used it to produce goods and services in such abundance that ability to consume for the first time became a social concern.

Until the early years of the present century, population growth rates were highest in those nations and regions where growth rates in energy consumption were highest. Without exception, these were the nations and areas peopled by western Europeans or their descendants. Then a curious change started. Population growth rates in the nations consuming much energy started to decline, while population growth rates in the nations consuming relatively little energy started to rise. One consequence was that per capita energy consumption rates in the developed countries rose even faster while per capita energy consumption rates in the underdeveloped countries increased at a lesser rate or actually started to decline. In other words, the energy gap between individuals in the industrialized and the nonindustrialized countries was increased by the different rates of population increase.

THE NATURAL SUBSIDY

Whenever man can recover more useful energy or work from any part of the natural energy system than he has to expend to recover it, he is the beneficiary of a *natural subsidy*. The natural subsidy inherent in streams of falling water, forests, and deposits of coal, petroleum, natural gas, and uranium is unevenly distributed over the surface and within the crust of the earth.

The natural energy subsidy is possible only because the energy resources that man uses are the result of an enormous amount of work that he has neither to perform nor to pay for. In nature work is done in the conversion of solar energy into the chemical energy of plant food, in soil formation, in the continuous operation of the hydrologic cycle and the atmospheric wind system, in the concentration, preservation and transformation of initially scattered, minute, organic packets of chemical energy into deposits of the fossil fuels, in the magmatic, hydrothermal, and sedimentary concentration of radioactive minerals, and in the crustal uplift and erosion of overburden that have made many fossil fuel and uranium deposits more readily accessible than they otherwise would be. All geochemical concentrations of minerals useful to man that can be recovered at a net work profit represent natural subsidy. The subsidy is available only to those who possess the appropriate technological keys and its size depends on the size of a nonrenewable resource or on the sustainable yield rate of a renewable resource, as well as on the technical efficiency of the exploitation system.

The market value of goods and services enjoyed by man is related to the work done to produce them and to the demand made for them. However, the contribution to the wealth and capital of the nations benefiting from the natural subsidy of fossil and nuclear energy resources can be enormously greater than the work required to produce the energy materials and to convert them to usable form. The measure of the subsidy and of the contribution to national wealth is the net work profit achieved by the exploitation system.

BENEFITS AND COSTS OF ENERGY

Increasing control over inanimate sources of energy has brought great benefits and entailed substantial costs. The benefits have been for the most part tangible and immediate. The costs, except for the direct labor and capital costs of exploitation, have generally been intangible, or at least not measurable by market values, and intermediate to long-term in their rates of accrual. Moreover, the benefits and costs of energy use have not been shared equally. Both within and among nations the costs of energy use frequently have been borne most heavily by those who benefited the least. The owner of a factory that obscured the sun with its smoke could move his residence to a clean suburb, whereas his employees could not afford to move away. The use of coal benefited many people more than the miner. In many regions of the world, primitive agricultural man found his leisure, the currency in which he spent the energy surplus from his farming, abruptly taken away, along with most of the satisfying ritual and value of his life, to be replaced by a life of plantation toil and indenture under a price system that bewildered him while it enslaved him. The fact that his new masters were also economic slaves did not lessen the social cost of the new system to the laborer.

Even where the benefits of abundant energy are most evident and most praised, where heat, light, motion, and impact come from the gentle pressing of buttons and pedals, accrued costs are beginning to be perceived. Such costs include environmental degradation, damages to health, and the economic and social costs of the depletion of fund resources. The benefits and costs of high energy use will be discussed more fully later in this book.

Energy, Environment, and Evolution

SUPPLEMENTAL READING

Carter, George F. 1975. *Man and the Land,* 3rd ed. New York: Holt, Rinehart and Winston.
A useful introduction to the origin of man and the role of environment in the development of culture.

Clark, W. E. LeGros. 1966. *History of the Primates,* 5th ed. Chicago: University of Chicago Press.
A concise anatomical guidebook to the primate path through the Cenozoic era.

Cloud, Preston (editor). 1970. *Adventures in Earth History.* San Francisco: W. H. Freeman and Company. See especially Section X, "The Rise of Man, the Present, and the Future," p. 859–968, which contains articles on the Pleistocene epoch, its geology and temperatures, and articles on the nature and distribution of man.

Darnell, Rezneat M. 1973. *Ecology and Man.* Dubuque, Iowa: Wm. C. Brown Co.
Excellent introduction to ecological concepts: environment, population, community, ecosystem.

Holloway, Ralph L. 1974. The casts of fossil hominid brains. *Scientific American,* v. 231, July, p. 106–115. Available as *Scientific American* Offprint 686.
Man's brain began to differ from that of other primates three million years ago.

Hubbert, M. King. 1962. *Energy Resources.* Washington, D.C.: National Academy of Sciences.
Terse but classic review of energy resources, of the evolution of man's ability to control energy, and of the fundamental relations between human population growth and the availability of energy.

Jorgensen, Joseph G. 1972. *Biology and Culture in Modern Perspective.* San Francisco: W. H. Freeman and Company.
Contains forty-one articles on the physical and cultural evolution of man, with emphasis on contemporary concerns.

Krauskopf, Konrad, and Arthur Beiser. 1966. *Fundamentals of Physical Science,* 5th ed. New York: McGraw-Hill.
Useful chapters on earth history, processes, and materials, as well as on energy, heat, electricity, and nuclear energy.

Menard, H. W. 1974. *Geology, Resources, and Society.* San Francisco: W. H. Freeman and Company.
Much of the eminently readable text deals with manifestations of energy within and on the surface of the earth—plate tectonics, earthquakes, volcanism, erosion.

Newell, Reginald E. 1974. "The Earth's Climatic History." *Technology Review,* v. 77, n. 2, p. 31–45.
Fine summary of what is known about the causes of climate change and its relations to food supply. Another Ice Age is coming.

Odum, Howard T. 1971. *Environment, Power, and Society.* New York: John Wiley and Sons.
Book of ambitious scope that attempts to analyze the energy of ecosystems, social orders, economics, politics, and religion—for the most part successfully.

Sears, Francis W., 1945. *Principles of Physics.* Cambridge, Mass.: Addison-Wesley.
One of the better physics texts for lucid explanation of thermodynamic concepts and units.

Wilson, Mitchell, et al., 1967. *Energy.* New York: Time, Inc. (Life Science Library).
Valuable for its excellent illustrations of energy sources, guises, processing, processes, and uses.

Chapter 2

MAN AND OTHER ENGINES

In the high energy society a machine which will deliver energy equivalent to that of a fully grown man can frequently be secured for less than the fee of the obstetrician who delivers a baby.

—Fred Cottrell, 1955

ANIMALS AS ENGINES

An *engine* is a device for converting some form of energy into work. For most of his evolutionary history, his own body was the only engine man had. He developed simple *machines* such as the bow and the metate, which is a concave stone for grinding maize or cocoa. With his body, the engine, as a *prime mover,* he powered these simple machines (see figure 2.1). Later, he tamed more powerful animals and persuaded them to act as prime movers for other machines, such as the plow, the chariot, and the geared mill (see figure 2.2). Until very recently, men and other animals provided a substantial amount of the power used by mankind. Their energy came from food and feed, most of it derived from plants. Although food energy in the modern high energy society (that in which the average daily per capita consumption of energy from all sources exceeds fifty thousand kcal) forms only a minor fraction of the gross energy consumption, it is still the fundamental part of a high energy system, since no community can exist without it. The computer expert does need less food energy than the worker in a rice paddy, but not much less. To both, it is vital.

While steam was king, the power of the industrializing world came from heat engines. They were analyzed carefully and they became well understood, both theoretically and mechanically. A heat engine is a device by means of which heat can be used to do work.

Strictly speaking, a heat engine is one in which thermal energy is transmitted to a fluid such as steam or ammonia that does work. In this sense, neither the gasoline engine of an automobile nor a man is a heat engine; in both the automobile and the man, however, a combustible material is oxidized, producing energy, and the energy of combustion (rapid oxidation) is what gives them the ability to perform work. The

FIGURE 2.1
Cranked water pump of the eighteenth century was operated by prisoners. Men at the cranks were spelled after an hour (note hourglass on the platform).
(Courtesy of The Hagley Museum Wilmington, Delaware.)

energy provided by the automobile's fuel may be measured in British thermal units (Btu) and the energy provided by the man's food in kilocalories (kcal), but it is the same kind of energy being converted to work in different ways. Both might be called combustion engines. Today we restrict the term *heat engine* to a particular type of combustion engine, one in which energy is transmitted as heat and not as kinetic energy (as in automobile engines) or muscular energy (as in animals).

Food is the fuel for the man-engine. Plants convert solar energy, carbon dioxide, and water into plant tissue and oxygen; in this process, called photosynthesis, which no animal, including man, has yet been able to duplicate, energy is stored in chemical form in the plant tissue. In animals and other engines that use plant material as fuel, this chemical energy, released through combustion, is converted into work (including respiration and circulation), heat, flesh, bone, and waste materials. The portion of plant energy that is useful to animals is called

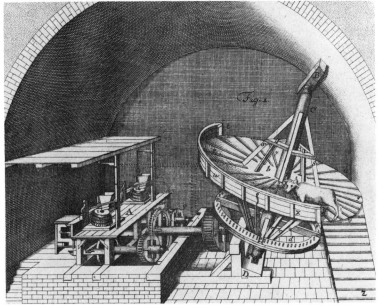

FIGURE 2.2
(*Top*) *The horse engine, a major power source until it was replaced by the steam engine, was the model for the unit of power called horsepower. James Watt, acting in 1783 on figures supplied by a millwright, stated that a horse could pull 150 pounds at a velocity of 2.5 miles per hour, thereby working at a rate of 3,000 foot-pounds per minute. In reality a horse cannot do that much work continuously. (Bottom) The inclined treadwheel could be driven by men or animals. In this representation, which appeared in* Theatrum Machinarum Molarium *(1737), a single animal is providing the power. Treadmills of this type were frequently illustrated in books on machinery throughout the seventeenth and eighteenth centuries. The first known portrayal of the device was published in the latter part of the sixteenth century.*
(*Courtesy of The Hagley Museum, Wilmington, Delaware.*)

food if man eats it and feed if other animals eat it. Animals can convert only a fraction of the available plant food into living tissue and work, that fraction determining their efficiency as engines. A man is about as efficient as a horse and considerably more efficient than a modern

Animals as Engines

automobile, in converting chemical energy into work. The horse and the automobile have more power and speed and the automobile better endurance.

The energy in food comes from carbohydrates (mainly sugar and starch), fats, and proteins. Proteins contain less than half the energy of carbohydrates and fats. All three may be of either plant or animal origin, although most of the carbohydrate intake is of plant origin. Not all carbohydrates can be digested. Cellulose, which constitutes the tissue of many large plants, is a notable illustration of indigestible food. About 70 percent of the world's population of men, women, and children get almost all their energy input from plant food, mainly the cereal grasses (principally wheat and rice) and the pulses (beans and peas). In addition to food energy, the human machine requires protein for body building and repair, as well as certain vitamins and minerals to keep the system functioning properly.

The total life of an individual man, below a theoretical maximum durability of perhaps 150 years, depends on the quality of the maintenance he has received. In the high energy societies, the average lifespan is about 70 years, considerably higher than in low energy societies. The working life at full load (between 8 and 10 hours of manual labor, daily) is closer to 40 years.

The physical principles governing combustion engines are so important to any discussion of man's energy problems that they need to be elaborated.

THE PRINCIPLES OF OXIDATION

Rapid oxidation is called combustion. Oxidation yields heat as well as new compounds and, for each specific oxidation reaction, a known amount of heat will be produced (see table 2.1). For example, oxidation of carbon according to the equation

$$C + O_2 \rightarrow CO_2 + 94.03 \text{ kcal}$$

(in which the components are assumed to have their molecular weights expressed in grams) yields 94.03 divided by 12 (the molecular weight of carbon) or 7.8 kcal.

Oxidation of a gram of hydrogen according to the equation

$$H_2 + \frac{1}{2}O_2 \rightarrow H_2O + 68.37 \text{ kcal}$$

yields 68.37 divided by 2 (the molecular weight of hydrogen) or 34.2 kcal.

Oxidation of a gram of methane (CH_4) according to the equation

$$CH_4 + 2O_2 \rightarrow CO_2 + 2H_2O + 210.8 \text{ kcal}$$

yields 210.8 divided by 16 or 13.2 kcal.

Table 2.1 Combustion Heat of Common Fuels

Substance	Formula	Heat per gram
Hydrogen	H_2	34.2 kcal
Methane (natural gas)	CH_4	13.2 kcal
Lipid (fat)	$C_{57}H_{104}O_6$	9.1 kcal
Carbon (coal)	C	7.8 kcal
Ethyl alcohol	C_2H_6O	7.1 kcal
Protein	$C_{1864}H_{3012}O_{576}N_{468}S_{21}$	5.7 kcal
Glucose (sugar)	$C_6H_{12}O_6$	4.1 kcal

All the important fuels, as well as the principal foods, contain carbon; most of them also contain hydrogen. Those that contain oxygen already are partially oxidized and have lost some of their energy potential. When we eat carbohydrates and burn or oxidize them in our bodies, the carbohydrates are being converted to carbon dioxide and water and releasing energy. The minimum average energy requirement for an adult human to exist is about 1,000 kcal a day. For an adult engaged in normal activities the requirement doubles, to about 2,000 kcal a day, and for a man engaged in manual labor much of the day, 4,000 kcal or more are needed. See table 2.2. The minimal energy

Table 2.2 Energy Requirements for Various Activities (in kilocalories/hour)

Light work		Moderate work	
Sitting	19	Shoemaking	80–115
Writing	20	Sweeping	85–110
Standing relaxed	20	Dusting	110
Typing	16–40	Washing	125–215
Typing quickly	55	Charring	80–160
Sewing	30–90	Metal working	120–140
Dressing and undressing	33	Carpentering	150–180
Drawing	40–50	House painting	145–160
Lithography	40–50	Walking	130–240
Violin playing	40–50		
Tailoring	50–85		
Washing dishes	60		
Ironing	60		
Bookbinding	45–90		

Hard work		Very hard work	
Polishing	175	Stonemasonry	350
Joiner work	195	Sawing wood	420
Blacksmithing	275–350	Coal mining (average for shift)	320
Riveting	275	Running	800–1000
Marching	280–400	Climbing	400–900
Cycling	180–600	Walking very quickly	570
Rowing	120–600	Rowing very quickly	1240
Swimming	200–700	Running very quickly	1240
		Skiing	500–950
		Wrestling	1000
		Walking upstairs	1000

SOURCE: *Man and Food* by Magnus Pyke, 1970, McGraw-Hill, p. 100, Table 6.1.

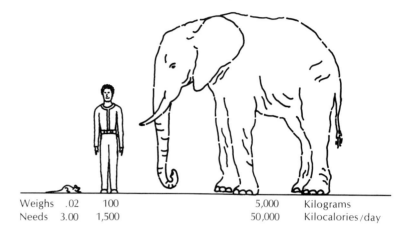

FIGURE 2.3
Animals do not require food energy in direct proportion to their weight. Smaller animals need more for each pound than do larger ones. An elephant needs 50,000 kilocalories a day to stay alive, while a mouse needs but 3. The difference—about 17,000 times—is not as great, however, as their difference in weight. The elephant weighs 263,000 times more than the mouse. One kilogram of elephant requires 10 kilocalories a day for bare subsistence; one kilogram of man takes 15 kilocalories; and a kilogram of mouse would require 158 kilocalories a day. The smaller the animal, the greater the proportion of its food energy that is burned up in body maintenance.

| Weighs | .02 | 100 | 5,000 | Kilograms |
| Needs | 3.00 | 1,500 | 50,000 | Kilocalories/day |

requirement to maintain life varies with the weight of the animal: a mouse needs about 3 kcal a day, a man about 1,000 kcal, and an elephant about 50,000 kcal.

EFFICIENCY OF ANIMALS AS ENERGY CONVERTERS

The efficiency of an animal as an energy converter has two distinct elements: the efficiency with which plant-food energy is converted into living tissue or into work. We are interested in the former when we feed animals for meat, milk, and egg production. We are interested in the latter when we feed an animal that is to do work for us. In either case, we must use units of energy in our calculation. Pounds of grain feed may be compared to pounds of edible product but, since the energy content of a pound of feed is almost never the same as the energy content of a pound of animal product, conversion efficiencies cannot be calculated by a comparison of weights. As we shall see in chapter 6, animal products represent a wide range of food-energy conversion efficiencies—from almost 20 percent in milk cows (in the United States) down to less than 2 percent in lamb. Conversion efficiencies vary with the type of feed as well as with the variety, age, and condition of the animal.

The work efficiencies of men and draught animals range from 20 to 25 percent on a sustained basis; for short periods, their efficiency may exceed 30 percent. Such low efficiencies (rather typical of combustion engines) come about because so much of the chemical energy in food and feed must replace tissues and sustain the normal physiological processes, such as the maintenance of a constant body temperature.

THE SEARCH FOR POWER

Man has long sought to control power as well as energy. His own body is very limited in power, the rate at which it can do work, either in brief bursts or for a sustained period of time. A man can lift and move forty rocks, each weighing fifty pounds and together weighing two thousand pounds, but he cannot lift, with his unaided muscles, a boulder weighing one ton. The early development of power technology was designed to augment the low power capacity of the human body (see table 2.3). When he discovered the principles of momentum, leverage, tension, and the wedge, a man could throw, pry, propel, and split with a directed force far greater than the capacity of his own muscles. The concept of concentrating mechanical or kinetic energy through the sharpened point or edge of a tool led to the flaking of stone implements, the scraping of hides, and the shaping of wood, as well as to the development of arrows, spears, and axes. It was discovered that energy could be concentrated along the line of contact of one hard object rolled over another, and thus the metate and the mortar and pestle were born. For those people who lived near lakes and streams, the advantages of transport by flotation must have been perceived early. Rafts and shaped logs represented a major step in power technology because they made use of natural forces outside man's body. The pole and the paddle probably followed close behind.

The technology of moving very heavy objects on rollers was well developed by the time of the construction of the great pyramids of

Table 2.3 Chronological Advances in Heat Engine Technology

Prime mover	Date	Output in horsepower
Man pushing a lever	3000 B.C.	0.05
Ox pulling a load	3000 B.C.	0.5
Water turbine	1000 B.C.	0.4
Donkey mill	500 B.C.	0.5
Vertical water wheel	350 B.C.	3
Vitruvian water mill	50 B.C.	3
Eighteen-foot overshot water wheel	1200 A.D.	5
Post windmill	1400 A.D.	8
Turret windmill	1600 A.D.	14
Versailles waterworks	1600 A.D.	75
Savery's steam pump	1697 A.D.	1
Newcomen's steam engine	1712 A.D.	5.5
Watt's steam engine (land)	1800 A.D.	40
Steam engine (marine)	1837 A.D.	750
Steam engine (marine)	1843 A.D.	1,500
Water turbine	1854 A.D.	800
Steam engine (marine)	1900 A.D.	8,000
Steam engine (land)	1900 A.D.	12,000
Steam turbine	1906 A.D.	17,500
Steam turbine	1921 A.D.	40,000
Steam turbine	1943 A.D.	288,000
Coal-fired steam power plant	1973 A.D.	1,465,000
Nuclear power plant	1974 A.D.	1,520,000

Egypt. Then came more sophisticated ideas such as the plow, the screw, the windlass, the wheel, the gear, and the pulley, all of which extended the *power* of man's body, and of his domesticated draft animals, but each of which was limited by the inherently low capacity for sustained power of the animate prime movers. Large amounts of work could be concentrated on one job, such as building a pyramid or propelling a sizeable ship, only by massing together great numbers of men pulling ropes or manning oars, and such concentrations of manpower quickly ran into both engineering and economic constraints.

The harnessing of the wind by sails and of water by specially designed wheels broke the limitations of power and created, for the first time, inanimate prime movers that did not need long pauses for food and rest, and that could produce more power with far less investment of men and materials. Western industrial and technological civilization received a great and vital impetus from these latter developments, as well as from efficient use of the horse on the fertile fields of western Europe.

It was not, however, until the development and proliferation of *inanimate* combustion engines that the Industrial Revolution, the voracious depletion of the world's stock of nonrenewable fuel resources that characterizes our contemporary economy, and an explosion in human population could take place.

RISE OF THE HEAT ENGINE

Newcomen's improvement on the Savery steam pump spread rapidly throughout western Europe. The Newcomen engine (see figure 2.4), however, was slow and very inefficient, wasting more than 99 percent of the energy in its fuel. Not until more than 50 years had passed did the full potential of the steam engine begin to be developed, when, between 1765 and 1769, James Watt greatly improved the Newcomen engine. By 1800, about five hundred Watt engines were working in England and had completely replaced the Newcomen engine. In the 1780s, Watt devised a mechanical means of converting the reciprocating motion of a steam-driven piston into the rotary motion of a wheel. Only then did the steam engine become capable of replacing water and windpower to drive mills, capable of replacing filled sails to move ships, and capable of creating a revolution in transportation by allowing the development of the steam locomotive.

In 1802 the steam engine was successfully placed on wheels by Richard Trevithick. As in the case of Savery's steam pump, needs of the coal miners encouraged the development of the locomotive. The sudden great expansion of the market for coal caused by Abraham Darby's perfection in 1709 of a method for making coke from coal

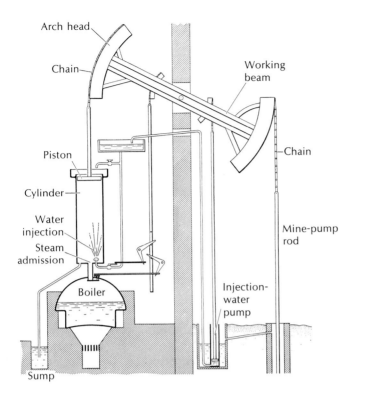

Arch head

Chain

Working beam

Chain

Piston

Cylinder

Water injection

Steam admission

Mine-pump rod

Boiler

Injection-water pump

Sump

FIGURE 2.4
The Newcomen engine was the first practical source of motive power independent of muscle, water, or wind. Its first major application was for the removal of water from coal mines; it was vital to the growth of coal mining in England. Steam formed in a boiler was admitted at low pressure to a cylinder, which was then cooled by a jet of cold water. As the steam condensed to form a partial vacuum, a piston was pushed into the cylinder by atmospheric pressure, thus operating a pump rod attached to a working beam.
(From Eugene A. Ferguson, "The Origins of the Steam Engine." Copyright © 1964 by Scientific American, Inc. All rights reserved.)

FIGURE 2.5
Model of the first practical steam locomotive, built in 1804 by Richard Trevithick, a Cornish mining engineer. At Penydarran in Wales, this locomotive won a prize by pulling 10 tons of iron and 5 coal-wagons with 70 men riding in them over 9 miles of iron-flanged tramroad.
(British Crown Copyright, Science Museum, London.)

(thereby saving the British metal industry from what would have been fatal economic erosion caused by the increasing scarcity and cost of charcoal) had thrown into sharp focus a weak link in the coal-supply system, namely the slow and costly haulage of coal from the mines to the docks in horse-drawn wagons of limited capacity that cut great ruts in dirt roads or wore through the planks of planked roads. The invention of coke and the invention of the steam engine had synergistic consequences: from them sprang the complex of coal pits, railroads, and steel mills that were the structural framework of the Industrial Revolution.

HEAT SCIENCE

Heat theory as it developed was a marriage of Newtonian mechanics and Democritian atomism. Here it should be pointed out that scientific principles or laws do not really explain, they simply provide models that appear to be able to predict events in the physical world. The model of Newton's apples dropping from the tree *because* of a gravitational force that can be shown to exist between any two bodies does not *explain* gravitation, no more than the facts that such a force is measurable and that the speed and path of the falling object are predictable tell us *why* the apple is falling. Through the efforts of Newton and many of his successors, we know a great deal about how the forces of nature act, how to measure them, and how to predict what will happen if a certain conjunction of phenomena occurs. That is all we need to know to construct an industrial society and a technological world.

A French artillery officer, an expatriate American Tory, two German doctors, and an English brewer are largely responsible for our knowledge of the scientific relations of heat and work. The artillery officer, Sadi Carnot, in 1824 at the age of twenty-eight published the basic principles that govern the efficiency of heat engines. Unfortunately young Carnot was struck down by a cholera epidemic and most of his ideas did not become widely known until thirty years later, by which time many had been rediscovered independently by others. Had he lived, Carnot, the founder of the science of thermodynamics who anticipated almost the entire energy theory, would have become one of the greatest scientists of all time. In his journal he set down all the critical experiments that needed to be done to test his theory.

The first of the German doctors was Julius Robert Mayer who, while practicing medicine in Java in 1840, noted that the blood in the *veins* of his Javanese patients was bright red, a color found only in the highly oxygenated blood of the *arteries* of persons living in Germany and other countries outside the tropics. Mayer's inspired guess that in hot climates the body requires less oxidation to maintain body temperature than it does in cooler climates led him into a new and disturbed career as a theoretical physicist. Rebuffed by the scientific establishment of his day and derided by others, he was finally committed to an asylum for the insane. To Mayer eventually went the credit for first establishing, by scientific argument, the principle of the conservation of energy based on the equivalence of heat and work, although he never conducted any experiments of his own. One of Mayer's most prescient perceptions was that the law of conservation of energy applied to living as well as inanimate objects. At that time, as Isaac Asimov has pointed out, the laws of nature applicable to the inanimate were believed not to apply to the animate and certainly not to human beings.

Well before Mayer, an American who left his native Massachusetts in 1776, Benjamin Thompson (Count Rumford), already had come very close to making an accurate statement about the relations between heat and work. Count Rumford, supervising the boring of cannon barrels in Bavaria in 1798, measured the heat produced and concluded that the heat was produced by the two horses who provided the power for the boring machine and ultimately by the oxidation of the fodder they consumed. He recognized a loss in efficiency in the conversions from chemical to mechanical to heat energy, for he noted that more heat could have been produced by burning the fodder. But Rumford proposed no theory and no equation, and his great chance passed.

The second German doctor in this history was Hermann von Helmholtz, a Prussian army surgeon. In 1847 he first identified the two forms of heat, latent and sensible, with the two forms of energy, potential and kinetic. Both Helmholtz and Mayer explained the heat developed by combustion in the same way, namely, as being due to the kinetic energy developed by the atoms as they are liberated from the orderly structure of the solid state to the disorder of the gaseous state. To Helmholtz goes the credit for making the conservation law a tool of research, by showing how other laws or hypotheses could be derived from it.

The English brewer was James Prescott Joule, who set out to determine just how much heat was produced by a given amount of work. He began his experiments in 1840 and continued them for twenty years. In every way he could devise, Joule produced heat from work and he found that the same amount of work, no matter how performed, always yielded an equal amount of heat. For many, Joule's measurements confirmed the hypotheses of Mayer and Helmholtz. From the experimental and theoretical base so well laid by Joule, Mayer, and Helmholtz, Rudolf Clausius, professor of physics at the Swiss Polytechnic Institute (1855–67), went on to state the first law of thermodynamics: that energy can neither be created nor destroyed in any observable process.

Hans Thirring remarks that

> when this discovery [implying that the original potential energy of a dropped stone is converted into kinetic energy during the fall and then partly into mechanical work by making a depression in the ground and partly into heat] was made by Robert Mayer and J. P. Joule . . . it was greeted with great scepticism, because nobody had noticed any temperature rise in a fallen stone or in the water at the bottom of a waterfall. The heat equivalent of mechanical energy is small; the water at the bottom of Niagara Falls is only $\frac{1}{8}$° C warmer; a 1° C rise would require a drop of 1,400 feet. [1968, p. 18]

STEAM ENGINES AND THEIR EFFICIENCY

Carnot saw that steam would be an excellent medium for the transport of heat because it carries a large amount as latent heat of vaporization—the heat that must be added to convert a liquid to its vapor. To condense steam at 100° C to water at the same temperature requires the abstraction of this latent heat. Although any hot vapor or gas can be used to transport heat in a heat engine, water vapor or steam is generally the most economical, not only because steam carries so much heat, but also because water usually is available at little or no cost.

Not all of the heat in steam, however, can be recovered as work in any practical steam engine. Here the second law of thermodynamics, which Clausius put simply as the observation that "heat cannot of itself pass from a colder to a hotter body" comes into play (see figure 2.6). Temperature is the intensity of heat. Heat flows down temperature gradients, ignoring differences in heat content and may be likened to water flowing downhill, with temperature resembling elevation. (Electricity follows a similar rule, flowing from a high potential to a low potential.) The raising of heat to a higher temperature is an uphill job that requires more work (or heat equivalent) than can be recovered from the high-temperature heat. Only part of the heat produced by combustion under a boiler can be transmitted to the steam produced and in turn only part of the energy in the steam can be converted to work in driving a piston or a turbine; most of the chemical energy in the boiler fuel is lost as heat, some up the furnace stack, the rest in the condenser.

Both Carnot and Clausius showed that the efficiency of a heat engine, or the amount of power obtainable from a given quantity of heat, depends only on the temperature drop, according to a formula

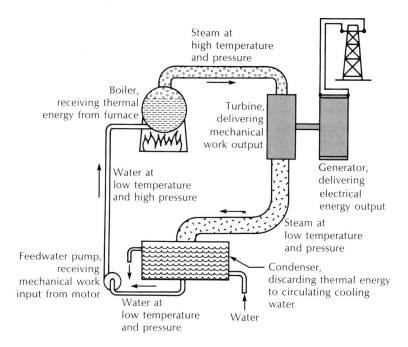

Steam at
high temperature
and pressure

Boiler,
receiving thermal
energy from furnace

Turbine,
delivering
mechanical
work output

Water at
low temperature
and high pressure

Generator,
delivering
electrical
energy output

Steam at
low temperature
and pressure

Feedwater pump,
receiving
mechanical work
input from motor

Condenser,
discarding thermal energy
to circulating cooling
water

Water at
low temperature
and pressure

Water

FIGURE 2.6
Elements of a steam power plant producing electricity.

where T_2 is the top temperature of the heat engine (the temperature of the steam in a steam engine) and T_1 is its bottom temperature (the temperature of the used steam), both stated in terms of an absolute temperature scale, such as the Kelvin, on which zero is absolute zero ($-273°$ C). In looking at the formula below one can see at once that, with any finite positive numbers substituted for T_2 and T_1, the efficiency of the engine will be less than 100 percent.

$$\frac{T_2 - T_1}{T_2}$$

Watt's best steam engine had an efficiency of only 5 percent. Carnot had demonstrated that the most effective way to increase the efficiency of a heat engine is to widen the temperature limits between which it works. If a steam engine exhausts to the atmosphere, as most steam locomotives did, the bottom temperature is usually well above 100° C, the temperature at which steam condenses at atmospheric pressure. This fact accounts for the low thermal efficiency of the steam locomotive, which never rose above 10 percent even in double-expansion or compound engines that used the same steam twice in two sets of cylinders, one high pressure, the other low pressure. If the engine exhausts to a condenser cooled by circulating water, a much lower bottom temperature can be achieved. A low bottom temperature does require greatly enlarged cylinders (in a reciprocating steam engine) to take care of the expansion of the steam necessary for it to cool. Practical limits were reached about 1900 with the huge, triple-expansion, reciprocating steam engines that powered some steamships and electric generating plants. The subsequent development of the steam turbine allowed a great increase in efficiency, because for the first time an enormous expansion of steam at low temperature could be accommodated, and condensers could be operated in a vacuum in order to reach bottom temperatures below 100° C. Today, the condensers of large fossil-fueled power plants operate at temperatures as low as 30° C. Now the only alternatives available to increase the efficiency of the steam cycle are to raise further the boiler pressure and to superheat the steam to still higher temperatures. Boiler pressures are limited by the strength of economically available materials and the design of the boiler; superheat (a higher temperature than that corresponding to the pressure the steam is under) is obtained at the expense of some of the heat in the system and therefore is self-limiting. Modern steam-turbine power plants have an overall thermal efficiency of about 40 percent.

A modern steam-generating plant (see figure 2.7) consists of a source of feedwater, a feedwater heater, a boiler in which the water is converted to steam by the heat from a furnace, superheaters that raise the temperature of the steam, and condensers that cool the steam after it has been used and convert it back to water.

Steam engines can use as fuel almost any material that will burn. Fuel wood was much used in the nineteenth century in forested lands. Then coal became the dominant fuel for steam engines. About 1910 fuel oil began to replace coal, especially for marine engines. Today, most of the steam horsepower in the world is provided by steam turbines in electric power plants; the furnace and boiler units in such plants mainly burn coal and fuel oil, unless natural gas is available. A few countries in the world, mainly in central Europe and mainland China, still maintain coal-burning steam locomotives. The power performance of steam has advanced marvelously; Savery's pump produced about 1 horsepower, whereas a modern coal-fired steam power plant can produce in electrical power the equivalent of almost 1.5 million times the power output of Savery's "miner's friend."

The working life of a steam power plant depends not only on wear and tear but also on obsolescence. For amortization purposes, large steam power plants are expected to have a useful life of between thirty and forty years. It has been found, during the past seventy years, that a thirty-year-old power plant, although perhaps in excellent physical condition, probably will be substantially less efficient than a new plant; in short, will be obsolete.

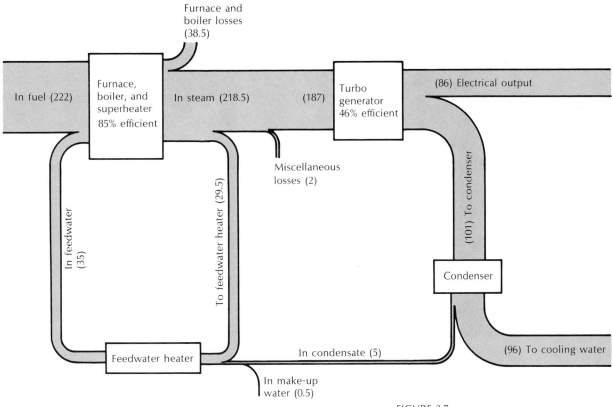

Furnace and boiler losses (38.5)

In fuel (222)

Furnace, boiler, and superheater 85% efficient

In steam (218.5)

(187)

Turbo generator 46% efficient

(86) Electrical output

In feedwater (35)

To feedwater heater (29.5)

Miscellaneous losses (2)

(101) To condenser

Condenser

Feedwater heater

In condensate (5)

In make-up water (0.5)

(96) To cooling water

FIGURE 2.7
Simplified diagram of modern 100-megawatt steam power plant in energy-balance form. Overall thermal efficiency is calculated by dividing the electrical output (86) by the heat content of the fuel (222); in this case, it is 0.387 or 38.7 percent. All units shown are millions of kilocalories per hour.

THE NUCLEAR STEAM PLANT

The nuclear power plant is a special kind of steam engine, in which heat is transferred to steam that drives a steam turbine attached to an electric generator (see figure 2.8). The disintegration of uranium 235 and plutonium contained in the fuel elements in the core of the nuclear reactor produces high-velocity particles with a great deal of kinetic energy. In colliding with one another and with the structural elements of the core, these energetic particles convert a large part of their kinetic energy into heat. The heat is transferred to a fluid coolant that circulates through the reactor core.

The coolant in the boiling-water reactor is water. The addition of heat produces steam and, as in a conventional power plant, the steam is used to power one or more steam turbines, the spent steam being condensed and returned as hot water to the reactor core.

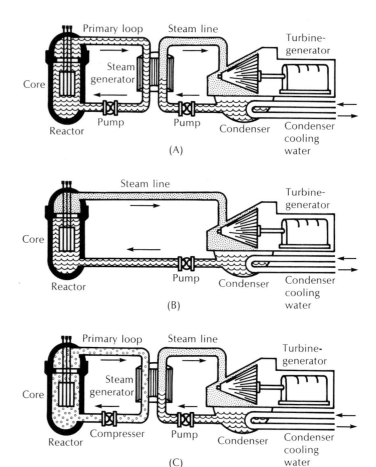

FIGURE 2.8
Three types of fission-reactor power plants are represented in these diagrams. At the top (A) is a pressurized-water reactor, in which the heat generated by fission of uranium 235 in the reactor core is transmitted to water circulating under pressure in the primary loop; heat from this water is transferred to to water in a separate loop, turning it into steam that drives a turbine-generator set. The boiling-water reactor (B) eliminates a heat transfer step at the possible cost of the environmental protection from radiation inherent in the pressurized water design. Gas-cooled reactors (C) also employ a closed primary loop but in this case filled with a gas that conducts heat from the core to the steam generator.
(From Cook, 1975.)

In other types of nuclear power plants, the coolant is a fluid other than common water and the heat from the coolant is transferred by means of a heat exchanger to a water-steam-turbine-generator system. The coolant may be heavy water, an organic liquid, a liquid metal, or a gas, such as air, carbon dioxide, or helium. Liquid sodium is the coolant in the breeder reactor now being developed in the United States.

The fuel can be either natural or enriched uranium. Enriched uranium contains more than the natural proportion of the fissile isotope uranium 235, or it contains the man-made isotope plutonium 239, or both. Only uranium 235, which constitutes but 0.71 percent by weight of natural uranium, is fissile or naturally fissionable; consequently, it is the only natural material that can be used to produce nuclear energy

Man and Other Engines

directly. The breeder reactor, subject of intense contemporary research and development efforts, would convert much of the abundant isotope uranium 238 (it constitutes 99.3 percent of natural uranium) as well as thorium 232 into fissile material. Uranium enrichment is an expensive process, the details of which are secret. Until recently, the United States was the sole producer of enriched uranium. A peculiarity of the nuclear power plant is the requirement that the reactor contain enough fuel to sustain a chain reaction. In the first successful controlled chain reaction, at the University of Chicago in 1942, the reactor core contained forty tons of uranium; it produced two hundred watts of power. Only a small fraction of the fuel is expended or burned in a day. To produce one thousand megawatts of power, only one kilogram of uranium 235 needs to be fissioned each day, or about 800 pounds a year. The fuel "inventory" in the reactor core, however, is very much larger, and it is forecast that the largest single factor in the cost of electricity from breeder-reactor power plants will represent interest charges on the inventory of fuel.

Nuclear reactors have a potential for high thermal efficiency because of the very high temperatures that can easily be generated in the reactor core. But the materials used must meet very exacting standards: they must be able to withstand high temperatures and intense neutron bombardment without becoming quickly so highly radioactive that they lose strength and toughness. With available materials, severe limitations are imposed on maximum temperatures, so that commercial nuclear power plants operate at substantially lower thermal efficiencies than do modern conventional (fossil-fueled) power plants. As a result there is more waste heat discharged per unit of power produced. The waste heat of a nuclear power plant is all transferred to the condenser's cooling water; in the conventional power plant, much of the thermal effluent goes up the stack from the furnace and boiler unit and into the atmosphere.

THE IMPORTANCE OF THE CONDENSER

Watt's major contribution to the steam engine was to introduce a water-cooled condenser chamber that allowed the steam to expand more and do more work before being exhausted from the engine. Modern condensers are equipped with pumps to maintain a partial vacuum. It is important to understand the full significance of the condenser, which plays a major role in the efficiency of a steam power plant. The vapor pressure in the condenser is the same as that of the steam exhausted from the turbine; this pressure controls the bottom temperature; the lower that temperature, the greater the efficiency of the turbine and the power plant.

When steam is allowed to expand, it cools and its internal energy is thereby decreased. The reduction is converted into mechanical energy in the form of an acceleration of the steam particles, thus making mechanical energy directly available. In a steam turbine, the steam flows through nozzles or jets, abruptly expands, and gains kinetic energy that is used against the blades of the turbine.

Normally condensers are operated at between 5° and 10° C above the temperature of the cooling water available. If condensation takes place at 32° C, the steam pressure will be only 0.7 pounds per square inch (psi), or 14 pounds per square inch below atmospheric pressure. At such low pressure, the volume of steam is about thirty thousand times the volume of the same amount of water, and very large condensers are needed to handle the volume of steam produced in a large power plant. The steam is condensed as it flows down over hundreds of thousands of copper (or copper-alloy) tubes through which the cooling water is circulated. The condenser is a heat exchanger, taking some 500 kcal of heat out of every kilogram of steam converted to water. Water leaving the condenser, although a great deal of heat has been extracted from it, will probably be hotter than the water of any nearby natural body such as a river or bay; if the ecological impact of the heated water is deemed unacceptable, the condenser effluent will be diverted to a cooling pond or to cooling towers.

Modern coal-fired power plants with superheaters produce steam at 540° C and discharge it at 32° C. The theoretical efficiency of such a plant is

$$(100) \frac{813 - 305}{813} = 62.48 \text{ percent}$$

but the actual efficiency usually is about 40 percent because of heat losses in the furnace and boiler and the need to use some of the system energy for heating and pumping the feedwater. Raising the bottom temperature by about two degrees would lower the ideal efficiency to

$$(100) \frac{813 - 307}{813} = 62.24 \text{ percent.}$$

If the plant has a capacity of 1,000 megawatts and operates at full load throughout the year, this seemingly insignificant difference would increase the plant's fuel requirement by 19,200 tons of coal annually, an amount that would take a train two miles long to carry. In plants that use water recycled through cooling towers instead of using water drawn from a stream or lake only once before discharging it, the condenser temperatures are usually higher. Thus the initial cost of the

Man and Other Engines

cooling towers is not the only cost of protecting the environment from the effects of hot water discharged from large power plants.

INTERNAL-COMBUSTION ENGINES

The proliferation of the small, mobile power plants that propel cars, trucks, and buses required the development of the internal-combustion engine of which there are two main varieties: the Otto, which powers most present-day automobiles, and the Diesel. Both were conceived and first built in Germany in the last quarter of the nineteenth century. The first commercially successful motor car of modern design, the Panhard, introduced in 1894, used an Otto-cycle engine developed by Gottlieb Daimler.

Although some early automobiles were powered by steam engines and others by batteries, the gasoline-fueled internal-combustion (Otto) engine won out because it had a higher power-to-weight ratio and therefore faster acceleration and more economical operation and because it lent itself to mass production and consequent lower cost much better than did the others. The speed and durability of the gasoline engine were demonstrated in the famous 1905 automobile race from Paris to Bordeaux and back, in which all three types of engine were represented and in which the winning car by six hours, as well as the next three finishers, were powered by the Daimler engine.

The efficiency of the ideal Otto cycle depends solely on the degree to which the mixture of fuel and air in the cylinders of the motor is compressed immediately before ignition. Compression in an Otto or Diesel engine commonly is measured as a ratio between the volume that the mixture of air and vaporized fuel would have at atmospheric pressure (at sea level) and the volume to which it is reduced by the pistons moving in the cylinders. The compression ratio should be as high as possible, but in the Otto engine a limit is imposed by the fact that raising the pressure raises the temperature, and the danger arises of igniting the fuel before the piston stroke is completed; this is called preignition and entails a loss of power and energy as well as damage to the engine. The Diesel engine gets around this limit by compressing the fuel and raising its temperature above the ignition point before it is injected into the cyclinder where it ignites spontaneously and immediately without need of an ignition system. The Diesel thereby attains compression ratios between 30:1 and 35:1, in contrast to the Otto engine, which uses compression ratios between 5:1 and 9:1. The efficiency of the Diesel engine therefore is higher than that of the Otto, and the running is somewhat smoother, but the engine is heavier and more costly in relation to the units of power or work produced.

The rotary engine can use either the Otto or the Diesel cycle; if it is to be small or light in relation to the unit of power produced, it must use the Otto cycle. There can be no gain in thermal efficiency, although there may be a gain in fuel economy because of a higher power-to-weight ratio.

We have seen that the operating efficiency of an electric power plant is considerably lower than the theoretical thermal efficiency of the steam turbine, mainly because of losses in the system outside the turbine and condenser. Likewise, an internal-combustion engine's thermal efficiency of between 20 and 30 percent may be reduced to 5 percent or less by losses in other parts of the machine. Unlike the losses in the power plant, the losses in the automobile are not limited to those that are unavoidable with available materials, fuels, and technology, a consideration we will return to in chapter 6.

Although internal-combustion engines can operate on a wide variety of fuels, those used today are mainly gasoline (for the Otto-cycle engines) and diesel oil (for the Diesel-cycle engines), both derived by refining crude oil. Power output ranges from a fraction of a horsepower to twenty-five hundred horsepower or more in locomotives and ships.

REFRIGERATORS AND HEAT PUMPS

The mechanical refrigerator is a heat engine working in reverse (see figure 2.9A). A gaseous refrigerant with a low boiling point, such as ammonia or Freon 12 (dichlorodifluoromethane), compressed to the liquid state and allowed to pass into a region of lower pressure, becomes chilled and draws heat out of any warmer object with which it is in contact. The heat drawn into the refrigerant causes it to evaporate; it is then recompressed to a temperature above atmospheric after which it is condensed, heat being drawn from the refrigerant into circulating cooling water.

A heat pump operates in similar fashion (see figure 2.9B), except that its objective is to *raise* the temperature of objects or of air in a building rather than to lower it. The heat pump uses a refrigerant gas, which being cooled by lowered pressure, draws heat from a river, a lake, or the earth's crust. This heat, upon the compression and condensation of the refrigerant, is transferred to circulating water at a temperature sufficiently high to heat the air in a building. Because heat drawn from the environment is essentially free and renewable, the energy-saving potential of heat pumps is high.

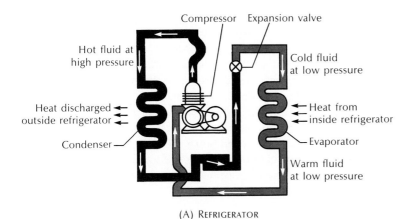

(A) REFRIGERATOR

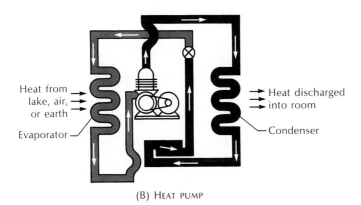

(B) HEAT PUMP

FIGURE 2.9
A refrigerator (A) and a heat pump (B) work the same way, except that a heat pump is a refrigerator working backward. In either case, the critical element is a fluid, commonly Freon, that remains a gas—or at least a vapor—at low temperatures. When such a fluid expands by lowering of pressure, it cools and is able to absorb heat from hotter surroundings. When its temperature is subsequently raised by compression, it can release heat to cooler surroundings.

Electrical or mechanical energy is required to run a heat pump and it is subject to the same entropy handicap as are all heat engines: it takes considerably more than the heat equivalent in work to extract any specified amount of heat from a body in a refrigerator or from the natural surroundings of a building. The total heat transferred, however, is the sum of the electrical or mechanical energy consumed and the heat taken from the lower temperature level and is therefore greater than the work input. The resulting thermal efficiency appears to exceed 100 percent thus violating thermodynamic law, but this is so only because we regard the environmental heat as a "free" input.

Because heat flows only from a hotter to a cooler body, a substance with a boiling point much lower than that of water must be used to transfer heat in a heat pump. Liquid ammonia, which has a boiling point of $-33.4°$ C, when enclosed in a pipe in a lake will start to boil or vaporize and in so doing will draw heat from the surrounding water. The ammonia gas then has its temperature raised by mechanical compression, whereupon some of its heat can be transferred to hot-water heaters in a building. What makes it better than an electric heater is that more useful heat can be extracted from a given amount of electrical energy, because not only can the work done in compressing the ammonia gas (and raising its temperature) be recovered but also the heat drawn from outside. The same device that acts as a heat pump in winter can become an air conditioner in the summer. Heat pumps using electric motors can supply three or four times as much heat to a room as can electric-resistance heaters consuming the same amount of electrical energy.

GAS TURBINES AND JET ENGINES

The transition from the steam turbine to a turbine impelled by exploding gas seems a simple step, but it took quite a while before problems of efficiency and structural materials were solved. During World War II, the gas turbine was developed into a practical power plant for fighter and bomber aircraft and the first operational aircraft powered by jet engines were put into the air.

The transition from turboprop to jet propulsion was rather simple (see figure 2.10). In the turboprop engine, gas formed in a combustion chamber (*B*) by burning a liquid fuel (kerosine) is allowed to expand to atmospheric pressure in a turbine (*C*), whose power is used to operate

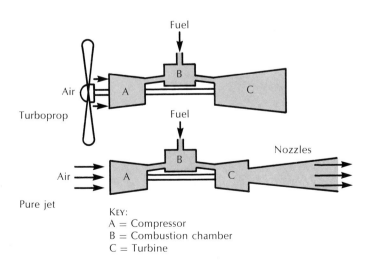

FIGURE 2.10
In the aircraft gas turbine, expanding hot gas from a combustion chamber (B) passes through a turbine (C) that turns a multistage air compressor (A). In turboprop aircraft, a shaft extends through the compressor to which a propellor is attached. In pure jet aircraft, hot gases leaving the turbine provide the kinetic energy for propulsion.

KEY:
A = Compressor
B = Combustion chamber
C = Turbine

Man and Other Engines

a compressor (A) that supplies compressed air to the combustion chamber and to drive a propeller. In the jet engine, the burnt gas expands sufficiently in a comparatively smaller turbine (C) to drive a compressor (A), but the remaining pressure drop and expansion takes place in a nozzle, producing a high-speed jet, which, by reaction, thrusts forward the engine and the aircraft to which it is attached. Jet engines are not as efficient as internal-combustion engines, but they are fast, quiet, durable, and do not vibrate.

On the ground, the gas turbine, because of its lower efficiency, still is not a strong competitor either of the large steam power plant or the small internal-combustion engine, although many improvements have been made and the fact that a turbine can accept any gaseous, liquid, or solid fuel, including coal, could become a major advantage in an era of costly substitute hydrocarbon fluids made from coal, oil shale, or tar sands.

THE FUTURE OF COMBUSTION ENGINES

Inanimate combustion engines (steam turbines, internal-combustion, diesel, and jet engines) are the world's main converters of chemical energy into mechanical and electrical power. They are the prime movers of a great revolution in the human condition that has brought the hope of abundance and material comfort to all mankind. But also they are the great degraders of energy, the genies of entropy, making high-grade energy into low-grade heat at a great pace and with an inefficiency that buys present affluence and comfort for some at the possible expense of a cold and powerless future for all. Prudence, not a cardinal virtue in the canon of growth, would seem to call for efforts to replace some of the present combustion engines by more efficient converters of stored chemical or nuclear energy, by converters that can use renewable stocks and can harness two enormous natural flows of energy, solar and geothermal. Fuel cells, for example, do not have to pay the heavy tribute exacted by the second law of thermodynamics and, despite the fact that experimental fuel cells have been disappointingly inefficient, may someday be able to use fossil hydrocarbons more efficiently than any combustion engine can. They might also use vegetable material and organic refuse as feedstock. The breeder reactor and the fusion reactor would unlock large stores of energy we are not now able to use.

Until such new converters are developed and made economical to use, the path of prudence would require that we consider improving both the technical and the social efficiency of existing engines and systems and replacing, wherever practicable, the use of nonrenewable resources, such as natural gas or petroleum, by the use of renewable resources, such as flowing water or solar radiation. Space heating by solar energy, now technically and economically possible, would be such a substitution and could greatly reduce demands on the fossil fuels and uranium, while diminishing the adverse environmental impact of increasingly intensive use of the nonrenewable energy resources.

Before considering such questions further, we shall need to study the nature of available energy resources and the physical economy of their use.

OTHER ENGINES OF THE FUTURE

There are good reasons for concluding that the steam turbine, which for half a century was the engine of the world's large land and marine power plants, has reached its limits of size and efficiency, for it is a victim of the exponential imperative (see chapter 5). More powerful plants of the future will be based on the nuclear reactor, whose economies of scale and ultimate efficiency as a power machine have not been reached. The electric motor has not been discussed in this chapter because it is not a primary engine. The problem of expanding the use of the electric motor is not one of design, but of generating and transmitting electricity to the motor without exhausting primary energy resources, without incurring disastrous environmental consequences, and without making inanimate power so expensive (in terms of the energy consumed) that industrial society will be forced back down the energy ladder, ultimately having to replace inanimate engines with human and animal engines. The technical efficiency of complete energy systems soon will begin to replace human convenience and desire for power as a criterion of design. Researchers will concentrate on engines that hold some promise of escape from the severe efficiency limitations that handicap all combustion engines. We have one such engine now, the water turbine, which drives the electric generators in hydroelectric plants but, as we shall see in the next chapter, the primary resource of falling water, although not yet fully developed throughout the world, has an ultimate development potential much too small to meet even a large fraction of present energy needs, let alone future demand.

The gas turbine-compressor combination, developed for aircraft, appears to have considerable potential for stationary power plants; efficiencies up to 50 percent are forecast. While this in itself, if cost

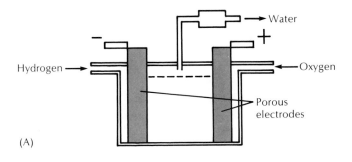

(A)

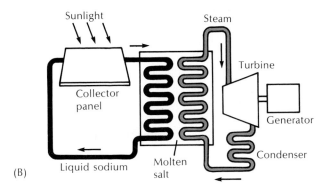

(B)

FIGURE 2.11
Two energy converters of great potential importance are the fuel cell (A) and the solar power plant (B). The fuel cell converts the energy in hydrogen or liquid fuels directly into electricity. The fuel is supplied to the anode (−) while oxygen is supplied to the cathode (+). Electrochemical reactions within the cell produce electric power when the cell is connected to a loaded circuit. In the solar power plant, sunlight falls on specially coated collectors and raises the temperature of liquid sodium to 1,000° F. A heat exchanger transfers the heat to water, turning it into steam whose energy drives a turbine-generator set.
(From Claude Summers, "The Conversion of Energy." Copyright © 1971 by Scientific American, Inc. All rights reserved.)

and performance were similar, would be sufficient to cause gas turbine-compressors to replace some conventional steam turbines, the possibility of solving the sulfur-oxide emission problem (by using clean gas made from coal as the fuel) internally adds great incentive to further development of the large stationary gas-turbine powerplant.

The fuel cell (see figure 2.11A), theoretically, should be able to achieve a conversion efficiency of almost 100 percent. Because it can operate on renewable resources, such as decaying refuse or vegetable matter, it would have very little adverse effect on the environment. The fuel cell is a device, much like a battery, that converts the chemical energy of fuel directly to electrical energy through electrochemical reactions that produce an electric current. Unlike a battery, in which the electrolyte changes composition and the electrodes are consumed, the fuel cell does not need to be recharged or replaced; it can operate as long as fuel is supplied. The only practical fuel cell yet developed uses hydrogen and oxygen to produce water and electric current, both of which it has supplied to Gemini and Apollo spacecraft. But so far fuel cells have been disappointing, in that their power output per pound of weight (important in vehicular propulsion) is low and their overall capital costs and conversion efficiencies do not allow them to compete with thermal power plants.

The very large engine of the next hundred years seems almost certain to be the nuclear breeder power reactor, or the nuclear catalytic burner reactor as some nuclear engineers prefer to call it. Breeders have been feasible technologically for about twenty years; the problem has been, and still is, to design a breeder that is efficient in the sense of using or "burning" most of the nuclear energy potential in its fuel and at the same time is "reasonably" safe to operate within the biosphere. These factors do not seem to be compatible: an efficient breeder is not very safe, and a reasonably safe breeder is not very efficient. Efficiency is the main factor in breeder economy; consequently, as the perceived need for the commercial introduction of the breeder grows, a compromise with safety may be suggested in order to rescue the potential of the nuclear age: the supply of abundant electricity for the industrialized nations.

Because controlled fusion has not been achieved and because there appears to be no reasonable way at present to calculate the economics of a fusion power plant, the fusion power reactor cannot now be regarded as a probable engine of the future, although it certainly is a potential one.

No engine, including man, can function without a supply of potential energy to be converted by that engine into mechanical work or its electrical equivalent. The availability of primary energy resources has always been the principal limiting factor of the ultimate, aggregate development of engines and power. Availability is a product of the existence of natural flows and stores of energy and man's technological ability to use them.

SUPPLEMENTAL READING

Ayres, R. U., and R. P. McKenna. 1972. *Alternatives to the Internal Combustion Engine.* Baltimore: Johns Hopkins University Press.

A comprehensive study of the internal-combustion engine and its alternatives.

Azimov, Isaac. 1962. *Life and Energy.* New York: Doubleday.

Lucid exposition of the role of energy in the growth and activities of man.

Cardwell, D. S. L. 1971. *From Watt to Clausius.* Ithaca, N.Y.: Cornell University Press.

Fine, explanatory description of the development of early heat engines and the evolution of thermodynamics in the early industrial age.

Carnot, S. N. L. 1943 [1824]. *Reflections on the Motive Power of Heat and on the Machines fitted to Develop This Power* (translated by R. H. Thurston). New York: American Society of Mechanical Engineers.

The marvelous hundred pages in which were laid down the principles of thermodynamics.

Glasstone, Samuel. 1967. *Sourcebook on Atomic Energy,* 3rd ed. New York: Van Nostrand Reinhold.

This standard reference contains readable explanations of the elements of atomic physics, reactor design and operation.

Hottel, H. C., and J. B. Howard. 1971. *New Energy Technology.* Cambridge, Mass.: MIT Press.

Good overview of research and development of alternate systems and converters.

Kleiber, Max. 1961. *The Fire of Life.* New York: John Wiley and Sons.

Included is an explanation of how the efficiency of men and horses as machines can be defined and measured.

Lehninger, A. L. 1965. *Bioenergetics.* New York: Benjamin.

Thorough treatment of energy flow through living things; good discussion of entropy and the concept of free energy.

Mott-Smith, Morton. 1934. *The Story of Energy.* New York: Appleton-Century.

Excellent summary of the development of energy theory and application.

Thirring, Hans, 1968. *Energy for Man.* New York: Greenwood Press.

Describes sources of energy in some detail, with estimates of potentials for supply; good review of engines used in power production.

Usher, Abbott Payson. 1954. *A History of Mechanical Inventions.* Cambridge, Mass.: Harvard University Press.

Well-illustrated and readable, this book not only describes in detail early machines, instruments, and engines, but also places them in social context.

Chapter 3

RENEWABLE ENERGY RESOURCES

The industrial revolution was made possible by the plowing-up of the great non-tropical grasslands of the world.

—Carl O. Sauer, 1956

RENEWABLE AND NONRENEWABLE RESOURCES

The terms *renewable* (income) and *nonrenewable* (fund) used to describe energy resources reflect man's time perception. A resource is renewable if it can be grown or replenished at a rate meaningful to man. Renewable energy resources include food, solar radiation, fuel wood, water power and wind power. A resource is nonrenewable if its rate of formation is so slow as to be meaningless in terms of the human lifespan. Coal, petroleum, and natural gas are formed over periods of hundreds of thousands of years and, although undoubtedly being formed today in special geologic environments, these resources are for all practical purposes not renewable. Natural concentrations of radioactive minerals such as uranium and thorium also are not renewable. Man has succeeded in creating fissile isotopes such as plutonium 239, which are energy resources in themselves, but he must start with natural materials and has as yet been unable to recreate those natural materials.

The distinction between the kinds of limitations of renewable and nonrenewable resources is vital. While there is no theoretical limit on the total quantity of energy man can get from a renewable resource, there is a limit on the *rate* at which he can draw energy from it. For examples, the available head (vertical distance down which water can fall) and discharge limit the rate at which energy can be drawn from a hydroelectric installation; the available arable land and its quality, the available technology of agricultural production and food processing, and the availability of labor and capital limit the rate at which food energy from the land can be secured; the capital and material required for construction, collection, storage, and distribution facilities, as well as the conversion efficiency of available technology, limit the rate at which solar, wind, and geothermal energy can be supplied as power.

There is no theoretical limit on the *rate* at which a nonrenewable resource may be used. Man chooses these rates according to his capacity to exploit them and his needs and desires. The limit on non-renewable resources is one of *quantity:* the total amount of a resource that can ever be put to economical use. Economical use is defined as the ability to get more work out of a resource or in exchange for it than is required to remove the resource from its natural environment, to transport it, and to process it into usable form. Note that the material itself may still exist in large amounts when it becomes no longer a resource. The economic limit to exploitation of a nonrenewable resource further implies a time limit, determined by the rate at which the resource is consumed.

Technology can and has greatly extended the limits of both renewable resources, but such extensions are temporary and tend to benefit only some of the earth's people. Most resource technology has been developed to increase the rates of production and consumption of resources, not to conserve them or even to produce a quasi equilibrium between man and his resources. Technology oriented to production and population growth persistently tends to destroy any incipient equilibrium.

It seems ironic that the only resources man has managed to exhaust completely are *renewable:* the bison, the blue whale (not extinct, but exhausted as a resource), and the passenger pigeon. The resources that seem in least danger of exhaustion belong in the *nonrenewable* category; they include aluminum, iron, and glass sand.

These paradoxes are explained by the energy relations involved. Energy drawn from living systems, although theoretically without any limit in time or total quantity, is subject to the vagaries of complex interactions of the multiple elements of the system as well as to the quality of man's management. Energy drawn directly from solar radiation or from the hydrologic cycle requires much less finesse in management and is therefore much more dependable. The resources already exhausted or in danger of exhaustion are those of hunted animals, whose population growth rates are limited by the wild food available in specialized environments, and who can hold on to their ecologic niches only if they maintain a certain minimal population against competitors, predators, and parasites. Such wild animals store energy only to a very limited extent, mainly in their own bodies. Abrupt changes in the availability of food, perhaps caused by man, can decimate their numbers as effectively as overharvesting. An ecologic system that includes wild food animals, although it has a certain energy storage capacity in the living tissue of the several layers of organisms that are included in the system, stores that energy for the benefit of the system, not for the benefit of an endangered species within the system; indeed, the species under attack by man will face increased problems of defense against its other enemies.

Renewable Energy Resources

In striking contrast to the dynamic, delicately adjusted energy system of the biosphere is the almost static energy system of rocks and minerals, where geochemical concentrations effected over millions of years await man's exploitation. The opening of large mines does change the energy system in a small area of the outermost part of the earth's crust, but generally little energy is transferred and the mineral resource is not destroyed.

THE EARTH'S ENERGY BALANCE

As we noted in chapter 1, energy enters the global ecosystem as solar radiation, as heat flowing from within the earth, and as tidal energy (see figure 3.1). By far the greatest of these three is solar radiation, amounting to about 7,700 times the other two combined, and to about 400,000 times the total installed electricity generation capacity in the United States in 1975.

About 30 percent of the solar radiation intercepted by the earth is directly reflected back into space from the clouds, water, and land surfaces of the earth; about 47 percent is absorbed by the atmosphere, and by the water and land bodies of the earth, becoming heat and

FIGURE 3.1
Flow of energy to and from the earth is depicted by means of bands and lines that suggest by their width the contribution of each to the earth's energy budget. The principal inputs are solar radiation, tidal energy, and the energy from nuclear, thermal, and gravitational sources. More than 99 percent of the input is solar radiation. The apportionment of incoming solar radiation is indicated by the horizontal bands beginning with "Direct reflection" and reading downward. The smallest portion goes to photosynthesis. Dead plants and animals buried in the earth give rise to fossil fuels, containing stored solar energy from millions of years past.
(Adapted from Hubbert, 1962.)

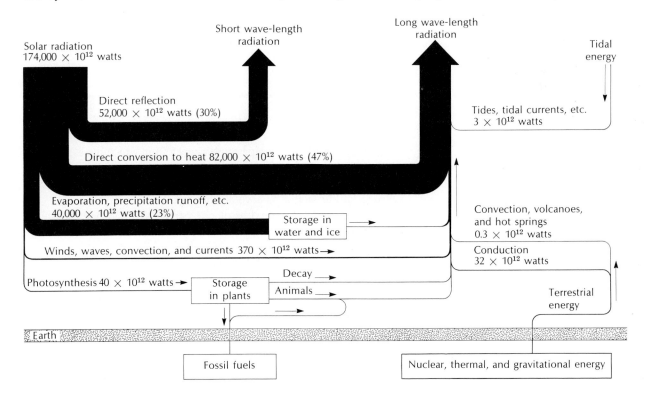

providing the ambient temperature range within which life as we know
it can exist; most of the remaining 23 percent powers the movements
of water and air on the earth (the hydrologic cycle), without which
terrestrial life would be severely limited.

About two one-hundredths of one percent (0.0225 percent) of the
incoming solar radiation is absorbed by the leaves of plants and used
in photosynthesis, which converts carbon dioxide and water into
organic carbohydrates and oxygen; this small portion of the solar
radiation intercepted by the earth provides the energy for the basic
biological requirements of all the earth's plants and animals.

Significant natural variations in the earth's energy balance are recorded in repeated episodes of glaciation in the higher latitudes as well
as in other worldwide climatic changes. These natural variations,
however, do not seem to have been large enough ever to have threatened the existence of life on the planet, although they have contributed

to changes in habitat that have been the downfall of many over-specialized living species.

Man has been making some contribution to the global energy system since his acquisition of fire, but a significant contribution only since the intensive use of fossil fuels began, a little over a century ago. Although for 99.9 percent of his existence as a genus, man used only renewable resources for his energy needs, now only about 10 percent of the world's energy consumption comes from renewable sources; about two-thirds of that is food for humans and feed for animals; about 25 percent is wood, refuse of vegetable origin, and excrement used as fuel; the remainder is hydroelectric, geothermal, tidal, and wind power.

SOLAR ENERGY

Although one could maintain that most forms of energy used by man (except nuclear, tidal, and geothermal) are of solar origin, the term *solar energy* usually refers only to useful heat and power obtained directly from the sun's radiation.

Solar radiation can be controlled by man for two purposes: to produce useful heat and to produce electric power. Controlled solar radiation has long been used to heat water and to provide warmth for sprouting seeds and raising flowers and certain vegetables; also it has been collected to heat houses and other buildings, for process-heat applications on a small scale, such as cooking, for heating water, and for refrigeration (see figure 3.2A). Harrison Brown once estimated that more than 20 percent of mankind's energy needs could be provided by solar heating in what he called "a properly designed world." Other than social inertia, there appear to be no serious barriers to much greater use of solar heating and cooling. Buildings need to be specially sited, designed, and constructed, and the solar-energy units need to be standardized and mass-produced to be economic alternatives to present space heating and cooling systems.

Solar power is quite a different matter. It is technically feasible (see figure 3.2B) to convert solar radiation to electricity, but existing technology is expensive and inefficient, and solar electricity is at present feasible only in applications, such as space satellites, where a high power cost is relatively unimportant. Solar-power enthusiasts are optimistic in their predictions of lowered costs and in their descriptions of the vast potential for solar power; they gain eager followers because of the appeal of a pollution-free, completely renewable source of electricity in face of the mounting problems of pollution, risk, and ecological disturbance that accompany our increased reliance on fossil and fissile fuels. The practical potential of solar power is impossible to calculate at present. With existing materials and technology, it has

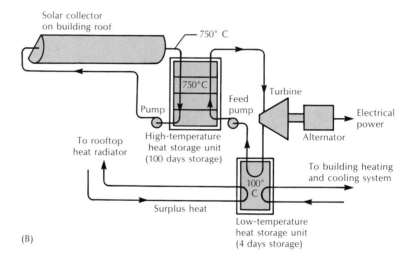

FIGURE 3.2
*(A) The simplest solar energy systems are those designed for domestic space and hot water heating. In such systems, an inclined, southerly-facing flat collector is typically coupled with a heat storage system from which heat can be drawn when required. This diagram shows a combination system that provides domestic hot water at 65°C. and hot water, as well, for space heating. (B) This solar-powered total-energy system is projected for use in suburban shopping centers or industrial plants where space for solar collection is available. One-axis-steerable concentrators would be situated on the roof; high-pressure dry nitrogen, steam, or sodium chloride would transfer heat at 750°C. to a high-temperature storage unit of insulated rocks or molten salt. Heat drawn from the storage unit would drive a turbine to produce electrical power and then, held in a low-temperature storage unit, would be available for space heating or air conditioning.
(From Morrow, 1973.)*

no large-scale potential. Substantial improvements in conversion efficiencies, and reductions in the amount of work required to construct the necessary materials and devices for collection, storage, and conversion, and in the amount of material (metals, glass, and other nonmetallics) needed per unit of power produced will have to be achieved before solar power will be as useful as its present backers claim that it will be.

FOOD AND FEED

Food (for man) and feed (for his animals) are basic energy resources. For a very long time, man hunted and gathered all his food. Only about ten thousand years ago did he start domesticating plants and animals. Agriculture increased the supportable population density and led to the development of towns and cities.

Different kinds of agriculture developed in different environments. The advantages of irrigating low land in arid and semiarid regions were perceived by some communities early in the history of agriculture and the so-called hydraulic civilizations grew up on the basis of irrigated agriculture. Although it would appear that irrigated agriculture could ensure a permanent equilibrium between man and the land, in practice it never has. Upper-basin deforestation with consequent siltation of water works, salt accumulation in the irrigated soils, conquest by invaders, or some combination of these has put an end to every hydraulic society man has yet erected.

Where vegetables could be grown readily and the ground could be tilled with rude implements, swidden agriculture developed. The forest was cleared and burned; crops were raised on the burned ground for one to three years, and then the village cleared a new area, moving in a circle that took between twenty and thirty years to complete. Properly done, swidden agriculture creates a closed ecological cycle within which man is in equilibrium with other species. Neither soil nor forest is permanently depleted, and the land retains forever its capacity to sustain the humans who draw their sustenance from it. This type of agriculture does tend to produce a protein deficiency if only vegetables are raised; therefore pigs are commonly domesticated in swidden communities, where they live off forage and vegetable waste and provide the protein insurance that all humans need.

In harsher climates where forests are meager if not absent and crops are not raised easily, more reliance for milk and meat is placed on grazing and browsing animals such as the goat, the llama, the water buffalo, the sheep, and the cow. To man a great advantage of these animals lies in their ability to gather and digest vegetable material that he cannot digest and to turn it into material that *is* digestible. They convert feed to food, making available to man more chemical energy from living tissue than he would otherwise have. This system of agriculture is more difficult to manage on one site than swidden agriculture. Soils and their vegetative cover are thin. Grazing and browsing animals, among which the goat is notorious, will, if not carefully managed, deplete the vegetative cover, and careless land cultivation may deplete the soil of nutrients; both cause soil erosion. Agricultural man has, under these circumstances, been a miner, moving on to new and unploughed lands when the old ones were exhausted.

A type of agriculture that has an element of the swidden as well as of the mining is that of the nomads, who rely on domesticated animals to supply milk, meat, transport, and hides for clothing and shelter and who move from place to place as their animals deplete the plant food available but who may return in a slow cycle to places they have used before, providing other nomads have not reached such places in the meantime and interrupted their cycle. Nomads are painfully susceptible to drought and animal diseases. There are a number of examples in history of nomads successfully invading societies based on sedentary agriculture and adopting the way of life of the conquered people, but no examples of sedentary populations choosing the nomadic life.

Agriculture carried out with the use of heavy, team-drawn, sod-turning plows already has been mentioned as giving rise to the industrial revolution and out of which emerged individual (as opposed to state) capitalism. In its early phase, heavy-plow agriculture, practiced in Germany on deep fertile soils not very susceptible to erosion and involving fertilization by the manure of the draft and other farm animals, must have been as ecologically harmonious as good swidden agriculture. But, whereas swidden agriculture appears to have natural limits of scale and its surplus cannot be readily stored, heavy-plow agriculture had no such natural limits and its surpluses could be easily stored or exchanged for goods that could be stored.

Although the hunting and gathering of wild land animals and plants now provides a negligible amount of the world's food supply, the hunting of marine animals is important and provides much of the protein for some countries such as Norway, Japan, Iceland, and Portugal. At one time whales were important as food, but today fish and shellfish are more important. (About one-third of the total edible marine food used in the United States of America is consumed by pets.) Fish is an excellent source of protein as well as of lipids. If a fishing society can obtain its essential vitamins and its carbohydrates from certain crops and perhaps from goats or cows, it has the possibility of achieving permanent equilibrium with its food supply.

In modern high energy societies, 20 percent or less of the labor force raises and processes food for the other 80 percent and for the dependents of all. That 20 percent of the labor force includes those engaged in the manufacture of farm machinery, chemical fertilizers, pesticides, and herbicides, as well as those who butcher, cure, can, freeze, or otherwise process agricultural products, and those engaged in agricultural research, education, and advertising. In a low energy society, these proportions tend to be reversed; 80 percent or more of the labor force works directly on the land or at sea to produce food for the entire society; very few persons are engaged in food processing, and even fewer in agricultural research, education, and advertising.

The great difference between the food consumption of high and low energy societies is not only in the total number of calories available to each person but also in the amount of protein available. High energy societies in general have more protein than they need and most of it is consumed as meat. Some medium energy societies obtain enough protein from fish and milk. Most low energy societies have diets deficient in protein, especially if they are in contact with high energy societies, which are able to buy protein foods raised or caught in the low energy societies. (There are some minor exceptions to this generalization: Eskimo communities, for example, have a high food calorie consumption because their bodies require more heat in their cold environment.) The sugars, starches, fats, and alcohols in the human diet are the main fuels of the body; protein is partly fuel, but is mainly the material vital for growth and replacement of tissue.

For life in a modern society, it has been said, an average daily consumption below 2,400 kcal will indicate that the population is suffering from the debilities of undernourishment, whereas a level above 3,000 indicates debilities caused or accentuated by overeating. In 1970, eighteen nations had daily per capita net food-energy supplies over 3,000 kcal while seventy-six fell below 2,400 kcal. Both undernourishment and overeating constitute malnutrition, but deficiencies of protein, amino acid and trace elements may also constitute malnutrition even where the total energy consumption is adequate.

Low energy societies tend to derive most of their food energy directly from plants; high energy societies divert more of their primary plant food to the raising of food animals, and the amount of meat and animal products in the individual diet may represent, in plant feed, five to ten times the amount of plant food eaten directly.

WOOD AND OTHER VEGETABLE FUELS

The human body can convert the chemical energy of only some plants into useful heat and work. The energy of some other plants can be converted by other animals into meat and milk and thus, indirectly, into food for man. Still other plants produce edible fruit or flowers and worms and insects may thrive on the main part of the plants, which cannot be digested by any large animal. A great deal of chemical energy stored in plants by photosynthesis is not available to man as food, either directly or indirectly. However, he may make use of such energy as fuel instead of food, by burning the vegetable matter—the wood, corn cobs, or dried vegetable refuse; by making alcohol or methane from it and burning those compounds; or by using the vegetable material or chemicals derived from it in a fuel cell, to obtain electrical instead of heat energy.

Wood for fuel was almost certainly the first source of energy, apart from food, used by man. Except in treeless regions, the vegetable refuse represented by the droppings of animals would have cost too much work to collect for use as fuel until the animals were domesticated and the droppings concentrated.

Where it was available, fuel wood was an important source of energy even to industrialized society until about 100 years ago. Before 1850, most space and process heating in the world was done by wood. In western Europe the use of wood for shipbuilding and for making charcoal, as well as for space heating and for burning heretics, had seriously depleted the forests by the time coal began to be used for making coke and as fuel; consequently, coal rapidly supplanted wood in the early years of the Industrial Revolution. In North America, where abundant forests had hardly felt the axe when Watt was patenting his steam engine, wood was widely available throughout the nineteenth and well into the twentieth centuries. Until about 1875 most locomotives in North America ran on wood and most space heating was done with wood.

Wood has some serious disadvantages when compared with coal. First is its lower heat content per unit of weight or volume. Second is the relatively low efficiency obtainable when wood is used in space heating or heat engines; coal gives a more concentrated, high-temperature heat. Third is the fact that the amount of human work required to cut and transport wood has remained about the same over the years while the amount of human labor expended per ton of coal mined in North America has kept dropping. As it became easier and cheaper to transport coal in the latter decades of the nineteenth century and early years of the twentieth, wood was rapidly replaced.

In high energy societies there is relatively little demand for fuel wood today. The pressure on the forests comes from the demand for lumber for construction and for woodpulp for making paper. In low energy societies the demand for anything—wood, dung, chaff, or other refuse—that will yield heat for cooking and warmth continues. Some of these societies, those that live more or less in equilibrium with a tropical or semitropical environment, have abundant wood for fuel. In Brazil, for example, it was estimated in 1952 that 85 percent of all energy used came from fuel wood. By far the largest number of people in low energy societies live in areas from which the forest cover has been stripped, for fuel and to clear land for cultivation. China, for instance, is said to have been half forest 200 years ago; today less than 10 percent of that great country is forested.

NIGHT SOIL AND OTHER ORGANIC WASTES

In China, India, and other countries whose forest resources are meager, animal excrement and vegetable waste from many sources is gathered to serve as fuel, as fertilizer, and as feed for pigs, ducks, and fish. Perhaps 40 percent of the gross energy consumption in India is supplied by such material and is not included in statistics on energy as gathered and reported to international agencies.

In high energy societies much vegetable waste is discarded. Only recently in the United States of America have corncobs been collected to be used as fuel. Considerable quantities of vegetable waste, such as some of that from potato, sugarbeet, and orange-juice processing plants goes to cattle feedlots, but the waste that comes out of the feedlots, some of which can hold one hundred thousand animals, is still regarded mainly as a disposal and pollution problem rather than as a source of energy or fertilizer.

Experiments have shown that energy can be recovered by burning municipal garbage, but it will not be economic to do so until the inorganic components can be separated and sold for recycling, and even then will require some subvention.

In some societies animal fats, apart from their use as food, have been sources of energy, mainly for light but also for heat. Candles made from animal fats are perhaps the oldest form of lighting that is still used. Whale oil was used in the industrializing countries for lighting during the nineteenth and part of the twentieth centuries, until "coal oil" (kerosine made from coal) displaced it. Fish oils are still used by the Eskimos for space heating. Apart from the animal fats collected in wartime for use in explosives, high energy societies waste animal energy in the form of discarded fat, bone, and blood.

ANIMAL AND HUMAN POWER

Nobody knows whether slaves or draft animals were first used by man as sources of power. The domesticated ass, today often called donkey, was being used in Mesopotamia and Egypt about 4000 B.C. Oxen were used to draw traction plows from about 3000 B.C. Horses probably were not domesticated until about 2500 B.C., by which time the nomads of central Asia were using them. The invention of the wheel led to the use of animals for pulling carts and wagons and for turning mills.

The origins of slavery are buried in antiquity. One school of historical thought holds that human slavery was unknown to hunting societies and that it was brought about when agriculture created a need for routine labor. But there is good evidence for slavery in some hunting societies, including those of the North American Indians. What probably happened was that agriculture greatly increased the usefulness of slaves and changed the climax of battle from butchery to capture. For many centuries agriculture was a major reason for slave raids and slave trading.

The early Egyptians used slaves for their great engineering works. Pyramid stones and obelisks were drawn over greased beams or rollers by thousands of slaves. One relief picture shows eight hundred rowers in twenty-seven boats pulling a barge laden with two obelisks. It is recorded that Rameses III presented 113,000 slaves to the temples during his reign.

Although horses and donkeys are more powerful than men and can deliver more work in a day, for about four thousand years men seem to have been preferred for agricultural work, for pulling heavy loads, and even for tasks in mills of various kinds where animal power could have been used. Most of these draft men were slaves. Slaves were impressed as soldiers, they pulled the wondrous battering rams and wheeled assault towers of the Assyrians and Greeks, they rowed the biremes of the Phoenicians, the triremes of the Romans, and the galleys of many lands, they labored in mines, worked in the fields, toiled in flour mills and at metal-working forges. A large river of slavery flowed under the civilizations of the Sumerians in Mesopotamia, of the Egyptian pharaohs, and of the Greeks of the Golden Age.

Where men were not used, the ox was the primary draft animal, pulling his burdens by means of clumsy yokes which strained his neck; the horse was used for riding, especially in battle; the donkey and the camel were the beasts of burden. Not until the need for large teams of

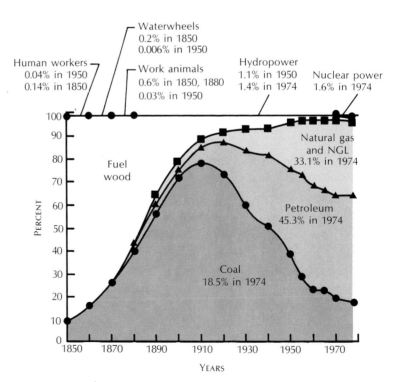

FIGURE 3.3
Energy input shares to the economy of the United States, 1850–1974.

oxen to pull the heavy plows of northwest Europe inspired the development of the shoulder-bearing harness in the early Middle Ages, did the horse become a premier draft animal. During the same period, the rapid proliferation of the water mill released both donkeys and men from work in the mill.

In the United States of America in the nineteenth century (see figure 3.3), horses did most of the work, especially in the first half of the century: in 1850, work animals (mainly horses) produced 53 percent of the work from all sources of energy, 13 percent came from human workers and an almost equal amount from the wind, 9 percent was derived from waterpower, 6 percent from fuel wood, and 6 percent from coal. In 1950, animals produced only 1 percent of the work output and human labor about the same. The fossil fuels had taken over, with petroleum and coal each accounting for 34 and natural gas 21 percent, a total of 91 percent. In 1973, the fossil fuel provided 95 percent.

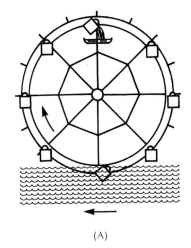

(A)

WATER POWER

Although the lifting of water was important in the early hydraulic civilizations, the power of flowing water to lift itself does not appear to have been harnessed until the invention of the noria (see figure 3.4A) shortly before the start of the Christian era. The noria is a waterwheel with buckets or paddles attached to raise water and it can be operated as an undershot waterwheel. Mechanical work other than lifting water could not be accomplished by the noria but depended on the development of the geared mill, which appeared in the first century before Christ. The Vitruvian mill (see figure 3.4B), as this new invention came to be called (after the Roman who first described it), consisted of horizontal millstones turned by a vertical shaft geared to a horizontal shaft turned by an undershot waterwheel and is the first known machine designed to transmit power through gearing. The design of a geared mill could be adapted readily to the stream that was to power it.

Historians have noted that established millers, using slave or animal power, tended to resist the introduction of water mills. They were reluctant to move these mills to sites where water power was available, to make the considerable investment in new facilities, to have to depend upon an aqueduct not under their control, and to sell the surplus slaves or other animals and the old mill. This pattern of resistance to new technology has been repeated many times in history.

The increased susceptibility of an industrial society to sabotage and acts of war because of its dependence on energy systems was illustrated very early in the history of Western technology. Barbarians laying siege to Rome in 536, thinking to starve out the city more quickly,

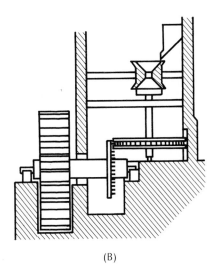

(B)

FIGURE 3.4
Early water-powered machines: (A) the noria used the power of a stream to lift water for irrigation; (B) the Vitruvian or Roman water mill was driven by an undershot waterwheel connected by gearing to a vertical shaft to which a grain mill was affixed.
(A. from Ewbank, 1849; B. from Beck, 1900.)

destroyed the aqueduct bringing water to the grain mills of the city. The Roman defenders quickly mounted undershot waterwheels and mills on barges anchored in the Tiber and in this way continued to supply flour for the city's bread.

Before and during the Middle Ages, the geared watermill spread from Rome throughout western Europe. By the end of the sixth century it was in general use in what are now France, Switzerland, and southern Germany. In the late Middle Ages it was used not only for grinding grain, but also for sawing wood and marble, for crushing wood, tanbark, and metallic ores, for fulling cloth, and for turning grindstones. The scarcity of slaves in northwest Europe and the value of draft animals for other uses may have accelerated its acceptance. Where water was lacking or not conveniently available, the geared mill could also be powered, at least for light uses, by men or animals. Water mills formed the power base for substantial industrial development before what historians call the Industrial Revolution. The network of streams and saltwater embayments of western Europe provided an almost ideal complement to the fertile fields, forests, and mineral deposits of the region. Agricultural, forest, and mineral products were milled at the streams and transported by barge and boat to places of use or of further processing. As marine technology and maritime prowess developed, the surplus became the basis for international trade. Advances in technology, the evolution of the merchant-burgher class, the growth of military power, the increasing wealth of the church, and the aggressive exploration and acquisition of distant lands and resources were sustained for centuries by power technologies based on horses, water, and wind as prime movers.

The economics of processing favored sites on streams rather than away from them. The economics of transport favored rivers, canals, and the sea rather than overland roads. Consequently, industry and commerce grew mainly on the banks of rivers and at places on or near the seacoast that provided good anchorage for sailing vessels. By 1700 a map of the population density of Europe would have outlined its waterways very nicely. Another advantage of situating industry and people on water bodies, one that was to grow important in modern times, was the great, and free, capacity of large streams and the sea to dispose of waste.

There were limits to the indefinite expansion of water power based on the geared mill. The power produced by such a mill could not be transported away from the mill; it had to be used there. To be of maximum economic value, the mill needed to be fairly large, process only one kind of material, and have adequate water transportation available. Sites for such mills, although numerous in western Europe, were far from infinite in number. When all the good sites were occupied, as they were by 1700, industrial expansion was limited, despite the existence of unexploited resources and an increasing demand for products. Not until electricity could be generated practically from falling water could this barrier to water power be broken, and by then most of the mechanical work of the industrializing world had been taken over by steam.

The world's first hydroelectric power plant was built at Godalming, England, in 1881. For seventy years thereafter, industrialized countries with good damsites, particularly those in regions where fossil fuels were expensive, developed their hydroelectric power rapidly. Even nations that had fossil fuels found hydroelectric power attractive. It did away with the necessity of mining coal, never regarded as glamorous or wholesome, it was clean from dam to kitchen, it was indefinitely renewable, and it provided power cheaply for the poor and the rural population. A hydroelectric cult grew up in the United States and the Soviet Union. In both countries, the development and dissemination of electricity derived from water became important politically. Lenin declared that Soviet power would be based on the collective state and hydroelectricity. In the United States, the proponents of publicly owned power battled the proponents of privately owned power in the political struggles over regional economic development that resulted in adventures in pragmatic democracy such as the Tennessee Valley Authority in Appalachia and the Bonneville Power Administration in the Pacific Northwest. In the 1920s and 1930s the development of hydroelectric power was regarded in both capitalist and communist worlds as a fundamental part of the wave of the future.

The development of hydroelectric power on a large scale depends on the availability of good sites. Good sites are those with high head (height of water fall), large discharge (the rate at which the river flows), large storage capacity, and nearness to load centers (where the power will be used). To a large degree the first two criteria are interdependent variables; that is, a low head can be compensated for by a large discharge and a small discharge can be compensated for by a high head. However, there are economic limits. Good sites are limited, they are unevenly distributed throughout the world, and, in general, they are not very near the major load centers. Although water power is theoretically a renewable resource, all hydroelectric power dams have finite lives ranging up to perhaps a maximum of four hundred years, because inevitably they will become silted up. This sedimentation is unfortunate, because hydroelectric power is the most efficient and the cleanest major source of power yet developed by man.

WIND POWER

The history of wind-driven boats explains much of the cultural evolution of man. We do not yet know how much culture traveled under sail before what we call the Age of Discovery, the age of Columbus,

Magellan, Balboa, Diaz, Cartier, Drake, and the other adventurers of the sea who came out of the Atlantic ports of Europe between 1490 and 1600. The Vikings certainly sailed to North America; the Chinese and others from the west may have reached both Americas well before the Vikings; and there is a possibility that one or more Mediterranean peoples crossed the Atlantic a millennium or more before the Vikings.

The Egyptians built and used sailing ships at least two thousand years before Christ was born. Chinese merchants had reached India under sail by the beginning of the Christian era. Cultural and commercial exchanges in the eastern Mediterranean during the millennium before Christ produced the "outlander" Greek philosophers who, from Anaximander through Democritus and Pythagoras to Epicurus, were the forerunners of modern scientific and political thought, in sharp contrast to the law-and-order home-fronters such as Aristotle, Socrates, and Plato. The Phoenicians in particular appear to have been responsible for the growing realization that novel and useful things could be obtained by maritime commerce from distances that made land trade impractical and to have provided the Romans with much of the geographic information they used to plan and build a colonial empire.

The Romans, although they sailed, had no genius for the sea. They seemed to have used the sea when they had to, but their faith was in roads. They used 100 galley slaves when three sails would have sufficed and the slaves could have been used elsewhere. The Romans were aggressive and adventuresome on land, but not at sea. In the latter days of the empire, they left the aggressive use of the seas entirely to the northern raiders, withdrawing to fortified ports, setting up land defenses against invasion, and childishly forbidding the export of ships' plans, as if their designs were by then even as good as those of the maritime barbarians.

While the water mill and the heavy plow were creating a new kind of energy base and a new society in western Europe, the sailing ship was the key machine of a corresponding and complementary revolution on the adjacent seas. The northern fishermen and sea rovers used both oar and sail, but sail provided the great power for seaborne raids and commerce. Development of the magnetic compass and the stern post rudder were as significant to marine-based society as the shoulder-bearing harness and the stirrup were to land-based society. Until the early tenth century, the two societies clashed as much as they cooperated. After another century of seeming indecision, they chose to cooperate, marched and sailed together in the Crusades, suffered and repented together through the Plague, drove the Moslem sword back across the Mediterranean, and came with a great burst of energy and enthusiasm into the Renaissance.

The Renaissance was a rebirth of hope for a better life and of interest in the real world, but it was more than that: it was the period in which nationalism, the idea of the temporal state as a good and efficacious social unit, began to take its modern form. Nowhere was this new idea more strikingly manifested than on the seas.

Rather surprisingly, in light of maritime history, Iberia led Europe into the age of aggressive nationalistic exploration and naval conflict. Briefly, in the late fifteenth century, Spain and Portugal were the major powers of the world's oceans. The Pope divided the non-Christian world between them, to have dominion over, to convert, and to exploit. They explored the Americas, rounded Africa to reach India, and circled the globe. These were not voyages of intellectual curiosity and only in part were they adventures in national pride. Plunder was the object and the plunder was to be returned to the royal establishments. As King Ferdinand instructed one of the Spanish conquistadors, in a letter dated July 25, 1511, "Get gold, humanely if you can, but at all hazards get gold." This was state piracy, unlike the private piracy of the northern mariners; whereas piracy was enthroned in the castles of Spain, it was enthroned in the hearts of Englishmen.

Columbus, perhaps a better salesman than a naval prospector, and not an Iberian, found neither gold nor silver, nor yet the rich spices of the Indies, but he pointed the way for some true Iberians, who followed their instructions well, ruthlessly serving God, country, and themselves (not necessarily in that order). High energy society, in which the energy machines were sailing ships and guns, overwhelmed the low energy societies that it invaded. Gold and silver flowed back from America, mainly to Spain, because the Pope had not divided the noble metal deposits of the world as equitably as he had the land and sea. The financially revived Spanish aristocracy put the new wealth into ostentation, pleasure, and the apparatus of power. The Spanish Armada came out of American mines.

In England, piratical adventures were not normally financed by the crown nor did the profits accrue to the crown. Money for ships and provisions was raised from private sources, and seamen did not have to be conscripted. Proceeds went in shares to the surviving direct participants and to those who had put up money for the venture. This also was the pattern for many of the commercial expeditions by which the English probed the defenses of a low energy society. Armed English pirates mainly attacked other high energy societies and thus played a role unknown in the Latin kingdoms; they often acted as "irregular" auxiliaries of the more formal British naval power, not infrequently subsidized and legitimized as privateers when a war was on, but not regarded as outlaws even in peacetime unless, of course, they attacked English property.

The vital difference between the English and the Spanish systems for exploiting low energy societies was in what happened to the profits. In Spain they went largely into unproductive goods and services. In England, they were reinvested in the English economy: more ships were built, mills constructed, land tilled, and textiles woven. The Spanish Empire was built on gold and silver, the British on rum and wool.

The defeat of the Spanish Armada by the English weather and Sir Francis Drake was symbolic of the oceans' passing from the hegemony of Spain to that of England, but the underlying reason was that England's power in the fundamental, physical sense was growing while Spain's was not. The Spanish empire settled into a long fading sunset, while England's empire rose out of the sea with the fresh scarlet tunic of morning.

Until about 1850 the sailing ship remained queen of the sea. Then the steamship came on with a rush (see figure 3.5). Early steamships were not as economical as sailing ships; consequently for many years they carried only perishable freight, the rest being still consigned to sailing vessels. For example, steamships were used in the California gold rush to transport passengers from the East to the West coasts by way of the isthmus of Panama or through the rivers and lake of Nicaragua, but during these same years Eastern coal was being shipped around Cape Horn to California in sailing vessels. By 1900, however, sailing vessels had been almost entirely supplanted, except for local voyages in unindustrialized regions, by steamships. There is now some experimentation going on with swivel-masted sailing vessels in expectation of the time when rising fuel costs may again make sail an economic alternative.

The use of wind on land has been much more limited than at sea. Wind power on land has been used only where other sources of power are not available or are more expensive, where the winds blow with some consistency, or where the inconstancy of the winds is not a serious handicap to the use of the power it can produce.

In most land areas of the world, the winds are erratic and undependable. The use of wind to produce mechanical or electrical power depends critically on whether or not it is necessary to have the power output during any specified time period be above some minimum level, in which case, it must be stored. Windmills used to pump water for farms or homes store the energy by raising the water into an elevated tank, whence it and its potential energy can be withdrawn as needed. In windy areas, windmills have been widely used for tasks that do not require the mills to turn constantly. Grinding grain and draining polders are two examples of such tasks. For small loads in isolated windy localities where any alternative would be more expensive, windmills are used to generate electric power and the surplus is stored in batteries. For large-scale power production, however, suitable batteries are much too expensive to be able to compete against available alternatives.

(A)

(B)

(C)

FIGURE 3.5
Sailing ships were more efficient than early steamships; the latter, however, were faster and more dependable. Shown are (A) the clipper HURRICANE, one of the swiftest sailing ships, (B) the sidewheeler GREAT EASTERN, 680 feet long, the largest steamship of its age, launched in 1853, and (C) early whaling vessels.
(A and B. Courtesy of Peabody Museum. C. Courtesy of Mariners Museum.)

TIDAL POWER

Places of high tidal range and constricted tidal bore have long fascinated engineers because of the possibility of developing substantial amounts of electrical power by harnessing the tidal flows. The basic engineering design problems are fairly simple, but construction and operation of such a facility are expensive.

First, it must be understood that a tidal power plant can hardly be a primary source of electrical energy, since there will be two or four times each day when there is no tidal flow and little or no power can be generated. Whether these times are momentary or last longer depends on the design and operation of the structures that dam and release the water. Very little tidal power is produced in the world and, if all the available sites were used, only a small fraction of the world's power production could be supplied.

TERRESTRIAL HEAT

The earth has a lot of heat in it (see figure 3.6). At fairly shallow depths temperatures are high enough to be used in heat engines. The average geothermal temperature gradient is about 1° F increase for each 100 feet of depth below the earth's surface. In some deep mines, rock temperatures are between 140° F and 150° F. In some deep oil wells, rock temperatures of 250° F have been encountered. At depths of between five and six miles, well within the capability of existing drilling technology, there are rocks containing truly enormous quantities of heat. We are not talking about the geologic heat traps, the accumulations near the surface of hot water and steam that constitute the basis for existing or planned geothermal power plants; these are exhaustible resources and will be treated in the next chapter. Here we are considering the normal heat distribution within the earth's crust.

The main barrier to the economic use of geothermal heat in its normal vertical temperature distribution is the very low thermal conductivity of dry rock. The hot rocks would need to be thoroughly shattered by explosives or by other methods, after which water would be introduced under pressure and withdrawn after absorbing heat from the shattered rock. It is not yet known whether or not the rate of heat transfer would be fast enough to repay the investment in drilling, fracturing, and forcing water through the ground. Pilot operations involving nuclear explosives or hydraulic fracturing of hot rock in the upper part of the earth's crust seem to justify the considerable effort because the amount of power that might be developed from a relatively small volume of rock is large. Any single dry geothermal power plant would deplete the heat in the accessible volume of rock faster than the constant flow of heat from the earth's interior could replenish it. It is possible to conceive of a very long-term continental system of geothermal power plants that would harvest heat in one area, move on to another of a series of sites, ultimately to return to the original site some thousands of years later when its heat content had been restored, a system that would resemble swidden agriculture.

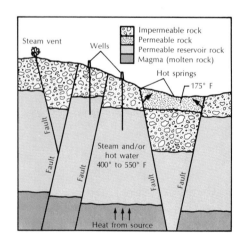

FIGURE 3.6
Schematic cross section of a geothermal reservoir capped by impermeable rock within a fault structure.
(From Bowen and Grotz, 1971.)

OCEAN POWER

Theoretically, two sources of power exist in the ocean. The vertical thermal gradient might be used to extract heat from the ocean, which might be raised in temperature sufficiently to drive a heat engine, or some of the enormous kinetic energy of large persistent ocean currents, such as the Gulf Stream, might be harnessed by large, slow turbines. The critical question is one of energy economics, in other words, whether or not a system using the energy of the ocean can be devised that will yield a net work profit when all the costs are included in the calculation.

FUSION POWER

The energy of nuclear fusion has been used so far only in hydrogen bombs, in which heavy hydrogen (^{2}H) or deuterium combines explosively with itself, releasing energy in large amounts.

If the energy of the uncontrolled explosive reaction could be controlled and released gradually, a vast new potential for power production would be created, one in which power production would have only waste heat and minimal leakage of tritium as environmental problems and which, if only deuterium were required, would be a quasi-income or almost renewable resource, because the deuterium available in seawater should provide fuel for tens of thousands of years of operation. One in every sixty-five hundred hydrogen atoms is a deuterium atom; there is about half a gram of deuterium in a gallon of seawater and that half gram has a fusion energy equivalent of three hundred gallons of gasoline. The separation of deuterium (^{2}H or D) from seawater is facilitated by the fact that the difference in mass or weight between H_2O (water without deuterium) and HDO (water with deuterium in it) exceeds 5 percent.

Fusion power research and development is being pursued in the two countries that can afford such expenditures and that might be able to afford to build and use the reactors. Predictions of ultimate success have no present value in planning; they reflect faith in provident technology, hope for the material future of certain portions of the human community, and vested interests in the research and development.

Although the potential quantity of fusion power that might be developed is so great that it could supply all the world's power needs for thousands of years, no one can say today what the unit cost of that power might be. There is little basis for assuming a very low cost that would allow everyone all the power he wanted.

THE NITROGEN CYCLE AS MODIFIED BY MAN

Nitrogen is an essential material in the construction of proteins and other biological compounds. Although nitrogen is the main component (79 percent) of air, it cannot be taken out of the air and used by most living systems, but must be combined with certain other elements, or "fixed," before it is biologically available (see figure 3.7). The natural supply of fixed nitrogen is limited. World agriculture now depends heavily on supplementation of the natural supply through nitrogen-bearing fertilizers. The element most commonly used to fix nitrogen is hydrogen, obtained either by the dissociation of a fossil hydrocarbon (natural gas in the United States of America, coal in China) or of water; in either case, there is an energy cost. The use of coal or natural gas as a source of hydrogen for ammoniated fertilizer, or the use of fossil or nuclear fuels to produce heat or electricity in order to dissociate water into hydrogen and oxygen (again to produce hydrogen for ammoniated fertilizers), links together the production of the world's fundamental energy resource, food, with the depletion of the nonrenewable energy resources on which the human population has become dependent. In the next chapter, we shall take a look at those nonrenewable resources, returning later to the difficult problem of freeing food production from its diminishing source of energy.

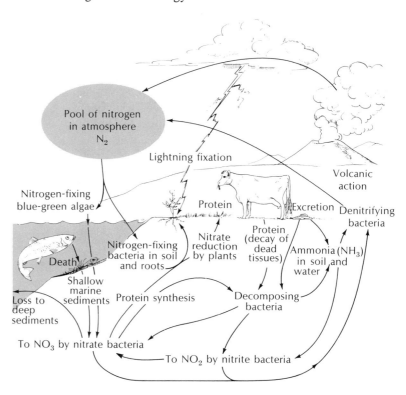

Figure 3.7
The nitrogen cycle.
(*From Paul Ehrlich and Anne Ehrlich,*
Population Resources Environment, *2d ed.*
Copyright © 1972. W. H. Freeman and
Company.)

SUPPLEMENTAL READING

Borgstrom, George. 1969. *Too Many*. New York: Macmillan.
One of several books by this author on world food problems: the unevenness of food distribution in the world, the pressure of population growth on food availability, and some "nutritional absurdities."

Brown, Harrison. 1954. *The Challenge of Man's Future*. New York: Viking.
A thoughtful book that spelled out the reasons for expecting world shortages of resources long before computers were called upon to emphasize the obvious.

Brown, Lester, and Gail Finsterbusch, 1971. "Man, food and environment." In *Environment*, ed. W. W. Murdoch, p. 53–70. Stamford: Sinauer Associates.
Review of the prospects of the availability of food, with emphasis on the environmental and social impact of expanded food production.

Ehrlich, Paul R., and Anne H. Ehrlich. 1972. *Population, Resources, Environment*, 2d ed. San Francisco: W. H. Freeman and Company.
Argues that the earth is grossly overpopulated, that limits to food production are near, that attempts to increase food production will decrease producing capacity, and that population growth and attitudes related to growth are the major barriers to the solution of "the population-food-environment crisis".

Harris, J. E. 1967. "The Employment of Steam Vapor in the Eighteenth Century." *History*, v. 52, n. 175, p. 133–148.
The great value of the steam engine was in its mobility and its power; the overwhelming use of the engine was in coal mines.

Hendricks, Sterling B. 1969. "Food from the land." In *Resources and Man* (National Academy of Science, National Research Council Committee on Resources and Man), p. 65–86. San Francisco: W. H. Freeman and Company.
Concise statement in which a four-fold classification of nations according to the presence or imminence of hunger is set up.

Karraber, Cyrus H. 1953. *Piracy Was a Business*. Rindge, N.H.: Richard R. Smith.
A brief summary of piracy, its history, the reasons for it, and its businesslike aspects, especially among nations on the North Sea.

Lewis, Archiblad R. 1958. *The Northern Seas*. Princeton, N.J.: Princeton University Press.
Fascinating history of the lands bordering the North, Baltic, and Irish seas during the first Christian millennium; emphasizes the dominant role of the sea and seafarers in mixing peoples and shaping national destinies.

Lidsky, Lawrence M. 1972. "The Quest for Fusion Power." *Technology Review*, v. 74, p. 10–21.
Succinct summary of possible fusion reactions and of the techniques through which it is hoped to achieve controlled fusion.

Newman, J. E., and Robert C. Pickett. 1974. "World Climates and Food Supply Variations." *Science*, v. 186, p. 877–881.
Suggests that most areas of famine could be reduced greatly with proper planning to take advantage of relatively favorable areas.

NSF/NASA Solar Energy Panel. 1972. *Solar Energy as a National Energy Resource*. College Park, Md.: University of Maryland, Department of Mechanical Engineering.
Optimistic estimates of solar energy technology.

Pyke, Magnus. 1968. *Food and Society*. London: John Murray.
Sprightly essays by an unstuffy nutritionist on human deterrents to the efficient use of food.

Ricker, William E. 1969. "Food from the sea." In *Resources and Man* (National Academy of Sciences, National Research Council Committee on Resources and Man), p. 87–108. San Francisco: W. H. Freeman and Company.
Critical analysis of the capability of the sea to add to the food supply.

Safrany, David R. 1974. "Nitrogen Fixation." *Scientific American*, v. 231, October, p. 64–80.
World agriculture depends on the continued augmentation of the natural supply of fixed nitrogen, and such augmentation entails an energy cost.

Sauer, Carl O. 1956. "The Agency of Man on Earth." In *Man's Role in Changing the Face of the Earth*, ed. by W. L. Thomas, Jr., p. 70–92. Chicago: The University of Chicago Press.
Evocative review of the origins and spread of agriculture and of the impacts on the environment of different kinds of agriculture.

Chapter 4

NONRENEWABLE ENERGY RESOURCES

. . . the first major penalty man will
have to pay for his rapid consumption
of the earth's nonrenewable resources
will be that of having to live in a world
where his thoughts and actions are
ever more strongly limited. . . .

—Harrison Brown, 1954

COAL

Photosynthetic conversion of solar energy to living tissue began on earth at least 2 billion years ago. Life, however, could not have been described as abundant or flourishing until some 600 million years ago, and then only in the seas. There was no extensive life on the continents until 300 million years ago. Since then, the remains of a small fraction of the world's plants and animals have been preserved under special nonoxidizing conditions and converted into coal, petroleum, and natural gas.

Coal is formed from the remains of trees. Most of the world's coal beds appear to represent accumulations of plant material in swamps near the shores of ancient seas that alternately transgressed and receded from the land. Such alternations, probably caused by the instability of the continental margins under crustal stress, allowed, first, the creation of swamps in which the fallen trees rotted slowly, if at all and, second, the burial of the matted remains by beach sands laid down during the ensuing transgression. In most coal fields there are many coal layers and seams, indicating repeated alternations of living swamp and sedimentary burial. Individual coal seams commonly lie on what are called underclays, the very fine material deposited by the sluggish streams of a subsiding coastal area immediately before the creation of a coastal swamp. Because of the rather unusual geological conditions required for their formation, coal fields are localized both in space and in geological age.

There is a continuous series from peat through lignite and bituminous coal to anthracite, reflecting a progressive metamorphism (change in form and composition) of the organic material. Peat is brown and porous; it contains visible plant remains and has a high

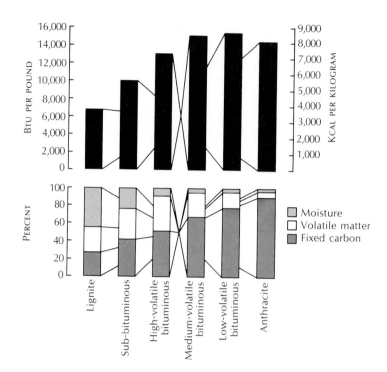

FIGURE 4.1
The rank of a coal has nothing to do with its ash or sulfur content, but is measured by heat content, calculated as if the coal had no ash or other impurities in it. The principal ranks of coal are compared above in terms of heat content and physical constituents, on an ash-free basis. Both fixed carbon and volatile matter are combustible, but moisture is a disadvantage.

moisture content. The weight of overlying rock and the earth's heat combine to drive moisture out of buried peat and change it into lignite. As the metamorphism of peat into lignite and lignite into bituminous coal proceeds, carbohydrates are converted to hydrocarbons and then to elemental carbon. The higher the percentage of elemental carbon and the lower the percentage of moisture and volatile hydrocarbons, the greater is the potential-heat content and the higher the *rank* of the fuel. The term *bituminous* is a term of rank, based on heating value. In addition to rank, these fuels have *grade*. The grade of a coal or lignite is determined by its content of waste materials, notably ash and sulfur. A low-grade bituminous coal is one that contains much ash, sulfur, or both. Clay or sand washed into the swamp where the trees grew account for the ash in coal; sulfur, generally in the form of iron sulfide, is a common chemical constituent of sediments deposited under anaerobic or reducing conditions.

A typical lignite, as analyzed by the United States Bureau of Mines, contains—after all ash is removed—33 percent fixed carbon, 26 percent volatile matter (hydrocarbons), and 41 percent moisture, whereas a typical anthracite—also on an ash-free basis—contains 92 percent fixed carbon, 5 percent volatile matter, and only 3 percent moisture. Lignite and coal may contain up to 15 percent ash (the limit is economic, not geologic) and sulfur to 4 percent. Low-sulfur coal has a sulfur content of less than 1 percent. Volatile matter and fixed carbon are the contributors to energy when these fuels are burned.

When exposed to air, lignite loses moisture and crumbles. It cannot be stored easily. Because of its high moisture content, it has a relatively low heating value and this, coupled with its propensity for crumbling, makes it uneconomic to ship long distances. Electricity generating plants using lignite are usually situated near the mine, the losses in transmission being taken in preference to losses in haulage costs and handling. The same considerations hold for sub-bituminous coal, a rank between lignite and bituminous coal.

Bituminous coal is the most abundant and widespread rank of coal in the United States of America. It is the coal used most commonly for industrial, electric power, and heating purposes. It may be either coking or noncoking—a property based not on rank but on whether the coal will produce a coke (a strong, porous block) when processed in a coke oven. Nearly all eastern bituminous coals are coking; western bituminous coals generally are not.

Coal is found in beds ranging in thickness from less than an inch to many feet. There may be many beds within a single sedimentary sequence or depositional basin; in West Virginia 117 coal beds are important enough to be named and there are hundreds of smaller beds. The lower boundaries of coal seams are commonly sharp contacts with the underclays; the upper boundary may be either well defined or gradational to sandstone or shale.

Certain geologic periods have seen conditions more favorable to coal formation than others; coal beds are so distinctive a feature of beds of late Paleozoic age in many parts of the world that this part of the geologic time scale is called the Carboniferous. In some parts of the world, coal beds are important among rocks of Cretaceous and early Tertiary age.

Coal beds may be buried deeply or exposed at the surface; they may be steeply tilted or essentially horizontal; they may be broken by faults (fractures that displace strata) or undisturbed by geologic forces. Whether or not a coal bed can be mined economically, in other words, whether or not it is a resource, depends upon its thickness, depth, continuity, inclination relative to the surface, rank, and ash content. Many coal beds exist that cannot be mined economically at present because they are too thin or too deep. As mining has become more mechanized, the minimum mining thickness has tended to increase because of the technological difficulties of mining thin seams. At the same time the minimum economical depth-to-thickness ratio has tended to increase for beds above a certain thickness: one can now afford to go deeper, in either an underground mine or a surface mine, to extract a coal bed that is, for argument's sake, more than five feet

(A)

(B)

(C)

FIGURE 4.2
Machines used in strip mining are large. A walking dragline (A) is used to strip the waste material or overburden from the coal seam. The dragline shown carries a 221-cubic yard bucket, capable of taking 330 tons in a single pass, on a 310-foot boom. Power shovels (B) are used both for stripping—in some mines—and for loading coal—all mines. The shovel shown moves 270 tons in each bite. Trucks (C) transport coal from the pit to a preparation plant or to a rail loading facility. The truck shown is 78 feet long, 21 feet wide, and 16½ feet high; it weighs 118 tons and carries 180 tons. (Photographs A and B courtesy of Bucyrus-Erie Co. C from Mining Engineering, October 1974.)

thick. The minimum minable thickness in surface or strip mining commonly is less than that in underground mining, being about eighteen inches in the former and thirty inches in the latter. This fact represents what may seem a paradox, that strip mining, in the sense of greater recovery of the resource material, is a more *conservative* method of mining than is underground mining.

The age of hunting for coal beds is over. Most of them have been located, and they only need to be explored further to determine their rank, ash content, continuity, thickness, and depth over an area where mining is contemplated.

Spectacular advances in the technology of mining and transporting coal have been made in recent years. Continuous mining machines are now used in most underground mines. Draglines with buckets that will hold more than 300 tons of rock, stripping shovels that can take 270 tons of rock at a bite, and trucks of 240-ton capacity are in use in mines in the United States, while in Europe and Turkey, bucket-wheel excavators, which can dig as much as 300,000 tons a day, are in use. Unit trains are used to haul bituminous coal 10,000 tons at a time from Montana and Wyoming to points of use as far away as Indiana and Texas, and coal is transported by pipeline (as a slurry) from northeast Arizona to the Mohave power plant near Las Vegas, Nevada.

Coal is mined by three methods. *Underground* mines are entered by a vertical or inclined shaft from the surface (in a few cases, where coal beds occur within mountains, the entrance may be horizontal). Where the shaft intersects a coal bed of minable thickness and quality, lateral openings are driven into it, and extraction begins. For many years, coal mining required much human labor. Holes were drilled into the face of the coal seam and loaded with black powder or dynamite, which was then exploded. The broken coal was shoveled by hand into a small car that, when full, was pushed by men or pulled by a mule along a track to a loading pocket adjacent to the shaft, where it was loaded into

FIGURE 4.3
The mechanization of underground coal mining in recent years has featured development of electric-powered continuous mining machines that break the coal from the advancing face of the seam and carry it back over the power unit to be loaded in trucks or cars for transport to the hoisting shaft. The square objects in the upper part of the photograph are roof bolts emplaced to reduce the danger of "rock bursts," one of the common hazards in deep mines. (Photograph courtesy of Joy Manufacturing Company.)

other cars or buckets and hoisted to the surface. New technology has greatly changed this labor-intensive system. Continuous mining machines, powered by electricity, cut out instead of blasting the coal, break it, and convey it continuously into waiting cars for transport to the hoisting shaft. Few underground coal mines are more than three thousand feet deep. They range in daily production from a few tons to ten thousand tons; from two to one thousand men are employed in individual mines.

Strip mines are surface mines created by stripping the overlying material, called overburden, from a minable coal seam. The overburden is removed by draglines or large electric shovels. In either case, the overburden is stacked behind the current operations, usually on the recently mined-out portion of the pit, whence it can be leveled or landscaped again. In strip mining, the coal bed is blasted to break it for shoveling, but the technology of drilling and blasting has evolved along with that of the giant shovels and huge trucks. Drill holes now may be a foot in diameter and the explosive material, ammonium nitrate, is chemically and physically more like fertilizer than dynamite. The broken coal is loaded by shovel or front-end loader into trucks.

FIGURE 4.4
A continuous mining machine for strip mining is the bucket-wheel excavator, developed in Germany. Used to excavate both overburden and coal, the bucket-wheel excavator discharges to a belt conveyor. The belts discharge overburden into stackers, which store the material for backfilling the pit when the coal has been removed. The belts discharge coal at a power plant or at rail loading docks.
(Sketches made by Dr. Ing. Otto Gold are reproduced by courtesy of the Mining Congress Journal.)

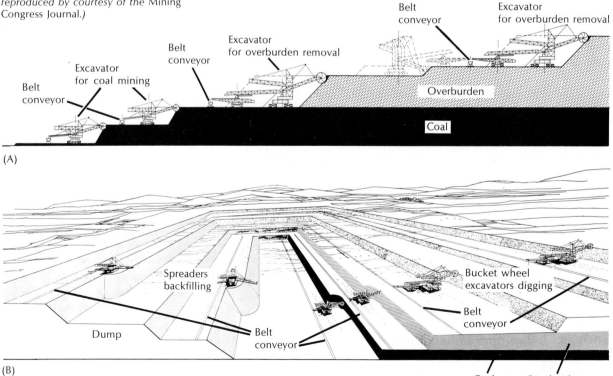

Most strip mines are less than two hundred feet deep. Strip mines produce as much as fifteen thousand tons of coal a day and employ as many as seven hundred men in a single mine.

The third method, which accounts for only a small portion of the coal mined in the United States of America, is *auger* mining, carried out by means of large horizontal augers that bore into a coal bed exposed on the side of a hill and push out the broken coal as they rotate.

More than half the coal produced in the United States is mined by stripping (as late as 1947, only 22 percent of our coal came from strip mines). Technological advances have been more effective in keeping down costs in surface mining than they have in underground mining. In addition, the application of new health and safety laws for underground miners have further increased the costs. The average ton of coal mined by underground methods now costs three or four times as much to produce as the average ton of coal mined in the new strip mines of Montana and Wyoming, even when complete reclamation of the land disturbed by stripping is required. *Reclamation* is used, not in the sense of restoring the land surface to its original shape and elevation, which is hardly possible after a seam of coal perhaps sixty feet thick has been removed, but in the sense of restoring the surface to or beyond its previous value for other uses. Another advantage of striping is that nearly 100 percent of a minable seam of coal can be recovered, whereas in underground mining almost half of the coal must be left behind in pillars that support the mine openings. Still, some coal beds can be mined only by underground methods, because they are too deeply buried for stripping, because they are steeply inclined, or because they underlie a surface too valuable to be disturbed by mining or impossible to reclaim. Stripping is best adapted to fairly shallow (less than 300 feet deep) horizontal coal seams in areas where the hillsides are not steep and where vegetation can readily be restored on the contoured waste material. The environmental problems of coal mining will be discussed further in chapters 10 and 11.

In the United States of America, most coal is transported by rail, unless it is used at or near the mine to fuel power plants. It is also transported by barge, but most of the routes between large coal deposits and centers of use rule out water transport.

Coal was being burned in China and in Britain (by the Romans) two thousand years ago. Englishmen of the eleventh century rediscovered the fact that coal would burn and, by 1234 the "sea coles" of northeast England were being shipped to London where they were used for heating. Also in the thirteenth century coal was being used by the Hopi Indians in what is now Arizona. As London grew and as wood for fuel in England became scarcer because so much was consumed in charcoal manufacture and in shipbuilding, the use of English coal for heating grew. The choking, foul-smelling smoke of the coal mined from the early shallow pits made its use even by smiths objectionable. In fact, according to Nef, the historian of coal, an intense and stubborn prejudice against the new fuel existed in England until the time of Elizabeth, when growing fuel scarcity brought a reluctant increase in the use of coal in English cities. Later, the perfection of a means of making coke from coal and the development of the steam engine greatly accelerated the growth in the use of coal. Then coal quickly became the fuel basis of the Industrial Revolution; it consolidated Great Britain's position as ruler of the world's oceans and undergirded the British Empire, the greatest single economic organization yet put together by man.

In the United States, coal was slower in supplanting fuel wood, which was much more abundant in the North American continent of the nineteenth century than it was in Europe. But in the fifty years between 1850 and 1900, coal increased its portion of the United States energy input from 9 percent to 71 percent. Then petroleum entered the scene, bulking large as automobiles, trucks, and buses proliferated. Petroleum took a large chunk of coal's market with the rapid replacement, after World War II, of the nation's steam locomotives by diesel-electric locomotives. Natural gas, in the same period, took a further bite from the coal market, by making it uneconomic for local gasification plants to produce "coal gas" or "city gas" from coal. The consumption of coal continued to decline until 1961, when an increasing utility market reversed the trend. With good hydroelectric sites growing scarce and with nuclear power costs and difficulties growing, electric utilities have been turning more and more to coal-fired power plants remote from urban centers as the most economical way of producing electricity over the lifetime of the plants, consistent with contemporary environmental constraints. This trend would have been much faster had it not been for the subsidy to consumers represented by wellhead-price control of interstate natural gas by the Federal Power Commission, the government subsidy of nuclear power through support of research, development, and accident insurance, and increasingly stringent ambient air-quality standards that made the high-sulfur coals of the Appalachians and Middle West unacceptable for urban power plants. In recent years, low-sulfur coal from the Rocky Mountain states has been hauled by unit trains to power plants hundreds of miles away, while new power plants have been built near the mines. Now the large coal deposits of the intermountain West are attracting coal-gasification plants. The gas produced from such a plant can be used either in new industry situated nearby or can be introduced into natural gas pipelines for distribution throughout the United States. Substitute crude oil also can be made from coal. For many years after James Young, a British chemist, took out the basic patent in 1850, kerosine or "coal oil" was distilled from coal; only later was it made, more cheaply, by refining crude oil. Methanol (methyl alcohol) and ethanol (ethyl alcohol) also can be made directly from coal and can be used as fuels.

In 1972, 57 percent of all coal mined in the United States of America was burned in electricity generating plants, 15 percent was used to make coke for steel blast furnaces, 10 percent was exported, mainly to Japan, and 12 percent was used in industry for process heat. Coal represented about 17 percent of the energy used in the economy of the United States in 1972. Coal also has other uses. Synthetic dyes were first produced from coal more than 100 years ago. Today, a number of organic chemical products, including one kind of nylon, are made from coal.

CRUDE OIL AND NATURAL GAS

Origin and geologic occurrence

Petroleum (rock oil) and natural gas are found in similar geological environments and often together; indeed some authorities classify natural gas as gaseous petroleum and, for that reason, the unambiguous term *crude oil* will be used in this book to mean liquid petroleum. Commercial accumulations of crude oil and natural gas are almost entirely limited to sedimentary rocks and to special geological circumstances. Such accumulations are often called *pools* although the fluids fill water-coated pore spaces in rocks rather than large open caverns. An oil field consists of one or more pools geographically isolated. Most oil and gas fields are a few square miles in extent; only a few giant fields underlie more than a hundred square miles of the earth's surface.

The geological requirements for an oil pool are a source rock, a sufficiently permeable reservoir rock, a sufficiently impermeable cap rock, a favorable trapping structure, and water in the formations. Because most oil and gas is found in sedimentary rocks, because most accumulations are close to thick deposits of marine or deltaic sediments that contain large volumes of shales having an appreciable organic content, and because this organic content contains compounds of carbon commonly manufactured by plants or animals, geologists believe that the *source rock* for most, if not all, petroleum is organic shale. As with coal, accumulation took place under conditions that prevented oxidation and promoted concentration of the organic remains and was followed by burial and subsidence. Subsequent metamorphism was physically and chemically unlike that that produces coal, which is marked by a progressive diminution of fluid and volatile constituents that ultimately yields pure carbon. The metamorphism of petroleum appears to yield increasing fluidity and volatility, and produces no elemental carbon. Crude oil consists almost entirely of liquid hydrocarbons with some gaseous hydrocarbons dissolved in them. Tar and asphalt are semisolid to solid forms of crude oil. In the United States, the standard unit of quantity for crude oil is the barrel, 42 U.S. gallons. In other countries, both the barrel and the metric ton (often indicated by the spelling *tonne*) are used. Because the density of crude oil has a considerable range, it is not easy to convert accurately from barrels to tonnes. The potential heat content (heating value) of crude oil likewise shows a range that reflects the natural range of composition; the average figure used by the United States Bureau of Mines in 1972 was 5,645,000 Btu per barrel. A barrel of oil weighs between 295 and 310 pounds (without the barrel).

Natural gas consists of hydrocarbons simpler and lighter than those of crude oil; the most abundant hydrocarbon in natural gas is methane (CH_4). Natural gas may contain liquid hydrocarbons in dispersed form; these are recovered in separation plants and called natural gas liquids (NGL). In addition, natural gas may contain water and gases that are not hydrocarbons.

Crude oil is lighter but more viscous than water. Natural gas is much lighter than crude oil and much less viscous than either crude oil or water. For various reasons, oil and gas are found in places other than the shale in which they were formed: they tend to seep through permeable rock to areas of lower pressure, they are buoyed up by the water contained in host rocks, they tend to rise toward the surface as the hydrostatic pressure lessens, and they are squeezed out of the shale by the weight of the overlying rocks and by earth-deforming processes that tend to compact very fine grained rocks, such as shale, more than coarser grained rock, such as sandstone.

Because sandstones tend to retain some of their original porosity and permeability better than do finer grained rocks, they form good *reservoir rocks* for oil and gas accumulations. The oil and gas move into the interstices of the sandstone as they are squeezed out of the compacting shale. Then they may migrate considerable distances upward within an inclined reservoir bed escaping to the earth's surface unless halted by a geologic trap. Buoyancy produces differential movement of gas and oil through the water, and of gas through the oil, so that, if a geologic trap is encountered, there will be a layering of the three materials—gas, oil, and water—according to their relative densities.

A *geologic trap* is formed wherever an impermeable formation exists in such a configuration that migration of oil and gas could be halted and contained sufficiently to form a pool. A trapping configuration is called a *structure* by oil geologists. A structure must contain an impermeable formation in order to be a trap and not all traps have oil or gas

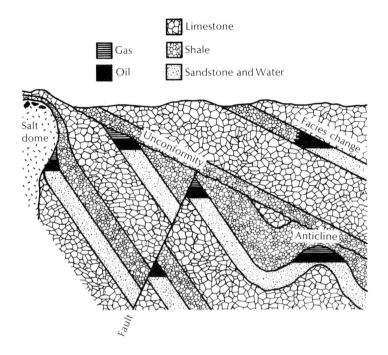

Limestone

Gas

Oil

Shale

Sandstone and Water

Salt dome

Facies change

Unconformity

Anticline

Fault

FIGURE 4.5
The principal geologic traps for oil and gas are shown in this hypothetical cross section of a portion of the earth's crust. Sandstone is more permeable than shale, limestone, or salt; consequently fluids can move through sandstone. Gas and oil commonly move upward, impelled by buoyancy when in contact with water, a common constituent of porous sandstone. If not trapped by a relatively impermeable cap or seal, oil and gas escape to the surface.

in them. Many oil fields have been found along the crests of upwardly convex folds in the rock strata, called anticlines, and especially in the high portions of an anticline that is longitudinally undulant, as many are. The high portions of such folds often are called *domes*. If the dome includes an impermeable formation, such as shale, as a *cap rock*, overlying a permeable formation, such as sandstone, as a reservoir rock, which in turn is underlain by an organic shale, as a possible source rock, a valuable accumulation of gas and oil may be found within the dome.

Some of the world's greatest oil wells and pools, in terms of rate of production, have been found in the cavernous caps of what are known as salt domes. These domes are not genetically similar to anticlinical domes. They have been formed by the upward movement of pear-shaped salt masses or plugs that have broken through some sedimentary layers and up-arched others in their push toward the surface. Because salt is highly soluble, the upper parts of salt plugs may develop cavities where groundwater has dissolved the salt. Should this cavernous cap rock underlie an impermeable up-domed sedimentary formation, an ideal geologic trap for crude oil and natural gas has been formed. The Spindletop oil field, near Beaumont, Texas, is a noted example of an accumulation under a salt dome cap rock. As Spindletop and other similar fields were developed, it was found that many salt domes have "flank" accumulations of oil or of oil and gas, below the top of the dome, in the upturned and truncated edges of the sedimentary formations through which the salt plugs pushed their way.

A third major kind of trap is the stratigraphic trap, in which a reservoir rock, truncated by erosion after tilting, is overlain by an impermeable cap rock, and the later upward movement of oil and gas is halted in the upper portion of the reservoir rock. It also happens that a permeable sandstone may grade laterally, because of differing depositional circumstances, into an impermeable siltstone or shale. Should this formation later be folded or tilted, even slightly, so that the impermeable part of the formation is higher than the permeable portion, a stratigraphic trap has been formed. The largest of all American oil fields, the East Texas field, was formed by a stratigraphic trap.

Lesser numbers of oil pools are associated with fault seals and tar seals, which are formed when a tilted reservoir rock has been sealed off in two other ways. Fault seals are the result of the emplacement of an impermeable formation against the truncated part of the reservoir rock by movement along a fault. Tar seals are the result of the coagulation of escaping oil in the upper part of a reservoir rock exposed at the surface of the earth.

Some traps contain oil but no gas, possibly because no gas was generated in the source rock, but more likely because gas, having a much lower viscosity than oil, can escape upward through rocks that petroleum cannot penetrate. (No rock is absolutely impermeable.)

Some traps contain gas but no oil. This phenomenon too is probably caused by earlier separation by differential movement of the two fluids through dense rocks, but it may result from natural distillation, under heat and pressure, of the light hydrocarbons of natural gas from the heavy hydrocarbons of crude oil. Below a depth of twenty thousand feet petroleum pools become scarce, yet gas accumulations are still found. The most likely explanation is that the conditions necessary for oil formation are obtained above such depths, and that only gas is formed under the pressures and temperatures common below twenty thousand feet. An important point is that gas, being much more compressible than oil, can exist in large quantities in the minute pore spaces of dense deep formations, whereas there would be room for little oil at such depths.

In all undisturbed crude oil and natural gas accumulations it has been noted that the fluids (gas, oil, and water; oil and water; or gas and water) are in density adjustment to the shape and nature of the trap. This means not only that the gas and oil have moved or migrated from their places of origin but also that they have become concentrated into pools that have sharp boundaries. It has also been found that oil must constitute about 15 percent of the mixture of oil and water in the rock pore spaces before it can be induced to move toward a well; consequently, it is probably true that there is a similar critical concentration, under given conditions of temperature, pressure, porosity, and permeability, at which crude oil (and gas) will start migrating. Once this concentration has been reached, the oil and gas will keep moving until trapped or leaked to the surface. Because they are constantly seeking places of lower pressure, oil and gas will adjust quickly to trap conditions and separate according to their densities. The transition zones between gas and oil and between oil and water commonly extend through a vertical distance of less than ten feet, even in large pools.

Exploration methods

The first petroleum put to human use oozed from natural oil seeps, and such seeps have been a guide to exploratory drilling in many parts of the world. The greatest single producing oil well in history, Cerro Azul Number 4, was drilled beside a seep near Tampico, Mexico. On 10 February 1916, it blew in, the column of oil rising almost six hundred feet into the air. The initial production rate was 260,000 barrels a day. The well probably penetrated a real pool of oil and gas floating on salt water in cavernous limestone.

Many oil fields have been discovered by random drilling, especially in or near areas of established production. As knowledge of the occurrence of oil increased, most wildcat wells began to be sited on the basis of geological and geophysical information and interpretation. Early random drilling in Pennsylvania and West Virginia showed that producing oil and gas wells tended to be grouped along anticlinal crests, and geological guides to oil have been used in more and more sophisticated ways since the "anticlinal theory" was repeatedly proven by drilling.

Several geophysical methods can be used to gather information about the structure and composition of buried rocks. Minute differences in the earth's gravity field, which reflect differences in the density of rocks beneath the surface, can be measured. Gravity surveys have been spectacularly successful in finding buried salt domes, because the salt plug has a lower density than the rocks it has pushed through and the resulting differences in gravity can be easily mapped by sensitive gravity meters.

Seismic methods have proved more generally useful than gravity methods. Reflection shooting has been most successful; this method involves the firing of small charges of explosive in shallow holes to produce minor artificial earthquakes. Elastic waves propagated by the explosions are reflected from the upper surfaces of the buried formations and sensitive instruments record the waves' return to the ground surface. The travel times of the waves are so plotted that the depth to

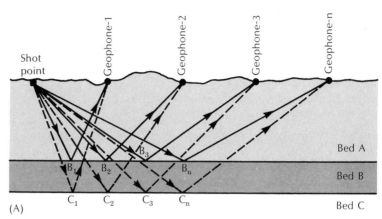

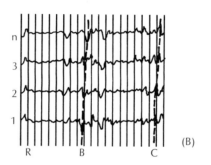

FIGURE 4.6
(A) One of the most common techniques in oil exploration is reflection seismic shooting. An explosive charge or other source propels shock waves downward from a surface shot point. Where rock types change, these waves will be reflected back to the surface. Sensitive recording instruments called geophones record the times of arrival of successively reflected seismic waves. (B) From a plot called a seismogram, upon which the grid lines represent milliseconds, the buried structure of the earth can be determined. The reflecting surfaces on the tops of beds B and C are shown as dashed lines on the seismogram.
(From Federal Power Commission, National Gas Survey, *v. 5, 1973, p. 209.)*

each formation from any point beneath the area of the seismic survey can be calculated; thus, anticlines and other significant features buried without surface expression may be located for exploratory drilling. Once drilling has begun in any area deemed worthy of the expense, seismic and gravity information obtained from the surface is checked and augmented by information obtained from drill holes. Solid core samples of the penetrated formations may be obtained by substituting for the usual rotary bit at the bottom of the drillstem a circular bit that cuts out a rock cylinder as it descends. The rock cylinder is encased in a core barrel mounted just above the cutting bit and, when the barrel, commonly about five feet long, is filled, it is pulled out of the drill hole and the core is removed for examination and tests of its porosity and permeability. Coring is slow and expensive and is usually done only within formations of special interest. Information about the physical characteristics of all the formations penetrated by the drill can be obtained by various well-logging techniques designed to measure differences in electrical properties, radioactivity, and resistance to drilling.

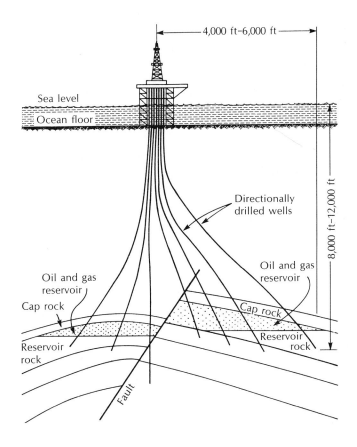

FIGURE 4.7
Directionally drilled wells from an offshore drilling platform. As many as 40 wells may be drilled from a single platform. The cost of drilling is lowered thereby, as is interference with other uses of the ocean.
(After Taylor, 1973).

Only the drill can prove the existence of petroleum in a trap found by geological and geophysical exploration. Recent research, however, suggests that it may soon be possible to detect and locate gas and oil, and possibly oil and water, interfaces by a new seismic technique. For the first time, it may soon be possible to tell, before drilling, whether or not there is oil and gas, or both, in the target trap.

A test well drilled in the hope of discovering a new pool is called a *wildcat well*, and the person who directs the venture, if it is not done by a large company, is known as a *wildcatter*. These terms are said to have originated because many such wells were drilled in remote, uninhabited country where wildcats seemed to be the ruling animal. If the test hole produces neither oil nor gas, it is called a *dry hole* or *duster* (even if there is water in it); it is plugged and abandoned. If a well yields oil or gas, it is equipped for production and becomes a *completed well*. Wells drilled in the same reservoir after a discovery are called *development wells*.

The deepest well so far (1975) drilled for oil or gas went down 30,050 feet, almost six miles below the surface. It is in the Anadarko

Basin in Oklahoma. Many wells along the Gulf Coast have gone down more than 20,000 feet. The average depth of producing oil and gas wells in the United States is about 5,000 feet. Offshore wells are drilled from fixed platforms, from floating drillships, or from structures that are towed to the drillsite and then partially submerged to provide stability. Fixed platforms have been erected in water almost 400 feet deep. Because offshore platforms are so expensive, as many as forty wells are drilled from the same platform, by slanting most of the holes outward from the platform as they are drilled. Such holes are curved and, near the bottom, may be as many as 40° from the horizontal.

Production and transport

If a pool of oil has natural gas dissolved in it under a confining pressure, the exsolving and expanding gas will drive the oil to the surface of a well drilled into it. This is what makes a gusher. Even the buoyancy of oil over water may push the oil to the surface. In either case, if the movement of gas and water can be controlled, the oil will not need to be pumped out, at least during the early life of the field. The terms *gas drive* and *water drive* are used to describe the action of gas and water in pushing oil out of a reservoir and into producing wells. In addition to pumping, recovery may be stimulated by hydraulic fracturing (forcing water by high pressure into minute cracks) or acidization (dissolving alkaline rocks by forcing acid into their pores) of the reservoir rock to increase its permeability near the well. In fields where the production of pumped wells is diminishing, gas or water may be injected into the reservoir in a process known as *secondary recovery*, an attempt to flush out the crude oil that clings tenaciously to the grains of the reservoir rock. In the United States of America, depending on the formation of the reservoir, 10 to 60 percent of the original oil in place is recovered, the present cumulative recovery average being about 31 percent. In other words, for each barrel of crude oil produced, more than two have had to be left in the ground.

Drilling derricks once were left in position to make it easier to rework the wells should improved technology or higher prices make it desirable to do so, but now they are dismantled and moved when a well is completed or abandoned. If the well is flowing freely, only a "Christmas tree" marks it; this is the arrangement of pipes, valves, and gauges required to regulate and monitor the flow of oil or gas from the well and to connect it with the network of pipes that gather the fluid and transport it to storage or processing facilities. When water and gas drive fail or are not economical, oil is pumped from the reservoir by the rocker-armed suction pump so familiar to those working or living near producing oil fields. The average daily production in 1972 from the 508,443 wells in the United States was only 18.4 barrels per well. Some famous wells have produced 100,000 barrels or more a day.

Natural gas does not need to be pumped from the ground; on the contrary, it is often difficult to keep it in the ground long enough for it to be of maximum assistance in driving out the oil with which it may be associated. In "tight" formations, which have a permeability too low even for natural gas to escape them freely, gas production has been stimulated by nuclear explosions (in New Mexico and Colorado) and large-scale hydraulic-fracturing tests are being considered.

Natural gas commonly contains water and small amounts of liquids similar to petroleum. These are uneconomic to transport and may corrode the pipelines; consequently, they are removed from the "wet gas" as near to the wellhead as possible, by means of a separator, and "dry" natural gas is then sent through the pipeline system. The natural gas liquids are added to petroleum storage tanks or put directly into crude oil pipelines. In the United States such natural gas liquids add about 10 percent to the domestic production of petroleum. Conversely, crude oil rising to the wellhead usually contains some natural gas, called *casinghead gas,* which is removed in a separator. Much of it is then used in the oil field: it may be pumped back into the reservoir in order to help drive or lift the remaining oil or be used as fuel for the engines required to run pumps and other devices. Formerly, casinghead gas that could not be used was burned off or flared in the field and this is still the practice in remote fields such as those of the Middle East. In the United States, the excess gas is now commonly put into natural gas pipelines. Most commercial gas, however, comes from wells that produce only natural gas.

Before World War II much petroleum in the United States was shipped overland in railroad tank cars; now most of it goes through pipelines to the refineries, which may be hundreds or even thousands of miles from the producing wells. Of the petroleum produced in Texas, for example, between 70 and 75 percent is refined within the state; the remainder flows as crude oil to refineries in Illinois, Ohio, Pennsylvania, and even New Jersey.

Petroleum is also transported by barge and tanker. The economies of scale achieved by enormous tankers for the intercontinental transport of petroleum have led to the construction of supertankers so large that they cannot pass through the Suez or Panama canals, that there are no ports on either the Gulf or Atlantic coasts of the United States that they can enter, and that, when fully loaded, take more than five miles to stop, even with the engines in full reverse.

After World War II, the rapid growth of a national grid of natural gas pipelines, the ease with which natural gas can be handled and used, its cleanliness, and its low price created a large market almost overnight. What had been a waste product became a valued fuel. In

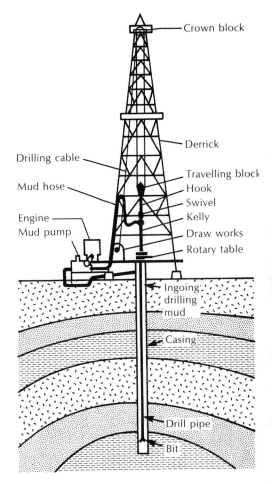

FIGURE 4.8
A simplified sketch of a drilling rig shows its principal features. The drill pipe is in 30-foot lengths, screwed together to form the drill string, which has a cutting bit at the bottom. The string and bit are rotated, their weight adding to the cutting power of the bit. Pressure in the bore hole is maintained by circulating drilling "mud," which also brings up the drill cuttings, keeping the hole clear. In weak formations or where groundwater must be protected, the hole is "cased" or lined with steel pipe cemented into place. When the bit wears out, the entire drill string must be hauled up and uncoupled, length by length; a "round trip" in a deep hole may take two days.

recent years, it has become even more desirable as stringent air-pollution standards have been implemented in the urban areas of North America.

Industrialized countries and countries where domestic production is small or declining have begun importing liquefied natural gas. It is not possible to lay pipelines under broad oceans—although they already crisscross the shallow North Sea—nor could much energy be transported as gas in even the largest tanker. Natural gas becomes liquid when refrigerated to $-259°$ F. Cryogenic tankers are being built to transport liquefied natural gas from areas of abundant production (North Africa, Indonesia, Soviet Union) to areas of great demand (the United States of America, Japan, West Europe). An alternative being given serious consideration is that of converting natural gas to methanol, which does not need to be refrigerated and can be transported in conventional tankers. The present enormous waste of natural gas in the Middle East may be reduced or eliminated when it can be liquefied economically or converted to methanol and transported to industrial countries. Another possibility is to use it as chemical feedstock for new petrochemical and fertilizer plants in the producing countries.

Sulfur is an undesirable impurity in both crude oil and natural gas. Crude oils contain between 0.07 and 5.55 percent sulfur by weight. Those containing less than 0.5 percent are called low-sulfur crudes; those with more are called high-sulfur crudes. High-sulfur crude oil and natural gas are also known as "sour" crude and "sour" gas, a reference to the odor of hydrogen sulphide, the gaseous form of sulfur in petroleum, an odor that can become almost painfully obtrusive in a refinery. Because they stink and may produce sulfuric acid when burned and because they corrode pipes, sour crudes and sour gas are rendered "sweet" before they are used, natural gas in desulfurizing plants near the producing wells, crude oil at or near the refineries. In refining crude oil, sulfur may remain concentrated in the heavier "fractions"; in recent years, plants have been built to desulfurize even these residual fuel oils so that they too may be burned without producing unacceptable amounts of sulphur dioxide.

Oil refining

The modern oil refinery is a plant in which various products are made to order out of one basic material, crude oil. The processes that characterize an oil refinery are distillation and "cracking." Progressive distillation or vaporization of the various components of crude oil results in an initial separation of the lighter petroleum products, such as gaseous hydrocarbons, gasoline, kerosine, and heating oil, from the

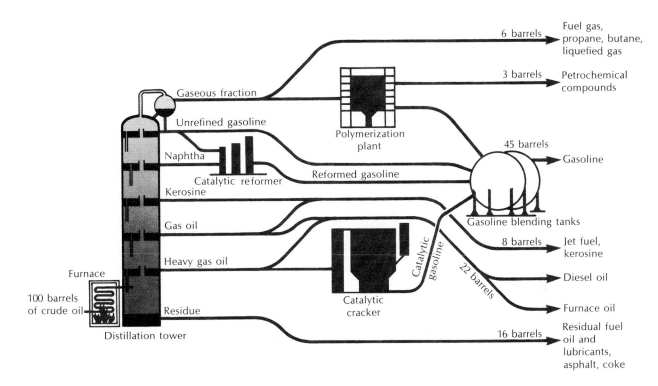

FIGURE 4.9
A highly generalized sketch of hydrocarbon flow through a modern oil refinery in the United States. The principal product is gasoline, which accounts for about 45 percent of the crude-oil input.

heavier products such as lubricating oil and residual fuel oil. "Cracking" is the production, through the application of heat and pressure, with or without catalysts, of lighter and more volatile hydrocarbon compounds from the heavier and less volatile compounds. By cracking the fuel oil product of the distillation tower, the proportion of gasoline or any desired lighter fraction can be increased over that obtainable by distillation. Through this process refineries in the United States of America make about 45 percent of each barrel of crude oil into gasoline. European refineries, responding to a different set of demands for fuel, transform only about 20 percent of their crude into gasoline. Other refinery processes such as hydrogenation, hydroforming, polymerization, and alkylation produce gasoline and other substances with particular physical and chemical properties. Seldom are the distillation fractions suitable as finished products; consequently, they are blended, cracked, re-formed, or changed in other ways to produce more than a thousand specialized products. Almost 15 percent of the crude oil consumed goes into products such as lubricants, waxes, solvents, paving materials, insulation, insecticides, fungicides, herbicides, and petrochemical products—none producing energy.

Natural gas and crude oil are the sources of about one-third of the chemicals produced in the United States, although petrochemical production (including nitrogen fertilizers) still accounts for less than

Table 4.1 Fractions Obtainable From The Distillation of Crude Oil

Boiling range (°F)	Fraction	Uses
−259° to −44°	Methane, ethane	Refinery fuel; petrochemical feedstock
−44° to 11°	Propane	Liquefied petroleum gas (LPG); petrochemical feedstock
11° to 30°	Butane	Blended with motor gasoline; petrochemical feedstock
30° to 300°	Light naphtha	Component of gasoline
300° to 400°	Heavy naphtha	Feedstock for catalytic cracking; blended with light gas oil to make jet fuel
400° to 500°	Kerosine	Component of jet fuel and heating oil; used as solvent and illuminant
400° to 600°	Light gas oil	Component of fuel oil and diesel fuels
600° to 800°	Heavy gas oil	Blended with vacuum gas oil into feedstock for catalytic cracking
800° to 1100°	Vacuum gas oils	Source of lubricating oils and distillates; feedstock for catalytic cracking
More than 1100°	Residue	Blended with gas oils to make heavy fuel oils; source of asphalts or waxes.

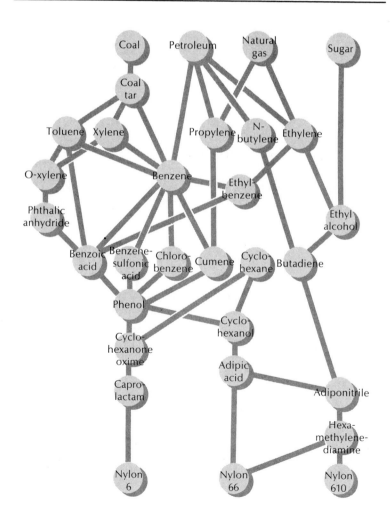

FIGURE 4.10
*Raw materials and chemical intermediates in
nylon production.*
(From Kneese et al., 1970.)

Table 4.2 Principal Petrochemical Products

Starter Material	Product	Use
Methane	Liquid anhydrous ammonia	Fertilizer
Methane	Carbon black	Tires
Methane, propane, butane	Methyl alcohol	Antifreeze
Methane	Formaldehyde	Preservative
Ethane	Ethylene glycol	Permanent antifreeze
Ethane, propane	Polyethylene	Squeeze bottles
Ethane, benzene	Styrene	Synthetic rubber, polystyrene
Ethane	Ethyl alcohol	Solvent, antifreeze
Propane	Epoxy resins	Adhesive
Propane	Acetone	Solvent
Butane	Acetic acid	Vinegar, chemical processing
Butane	Butadiene	Synthetic rubber
Propane	Plastics	Synthetic rubber
Xylene	Polyester	Synthetic fiber
Benzene	Nylon	Synthetic fiber
Toluene	Trinitrotoluene (TNT)	Explosives

10 percent of the consumption of the two resources. The starter materials in the petrochemical industry are methane, ethane, propane, and butane, all obtained from natural gas or the distillation of crude oil, and benzene, toluene, and xylene, produced by the catalytic cracking of crude oil. Some petrochemical products may result from more than one line of chemical descent from fossil-fuel ancestors. Nylon, for instance, may be derived from coal, petroleum, natural gas, or sugar.

Discovery and early use

No one knows when the highly viscous form of petroleum called asphalt was first used, but it was probably at least six thousand years ago. There are many references to petroleum in the Old Testament; the builders of the tower of Babel used "oil out of the flinty rock" (Deuteronomy 32:13). At various times and places petroleum was held to have medicinal properties. Natural flares of gas at seepages near Baku in the Caucasus were visited annually by thousands of fire-worshipers. Marco Polo wrote that he saw in that region in the thirteenth century, "a fountain from which oil springs in great abundance, inasmuch as a hundred shiploads might be taken from it at one time."

Natural gas was known and used in China in the twelfth century; wells were dug to depths of several hundred feet, and bamboo pipes were used to lead the gas to places of use. It is believed that the high and steady temperatures required to make the famous Chinese porcelain may have been obtained by the use of natural gas.

An oil well was completed on the island of Zante in the Ionian Sea about 400 B.C. Ayres and Scarlott (1952, p.18) note that a derivative of the oil from this well was used in lamps, the first recorded use of petroleum for lighting. By 1000 A.D. there were producing oil wells in Burma, and asphalt seeps in Mexico were being used by the Toltecs. In 1640 an oil well was completed near Modena in Italy; kerosine from this well was used to light the streets of Genoa and Parma as early as 1803, by which time Romanian oil also was being used for lighting.

The first substitute crude oil was manufactured in England in 1694 by retorting oil shale and cannel coal. In 1815, according to Ayres and Scarlott, the year of the Battle of Waterloo, commercial oil-shale retorting was started in New Brunswick, Canada and, between 1850 and 1860, more than fifty commercial plants were constructed in the United States to distill oil from shale imported from Canada as well as from cannel coal brought from Scotland. The substitute oil from such plants was very expensive; the selling price of between thirty and forty dollars a barrel restricted it to such uses as lighting the homes of the prosperous and the workbenchs of certain craftsmen.

The famous Drake oil well completed in 1859 in Pennsylvania was far from the world's first oil well, as it has often been called. Neither was it the first oil well in North America, nor even the first in the United States. In 1829 a successful oil well started producing in Kentucky and, in 1857, a well was completed in Ontario. The Drake well came in at a propitious time. Kerosine, introduced in 1850, was superior to whale oil for lighting, but kerosine made from oil shale or cannel coal was very expensive. By 1859, it was known that kerosine could be produced far more cheaply from crude oil; it was the growing demand for cheap kerosine that accounted for the world's first "oil rush," set off by the success of Drake's well. From then on, the use of whale oil and the production of substitute kerosine (coal oil) declined swiftly.

The first modern distribution system for natural gas was completed in 1884, bringing gas to Pittsburgh from a field fourteen miles away. For almost sixty years, however, only cities close to gas wells found it possible to use natural gas; most of the gas used for lighting, heating, and cooking during those years was manufactured from coal or from liquid petroleum products such as naphtha, and was called "coal gas" or "city gas." The Welsbach mantle, introduced in 1886, tripled the light output of both kerosine lamps and gas burners and greatly increased demand for these fuels.

The explosive proliferation of the automotive engine in the early years of the twentieth century, as well as the increasing use of fuel oil in steamships and locomotives and for space and process heating, brought about a phenomenal growth in the demand for and consumption of crude oil in the industrial nations. Because discovery and production of crude oil from 1859 to about 1915 outpaced demand and consumption, prices were low. There was no incentive for conservation and crude oil and natural gas were wasted. There was a great incentive and opportunity to create price-controlling monopolies to produce, refine, and market petroleum. By 1930, demand had so caught up with supply that pessimism about the adequacy of future supplies was growing. By this time, however, successful monopolism (oligopolism is the more accurate but less used term) had created abundant capital in the oil industry, some of which could be diverted to more intensive and more extensive search for new reserves. From 1930 to the present that search has been highly successful; the global discovery rate has kept well ahead of demand, although the latter has continued to rise steeply.

The first offshore oil wells were drilled at Summerland in southern California in 1912; they were shallow wells drilled from piers that ran out from the shore. The first modern offshore oil well, not connected to the shore by any structure, was drilled off Louisiana in 1947. At first, offshore wells were drilled only in water less than fifty feet deep. Then, as techniques and equipment improved, drilling moved into deeper water and today there is already at least one completed well where the ocean water is eleven hundred feet deep. The technology of offshore drilling has opened the world's continental shelves and the upper portion of the continental slopes to exploitation for oil and gas. Within the past decade another hostile environment, that of the Arctic, has been invaded successfully by oil explorers and drillers, both in North America and Siberia.

The fast growth of military and commercial aviation, especially in the jet age, has added to the demand for crude oil. For the automobiles, buses, trucks, diesel locomotives, and powered aircraft of today there is no practical substitute for crude oil as a primary energy source. Because transport consumes between 25 and 35 percent of the gross energy input to an industrial society, crude oil has become one of the most important natural resources of the modern world.

"Reserves" and "resources"

The different categories of energy *reserves* used in the oil and gas industry, as well as by the United States Department of the Interior and the Energy Research and Development Administration, have given rise to some confusion, which has been exacerbated by the use of the terms *resource base* or *oil-in-place* to mean all the oil that existed at the start of man's exploration for it. *Resource base* has a solid sound and when numbers are attached to it and histograms based on it, the effect can only be described as legerdemain, for the viewer sees, or thinks he sees, something that does not exist.

Category			Definition
POTENTIAL	PROVED	Can be extracted at a profit under existing economic conditions and with available technology	Have been measured within small margins of geologic error by properly spaced drillholes or other openings
	PROBABLE		Have been calculated by extrapolation, based on geologic information and judgment, from drillholes or other openings that have penetrated commercial concentrations
	POSSIBLE		Lie, in areas of established production, beyond the boundaries of reasonable projection of probable reserves
SPECULATIVE	Geologic		Some reason to believe they may exist in areas of no present production and little geologic information
	Economic — Discovered		May become reserves if prices rise or costs diminish but material is now too lean, too far from market, or in reservoirs too refractory for economic recovery
	Economic — Undiscovered		

FIGURE 4.11
A classification of reserves of nonrenewable energy resources.

There is an accepted hierarchy of designations for reserves in the oil and gas industry, which more or less parallels a similar hierarchy in metal mining. What the oil and gas industry calls proved reserves, the metals mining industry calls measured reserves. Potential reserves, in descending order of certainty, are called, by the former industry, probable, possible, and speculative reserves; by the latter, indicated and inferred reserves.

Proved reserves are those that can be calculated from information obtained from drill holes closely enough spaced in a given pool or field that there can be little doubt of the continuity between the holes and therefore, of amount and recoverability of the oil or gas.

Probable reserves are based on an extrapolation from drill holes into or through one or more producing horizons in individual fields; the calculations represent probable extensions of known productive accumulations of oil and gas.

Possible reserves may be found outside productive fields but within the formations known to be productive. Possible reserves may, for example, represent an estimate of the probability of finding undiscovered traps within a geologic basin where there is already some production from one or more formations of wide extent.

Speculative reserves are those that may be found in geologic basins or in terrain where no oil or gas has yet been found but where the geologic makeup of the earth's crust is similar to that of regions that have yielded oil or gas or both. They may also include known accumulations of too low a grade or too high a cost to warrant recovery under present conditions that might become recoverable with improved technology or higher prices.

Nonrenewable Energy Resources

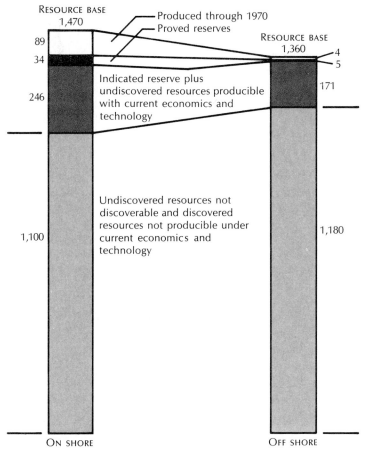

RESOURCE BASE
1,470

Produced through 1970
Proved reserves

RESOURCE BASE
1,360

89

34

246

Indicated reserve plus
undiscovered resources producible
with current economics and
technology

4

5

171

1,100

Undiscovered resources not
discoverable and discovered
resources not producible under
current economics and
technology

1,180

ON SHORE

OFF SHORE

U.S. Crude Oil Resources (billions of barrels)

FIGURE 4.12
*This sort of graphic presentation can be extremely misleading. Visually and arithmetically, knowns (production and proved reserves) are added to estimates that decrease in accuracy and probability of existence as they increase in amount. This specific figure has led to the inference that the United States has an adequate supply of oil resources, a conclusion that is not justified by the facts on which the drawing was based.
(From United States Department of the Interior, 1972.)*

These categories decrease exponentially in reliability from probable to speculative, and estimates for two categories should never be added together, but frequently are. From experience it has been found that measurements of even proved reserves can be in error by 25 percent; one can never know exactly how much oil is in a pool or can be economically extracted from a pool until that pool is exhausted. Exhaustion can mean that only water comes into the wells or it can mean that more oil can be extracted only by methods too costly for the returns obtainable.

A chart in which *resources* appear ready to be transformed into *reserves* by technology (through a lowering of the cost of production or a raising of the value of the use) is doubly misleading: it gives an appearance of concrete existence to material that may, in fact, not exist and it suggests strongly that technology always moves to make *reserves* out of *resources* when it may well move in the opposite direction by favoring a substitute material or may simply become stabilized, by the law of diminishing returns (or of diminishing marginal utility, as economists like to say).

A depiction of the *resource base* as large and discrete in relation to proved reserves and past production is bound to lead the unwary to the conclusion that there are vast underdeveloped oil and gas resources in North America and that, by our national energy strategy, we should subsidize exploration for those resources. That the size of the resource base portrayed is only an informed guess strongly flavored by hope and group interest, innocent of any definable relation to the physical economics of recovering the resource if it does exist, is hardly to be perceived, especially if one is influenced by a desire that the pretty picture be true.

The fact that ultimate recovery almost always exceeds the proved reserves measured at any time before a pool is exhausted is no argument for the hypothesis that reserves are continuously created. It is the rate of additions to proved reserves that is the sensitive barometer of depletion; when the curve of proved reserves starts to decline persistently, the handwriting is on the wall, and no hopeful estimates of the resource base or of speculative reserves are going to change the inevitable economic exhaustion of the resource.

TAR SANDS

Tar sands are sandstones in which the pore spaces are wholly or partially filled by congealed petroleum. When tar sand is heated to a relatively low temperature the tar liquifies and can be separated from the sand and sent to a refinery to be distilled and cracked into lighter and more fluid fractions. The Athabasca tar sands of Alberta constitute the largest known single deposit of petroleum in the world, estimated to contain 300 billion barrels of recoverable oil. This is an estimate based on technological judgment; how much of the oil may turn out to be economically recoverable is unknown.

Deposits of tar sand, because they originate as accumulations of petroleum trapped by tar seals within reservoir rocks, tend to have sharply defined boundaries; variations in tar content reflect variations in the original porosity and permeability of the sandstone reservoir. The tar sands of Alberta are now being developed on a relatively small scale by being mined by surface methods. Much of the estimated reserve lies at depths that will necessitate either underground mining or some in situ extraction process.

OIL SHALE

So-called oil shale, from which substitute crude oil can be made, does not contain petroleum in any form. It contains, instead, a solid bituminous material called kerogen, which, when heated to a fairly high temperature will yield a substitute crude oil that may be refined and treated like petroleum. Kerogen appears to consist entirely of plant remains, unlike crude oil, much or all of which appears to have been formed from animal remains. The oil shale deposits in Colorado, Wyoming, and Utah occur in old lake beds. Oil shale beds are found in a wide range of thickness and richness; a single bed may grade laterally from rich to lean. The richer oil shale in the United States is estimated to contain 80 billion barrels of substitute crude oil recoverable with available technology—about 0.8 barrel of oil per ton of shale. The major deterrents to the development of a large shale oil industry are the scarcity of water in the semiarid regions where the shale is found, and the adverse environmental effects of large-scale mining and waste disposal. Shale retorting requires considerable quantities of water, which cannot be reused and the small surplus of water available in upper Colorado River basin is coveted by other potential users. At present, more than a ton of shale must be mined for each barrel of oil produced, and the shale expands in volume during retorting. To put it back in the holes whence it came would be costly and probably impracticable; to dispose of it on the surface would ruin the scenery and invite pollution of both air and water because the spent shale is fine and light, easily picked up by winds, and readily leached of harmful chemicals by rainwater. Only the development of an economic process for retorting the shale in the ground, without mining it, and extracting the resulting oil through wells will obviate these environmental costs.

NUCLEAR-ENERGY RESOURCES

Although there are many radioactive isotopes, some natural and more that are manmade, only one natural radioactive isotope is fissionable: uranium 235 (U-235) which forms only 0.7 percent of natural uranium. It is the present basis of the nuclear-power industry. Other natural isotopes that are fertile, meaning that they can be made into fissionable isotopes not found in nature by being bombarded with neutrons in a reactor, are uranium 238, the abundant isotope (99.3 percent) of natural uranium, and thorium 232, which is at least as abundant in nature as uranium. The breeder reactor is designed to use the abundant energy of the fertile isotopes, which are largely wasted in the present generation of reactors.

Uranium occurs naturally in at least three distinctly different modes. Some uranium is found in vein deposits; much is found in tabular or nodular bodies within sandstone layers; and much is disseminated throughout large masses of sedimentary formations and igneous rocks.

Uranium does not occur in nature in a pure or elemental form. From the several minerals in which it is found, it is extracted as uranium oxide (U_3O_8), 84.8 percent of which, by weight, is pure uranium.

Vein uranium deposits have sharp boundaries; although veins may have pockets of high-grade ore containing more than 1 percent of uranium oxide, the average grade from the vein deposits of Marysvale, Utah has been about 0.20 percent. More than 70 percent of all uranium mined in the United States of America has come from deposits in sandstone formations in the Colorado Plateau region. These deposits tend to be uniform in grade, although ranging greatly in size. They are tabular or lenticular bodies, averaging only a few feet in thickness and a uranium oxide content of about 0.25 percent. The edges of the ore bodies are defined by an abrupt drop in grade; large halos of low-grade material are not common. Large amounts of uranium, but in concentrations too low to be economically recoverable at present (between 0.005 and 0.1 percent uranium oxide), are disseminated throughout certain sedimentary formations, such as the uraniferous lignites of North Dakota, the Chattanooga shale of the Ohio Valley, and the phosphate beds of Idaho, Utah, and Wyoming.

The geologic habitats of thorium are similar to those of uranium; it occurs also in veins, in sedimentary rocks (but less commonly than does uranium), and disseminated in large granitic bodies. Concentrations of thorium are similar to those of uranium, generally below 1 percent.

Uranium is mined by both underground and surface methods. Most of the uranium ore of the Colorado Plateau region has been taken from relatively shallow underground mines; the small ore bodies make this method preferable to surface mining because less waste rock has to be moved per unit of ore recovered. Certain shallow, sheetlike deposits in Wyoming and Texas are mined by stripping, in a manner similar to coal strip mining. If the very large, low-grade deposits of uranium and thorium, such as those in the Chattanooga shale and the Conway granite of New Hampshire ever become economical to mine—and they may if an efficient breeder reactor is introduced—they will probably be mined by surface methods.

Because of their low grade, uranium ores are concentrated in mills generally as close to the mines as technically and economically feasible. By a process requiring a copious supply of water, for which reason uranium mills are found along streams, a concentration of uranium oxide called yellow cake is produced. The yellow cake goes to a refinery where, by a chemical process, the metallic uranium is reduced from the concentrated ore. Then in the United States, an expensive operation called enrichment is undertaken to increase the portion of the fissionable isotope U-235 from its natural 0.7 percent to about 4 percent for the fuel elements for a nuclear power reactor. In Canada the uranium is not enriched but is used as natural uranium in the fuel elements. The process of enrichment packs much more energy into each fuel element and is said to result in a high conversion efficiency within the power reactor; proponents of the use of natural uranium claim a much superior overall system efficiency that offers a safety advantage, because of the design of reactor that can be used, and a political advantage in that Canada is not dependent upon the United States of America, until recently the sole source of enriched uranium.

For a number of years, all uranium produced in the United States was purchased by the Atomic Energy Commission (the AEC—now the Energy Research and Development Administration). Then some began to go to the infant commercial nuclear-power industry and now all newly mined domestic uranium goes into the open or commercial market. The AEC's uses were for weapons production, weapons experimentation, power research and other civilian-applications research. The agency ceased buying uranium, not because it stopped its research and development, but because it had accumulated a large stock of uranium.

GEOTHERMAL TRAPS

All geothermal energy so far exploited or known to be economically exploitable consists of heat contained in water and steam trapped in pockets within the upper part of the earth's crust. These pockets are of finite size and the heat is not renewable at any rate significant to man. Consequently, at present, geothermal energy is not a renewable resource. Only if heat can be extracted economically from the hot dry rocks that everywhere underlie the earth's surface at depths greater than a few miles and that derive their heat from the constant disintegration of radioactive materials within the earth, will geothermal energy become an income or renewable resource.

Geothermal traps, like oil and gas traps, require an impermeable caprock in a geometric configuration that prevents hot water and steam from rising to the surface. Such traps have been found in regions of recent volcanic activity in Italy, Iceland, New Zealand, Japan, Mexico, Siberia, and the western United States. Prospecting consists in a search for hot springs and for areas of unusually high heat flow. In recent years geophysical techniques based on changes in properties of the earth that depend on temperature have been developed.

Where natural hot springs exist, geothermal energy has long been used to heat buildings and greenhouses as well as to provide naturally warm and perhaps therapeutic baths. In 1904 at Larderello, Italy, the first electric power plant using geothermal steam to propel turbine blades was completed. More than half a century passed before the next geothermal power plant was constructed. Since 1958, about a dozen such plants have been built. The total installed geothermal generating capacity in the world in 1975 was about equal to that of a single large coal-fired power plant, about eleven hundred megawatts.

Geothermal heat traps are exploited by drilling wells into them, in a manner similar to that used in drilling for oil, and the completed wells are connected by large insulated pipes that lead the geothermal steam to the power plant. Geothermal steam commonly contains corrosive impurities hazardous to the turbine blades, which must be removed before the steam is introduced into the turbines.

LITHIUM

Fission is the disintegration of a heavy atomic nucleus into two lighter nuclei, fusion the integration of two lighter nuclei into one heavy nucleus. Both processes produce tremendous quantities of energy. Fusion power, if and when it becomes technically feasible and economically practical, may or may not require the isotope lithium 6 in addition to deuterium. Lithium, unlike deuterium, is not so abundant as to constitute a quasi-income resource. Lithium 6, which constitutes only 7.4 percent of natural lithium, is present at an average concentration of 7 parts per billion in seawater and between 1.5 and 2.4 parts per million in the crustal rocks of the earth.

Lithium is extracted from some natural brines (the water of the Great Salt Lake, for example) and from the mineral spodumene found in certain rather uncommon rocks. Dr. M. King Hubbert (1969, p. 222) has calculated that the known lithium reserves of the world, if all were used in production of fusion power, would allow the generation of fusion power approximately equivalent to the energy obtainable from the combustion of the world's initial recoverable supply of fossil fuels. That is a lot but, as we are beginning to realize about the fossil fuels, it has a limit.

SUPPLEMENTAL READINGS

Armstead, H. C. H., ed. 1973. *Geothermal Energy*. Paris: UNESCO.

Rather detailed review of research and development of the use of heat caught in geothermal traps.

Ayres, Eugene, and C. A. Scarlott. 1952. *Energy Sources—the Wealth of the World*. New York: McGraw-Hill. Emphasis on the finiteness of fossil fuel reserves and on the potential for their conservation and ultimate replacement by other energy sources.

Cloud, Preston. 1971. "Mineral resources in fact and fancy." In *Environment*, ed. W. W. Murdoch, p. 71–88. Stamford: Sinauer Associates.

Critical discussion of the limits to geologic resources.

Flawn, P. T. 1966. *Mineral Resources*. Chicago: Rand McNally.

Good discussions of the classes of reserves, kinds of mineral resources, and the various forms of ownership, exploitation, and taxation.

Harris, John R. 1974. "The rise of coal technology." *Scientific American*, v. 231, n. 2, p. 92–97.

The relation of the Industrial Revolution to the development of technology based on coal.

Hubbert, M. K. 1969. "Energy Resources." In *Resources and Man* (National Academy of Sciences, National Committee on Resource and Man), p.157–242. San Francisco: W. H. Freeman and Company.

This review includes an informative discussion of nuclear power, nuclear resources, and radioactive waste disposal.

Levorsen, A. I. 1954. *Geology of Petroleum*. San Francisco: W. H. Freeman and Company. A standard text, well illustrated.

Nef, J. U., 1932. *The Rise of the British Coal Industry* 2 vols. London: Frank Cass, reprinted, 1966.

A detailed and illuminating history that explores the complex relations of deforestation, the growth of British industry, and the rise of coal.

Tiratsoo, E. N., 1967. *Natural Gas*. New York: Plenum. An exhaustive treatment of the origin, composition, transport, and use of natural gas.

San Francisco Maritime Museum

Chapter 5

THE PHYSICAL ECONOMY OF ENERGY USE

The flow of energy should be the
primary concern of economics.

—Frederick Soddy, 1933

PHYSICAL ECONOMICS

Until man discovers how to make food out of inorganic materials, he, as well as all other animals, is dependent upon plants for the energy that sustains life. Man is a parasite in relation to plants; they can live without him, but he cannot live without them. Man burns and eats the carbohydrates of plant tissue and in so doing reduces those carbohydrates, with the aid of oxygen from the air, to carbon dioxide and water. In this process heat is liberated and cannot be recycled for it enters the entropy sink from which there is no escape. The carbon dioxide and water, however, can be recycled, but only by plants. In the presence of chlorophyll, the energy of sunlight, replacing the energy of combustion, is used by plants to reunite the dead products of combustion and metabolism into living carbohydrate tissue from which oxygen is released again into the air. This is the most fundamental and vital recycling operation in the biosphere; man destroys plants at his own peril.

A lesson of general application is that life is cyclic as far as material substances are concerned, with the same materials being used over and over in metabolism, but life is unidirectional as regards energy. The flow of energy is the primary concern of physical, as perhaps it should be of pecuniary economics.

Physical economics is the study of the benefits and costs of man's activities in terms of the physical laws that govern change, the laws of mechanics, chemistry, electricity, and heat and especially the laws of motion, thermodynamics, and work. Change always involves energy and usually work. There are several kinds of change. A football moving from a quarterback's hand to the arms of a wide receiver represents a change of *position*, as does a potato lifted from the ground. An ice cube melting represents a change of *phase*. A car accelerating on a curve represents a change in *velocity* and *direction.* A soft tire being inflated to hardness represents a change in *pressure,* and a pot of coffee being heated represents a change in *temperature.* In each of these examples, energy has been transferred and work has been done. The study of energy transfers, conversion efficiencies, and work is properly known as energetics. Physical economics deals with the energetic benefits and costs to man of energy transfers and conversions.

In the case of the thrown football, there is an energy cost but no energy benefit; the return is a service called recreation or entertainment, and energy must be diverted from some other system in which there is surplus to make up the deficit in the football system.

The lifted potato, from the viewpoint of energetics, is similar to the thrown football. Energy has been consumed in order to achieve a change in the position of an object. But from the viewpoint of physical economics, the two actions are very different since one is part of an energy-dissipating system called sport and the other is part of an energy-producing system called agriculture. The thrown football represents an unproductive discretionary expenditure of accumulated energy, whereas the lifted potato is an investment of accumulated energy in a process designed to produce more energy.

The ice cube represents a more complex system. The energy system whose end product is an ice cube very likely begins in a coal mine or a natural gas well. To make the ice cube requires an expenditure of electrical energy (to run the motor of a refrigerator) probably produced through the burning of a fossil fuel. An energetic analysis would involve determining how much energy in the fossil fuel was required, with all conversion and transmission losses accounted for, to produce the ice cube from water at or near room temperature, as well as how much heat would be required to melt it (because that heat is likely to have been produced for a purpose, such as space heating, other than melting ice cubes). Then, entering physical economics, we would calculate the amounts of nonrenewable, renewable, and human energy that had been used to make and melt the ice cube and compare these physical costs with the benefits obtained from the use of the ice. In an affluent industrial society, ice in the home has come to be regarded as a necessity; a careful analysis of the energetics will show how much household ice making is dependent on fossil fuels (in most of the United States) and why the ice cube is a luxury where fuel is scarce.

The car, the tire, and the coffee pot can be analyzed in the same way; each indicated change requires work to be done and, in each case, the energy required comes from two kinds of sources, inanimate and human. The inanimate sources (fossil fuels, waterpower, and so on) yield the energy required to run the car engine, the air compressor, and the hot plate or burner on which the coffee pot rests; they also are the sources of much of the energy required to fabricate the car, the compressor and the tire, the coffee pot and its heater, and all the conversion and transmission devices needed.

Human energy is involved in the entire picture. One of the marks of an industrial or high energy society is that most manpower is expended, not in physical effort, but in supervision, control, and guidance. The energy contribution from inanimate sources for mechanical, thermal, or electrical work in the high energy society is commonly much greater than that from humans. The more energy we can extract from coal, crude oil, or falling water by our converters and transmission devices, the less the unit cost of useful heat or power will be in terms of human effort.

Over the past hundred years the efficiency of the energy systems of industrial countries has increased. Still, the aggregate efficiency in the United States is only about 36 percent. In other words, almost two-thirds of the energy we put in does not come out the other end in a form we can use. This fact alone would make the study of physical economics seem worthwhile. But there are other matters of importance and perhaps we can get a new and useful view if we approach them in the light of the physical economy of energy and work.

Ultimately what is called economics (but is, rather, pecuniary economics) must take into account the physics of energy as well as the behavior of men. The concepts of energy, power, and work are keys to an understanding of what man can and cannot do with the earth's physical resources. Money, capital, and profit are concepts that relate to man's desires and values insofar as they can be realized, or are perceived to be realizable, through the use of natural resources, direct and indirect. In addition to the values he places on the resources that sustain life and make it comfortable, man assigns value to scarcity, to beauty, to speed, to leisure, to power over other men, and—in industrial society—to time. All these enter into pecuniary economics and each has an impact on how resources are used, how energy is spent, what power is used for, and to what ends work is expended. Gold and diamonds are not very useful, but man assigns great value to them, because of their scarcity and beauty. Modern man puts a great value on time, regarding a resource available this year as more valuable than that same resource would be if it were not available for ten years. When resources are abundant, they are valued mainly by the amount of work it takes to prepare them for use; when they become scarce, their value starts to rise above that base toward the utility or commodity value of the resource in the society. Petroleum is many times more valuable to an industrial society than its cost in work; recently its price has risen far above its work cost, toward its real value to the consuming society.

Food, the fossil fuels, and uranium are economic sources of energy only if they can be obtained at an energy or work cost that does not exceed the energy or work benefit that can be obtained from them.

This, the fundamental law of energy use, defines energy resources and provides the standard that all potential energy resources and energy technologies must meet. Unless the ratio of benefit to cost, measured in units of energy, is greater than 1:1, the potential resource will fail to become an actual resource, and the new technology would drain the economy rather than enlarge it.

Estimates of crude oil in the ground, calculations of the earth's total heat content or of the rate at which solar radiation is intercepted by the earth have little meaning in terms of energy resources. The moment that more energy is required to find, extract, process, transport, and use a barrel of crude oil than can be obtained from it or in exchange for it, there will be no more potential reserves of petroleum, and the amount of crude oil then remaining in the ground, no matter how large, will be of no significance to mankind. The heat content of the earth's crust and the rate of incoming solar energy, both demonstrably huge in terms of man's present energy needs, are also of little significance to man unless he finds means of tapping these two energy sources in ways that accord with the first law of energy use. At present, he has found no such way for either diffuse goethermal energy or diffuse solar radiation. The conversion of either kind of energy is technologically feasible, but neither is economically feasible. Technological feasibility neither equals nor forecasts economic feasibility. The prevalent concept that almost anything will become economically possible if enough money is spent on research and development is supported neither by history nor by the logic of physical economics.

THE CONCEPT OF NATURAL SUBSIDY

Man vainly seems to feel that he does or directs all the work done on earth. Rivers perform work in moving countless tons of sediment to the sea and in cutting canyons and valleys. The effect of the sun's radiation on the earth lifts water vapor out of the sea, causes winds to blow and transport that water over many miles, provides energy for plant growth, and breaks bare rock by repeated heating. Convection currents within the earth drag crustal plates around, raise continents, and cause mountains to form.

Nor can man pretend that he does all the *useful* work on earth. The physical and chemical processes of soil formation are useful, in fact vital, to him, as is the work done by natural forces in driving the hydrologic cycle. The work done by nature in forming deposits of the fossil fuels and the industrial minerals and in bringing many of these deposits to the earth's surface or close to it is of enormous value to industrial man. The work of photosynthesis, carried out entirely by plants, provides the world with all its food.

Western man in particular has been loath to admit his dependence on nature. The story of the Garden of Eden shows first his conviction that he really should not have to work at all, second his belief that he was condemned to a life of sorrowful toil because of his violation of the house rules, third his feeling that the punishment was far too severe for the crime, and fourth that the sentence will one day be annulled and followed by suitable compensation for at least the more deserving among the petitioners. Such mass self-pity, culturally transmitted for more than 2,000 years, leaves little room for recognition of the rather small amount of work man must do of all that needed to support him on this planet.

Man, in short, is the continuing beneficiary of a subsidy in the form of work performed by natural forces of which he takes advantage. How much subsidy he derives depends mainly on his ability to use the results of the work of nature; it depends, in other words, on his agricultural, mining, and power technologies. It may be thought that it depends too on population, but in this the relation is not as straightforward as it is with the stage of technological development; indeed it appears that very large populations may make less total use of the natural subsidy than do some smaller populations, a point to be returned to later in this chapter.

The natural subsidy is not uniformly distributed throughout the world. Some nations have abundant large streams that can be used to power water mills or to run the generators of hydroelectric power plants; other nations have few such streams. Some nations have broad areas of deep fertile soils; other nations have deserts instead, or the thin poor soils of the humid tropics. Some nations have abundant supplies of the fossil fuels; others have none, or almost none. Some nations have, or had, thick forests; others have virtually no timber or fuel wood. Some nations have regular seasons and a predictable rainfall with small variation from year to year; other nations have much less predictable climate and suffer from both floods and droughts. Some nations have large forests, good soils, many rivers, large deposits of coal, accumulations of oil and gas, *and* a predictable climate; other nations are endowed with few if any such good things.

MEASURING THE SUBSIDY

The amount of work that went into creating a thick coal bed cannot be determined accurately, although the amount of solar radiation falling on a field during a growing season can. Neither, however, gives us the measure of the subsidy to man, which is the amount of energy or work profit he is able to realize from the mining and use of coal, or from the growing of crops. Because the first law of thermodynamics prevents one from getting more energy from a system than has been put into it, any excess or profit man is able to obtain from his work must represent a contribution of energy from a source other than himself. It is this energy profit that is the measure of the subsidy.

Let us consider a few examples of natural energy subsidy. First, a very simple example, that of water falling through the turbines of a hydroelectric generating plant. Such turbines and the generators to which they are connected are about 95 percent efficient in converting the kinetic energy of falling water into electric energy. Man's work consists in the effort required to prepare the dam and reservoir site, to design and construct the dam, to mine and transport the cement and aggregate used in the dam, to mine and process the metals used in the water conduits, turbines, generators, wires, and transmission-line towers, and to install, maintain, operate, and replace the equipment and the transmission lines. The energy profit or subsidy is measured by subtracting the cost of all this work for the estimated life of the reservoir from the total electrical energy transmitted from the dam, measured at the points of use, over the life of the reservoir. For a large dam and reservoir, such as Hoover Dam on Lake Mead, the energy profit (or subsidy) is enormous.

Next, coal. In 1900, the average daily energy consumption required to sustain a coal miner and his family was probably about 200,000 kcal; the per capita daily energy consumption at that time was 90,000 kcal and if one assumes that more than half of that was going into physical plant and other stocks not directly benefiting coal miners and one takes 40,000 kcal as the share of each of the members of a miner's family of five, the total of 200,000 kcal probably is still a generous estimate. Interest, amortization, maintenance, and operation of mine equipment may have equaled another 15,000 kcal per miner, for a total of 215,000 kcal. He produced in one day about 32 million kcal or 150 times the energy required to maintain himself as a miner and to sustain his family. In 1900, in the United States, a ton of coal represented the average daily energy consumption of 70 persons. In a six-day work week, a miner's production represented the total weekly energy consumption of 200 persons. The rate of energy consumption at the turn of the twentieth century in the United States was four times that of an advanced agricultural society; the surplus came mainly from coal and was transformed rapidly into industrial capital, new technology, and cheap food and goods. Energy from coal (and later, oil) provided the tremendous subsidy required to transform an advanced agricultural nation quickly into an advanced industrial nation.

The energy profit or subsidy obtained from the winds that drove sailing ships was also large, so large that the early steamships, even with the large natural subsidy represented by the coal they burned, were not as efficient as well-designed sailing vessels. They were used

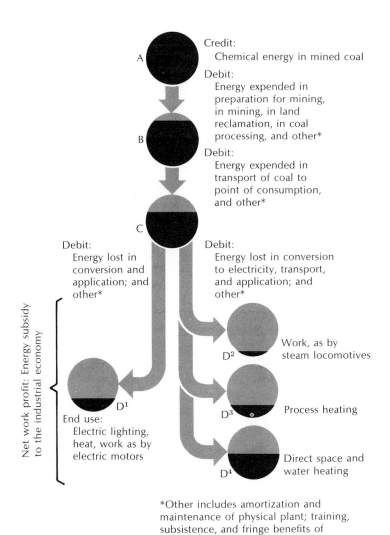

A

Credit:
　Chemical energy in mined coal

Debit:
　Energy expended in
　preparation for mining,
　in mining, in land
　reclamation, in coal
　processing, and other*

B

Debit:
　Energy expended in
　transport of coal to
　point of consumption,
　and other*

C

Debit:
　Energy lost in
　conversion and
　application; and
　other*

Debit:
　Energy lost in conversion
　to electricity, transport,
　and application; and
　other*

Net work profit: Energy subsidy
to the industrial economy

D¹

End use:
　Electric lighting,
　heat, work as by
　electric motors

D²   Work, as by
　　　steam locomotives

D³   Process heating

D⁴   Direct space and
　　　water heating

*Other includes amortization and
maintenance of physical plant; training,
subsistence, and fringe benefits of
workers; all expressed in equivalent energy

FIGURE 5.1
*The amount of energy profit derived from
the natural subsidy in coal depends upon the
end use.*

mainly for their speed and reliability of schedule, and the increased
speed was purchased by an increase in energy consumption, but more
importantly, by consumption of a nonrenewable resource (coal) in
place of a renewable resource (wind).

　Characteristic of industrialized or high energy societies is the use of
part of the fossil and nuclear fuel subsidy to finance agriculture. Those
primitive hunting and agricultural societies that have achieved a
culturally directed equilibrium with their environment harvest more
food energy than they use up in the annual labor of hunting and agriculture. The profit is taken, not in crop surpluses, but in leisure time

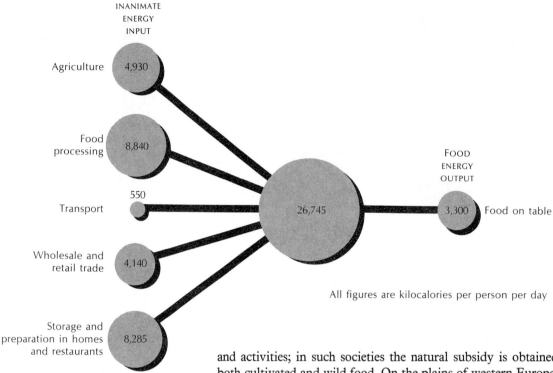

INANIMATE ENERGY INPUT

Agriculture 4,930

Food processing 8,840

Transport 550

Wholesale and retail trade 4,140

Storage and preparation in homes and restaurants 8,285

26,745

FOOD ENERGY OUTPUT

3,300 Food on table

All figures are kilocalories per person per day

FIGURE 5.2
About 3,300 kilocalories of food energy is set on the table each day for each American. It takes more than 8 times this amount of energy in fuel and electricity to get that amount of food to the table ready to eat. (Data used are for 1973 and are from Hirst, 1973.)

and activities; in such societies the natural subsidy is obtained from both cultivated and wild food. On the plains of western Europe in the Middle Ages cultivated food provided a surplus that was not taken in leisure, but was sold or traded for services and goods, some of which could be devoted to production of more energy from the soils and flowing waters. In parts of the world where there have been no cultural constraints on population, the natural subsidy from the land has been used, neither in leisure nor in surplus goods and services (a higher material standard of living), to support a higher population. In other words, population has enlarged to the limit of food available.

As soon as he could, western man began to mechanize agriculture and to subsidize it from other sources of energy in order to increase the food productivity of each agricultural worker or of the time spent working. In the industrialized world, tractors and other mechanical equipment have replaced horses and men, fertilizers have replaced manure, and herbicides, fungicides, and pesticides are used in large quantities. All these objects and substances are made, transported, operated, or applied by inanimate energy. About five units of fossil energy go into every unit of food energy produced at the farm gate in the United States. When one considers the energy used today in transporting, canning, freezing, refrigerating, storing, processing, and marketing farm products, almost all of which comes from the fossil fuels, one can see that food production and processing in a modern industrial society are heavily subsidized by the fossil fuels. What we do is to assign part of the natural subsidy we derive from coal, natural gas, and petroleum to the production of food, while other portions of the subsidy go to enlarge the physical plant of the nation (its factories, transport facilities, commercial, and residential buildings), to support of the growing service sector of the economy, to leisure, and to the collection of artifacts.

Because so much of the natural subsidy is used by the industrialized or high energy societies, the human being no longer is an economical source of power. A man working for a year can deliver only about three dollars worth of mechanical energy, calculated at a rate of two cents a kilowatt-hour.

The natural subsidy is available only to those who know how to exploit it. This is the realm of energy and power technology.

THE ROLE OF TECHNOLOGY

Without technology, man was limited in the amount of natural subsidy he could draw from nature. If we regard spears and hoes as implements of power technology, man as he defines himself could not exist without technology. With relatively simple tools and weapons, man gathered and grew, killed and harvested, and developed the capability of producing and even storing energy surpluses. Some nations, such as Egypt and Sumeria, obtained sufficient surplus energy from agriculture to enable them to subsidize the development of technologies for winning precious stones and metals from their native rocks, and for firing clays and metallic ores used to produce bricks and weapons and ornaments. For several thousands of years, mining and metallurgy would not have been economic activities in a physical sense, for much more work went into them than could be obtained from them. But work, or usefulness in terms of energy has never been the object of man's passionate quest for gold or diamonds or prestige, and this is one of the reasons that physical economics cannot easily take the place of social economics, where desire plays such a large part in demand.

Food surplus supported slaves and other work animals. The slaves could be employed in many ways. They could mine, smelt, and fabricate. They could be drafted into armies for defense or conquest. They could, especially in the offseason for field work, form great corvées for pyramid building and other stately endeavors. They could man galleys. They could walk treadmills. They could be used in spectator sports as gladiators or human mice for large cats to tease. But all this was supported by agricultural surplus. Unless the population, including that of the slaves and those other animals whose food came from cultivated crops, was kept within bounds, there would be no surplus. The fertility of the soil had to be maintained. In Egypt, the annual floods of the Nile took care of this potential problem, but elsewhere, soils became depleted and eroded, reducing or wiping out the energy surplus of the society. George Perkins Marsh, a pioneer of conservation in the United States of America, argued (1864; 1965, p. 9–13) that the fall of the Roman Empire was due to deforestation (overharvesting of timber because of population pressure), which led to soil erosion and siltation, which diminished harvests and weakened the nation.

But food surplus produced by manpower alone is limited, even under favorable environmental and cultural circumstances. It takes agricultural technology and the application of power greater than that of men to produce a substantial surplus. That first occurred in northwest Europe in the Middle Ages. There, on the plains of southern Germany and northern France, productivity was increased by use of the heavy plow, three-crop rotation, and manure, as well as by the increasingly efficient use of draft animals through such important inventions as the shoulder-bearing harness and the iron horseshoe. Laborers no longer needed on the farm were available to work in the water-driven mills, in the mines, or on the ships and barges of commerce.

Such *economic technology* improves the efficiency with which a resource is used. It may reduce waste of the basic material involved, it may reduce the energy cost of production, it may make the product more durable or more effective. Power and heating devices of the early

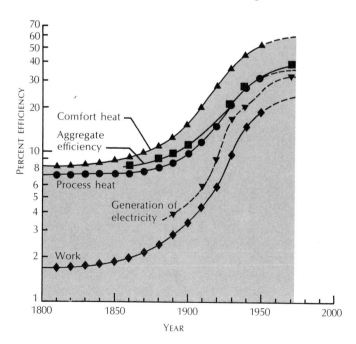

FIGURE 5.3
For about 100 years the efficiency of energy use in the U.S. rose: comfort or space heating, from 8 to almost 60 percent; process heat for steel mills, glass furnaces, and metal working, from 7 to about 35 percent; the mechanical work of motors and machines, from less than 2 to more than 20 percent. Electricity generation shows the sharpest rise of all. Now, all seem to have reached plateaus. Note the logarithmic vertical scale. (The data used for 1800–1950 are from Palmer C. Putnam, 1953, p. 90.)

stages of the Industrial Revolution were very inefficient by modern standards. As late as 1860, the aggregate efficiency of the energy system of the United States was only about 8 percent; in the following hundred years that figure rose almost fivefold. Economic technology improved the efficiency of space heating from the 5 percent of fireplaces to the 75 percent of a modern furnace, increased the efficiency of fossil-fueled electricity generating plants from 15 percent to 40 percent with fifty years, reduced the cutoff grade for minable copper ore from 5 percent to 0.5 percent in seventy years, and in forty years reduced by half the power required for aluminum smelting. For more than a century, increases of technical efficiency in the use of energy have kept ahead of increasingly adverse geological conditions; as a consequence, net work profit derived from the natural subsidy has increased and the unit cost of energy has decreased.

Economic technology expands a known resource. Because of advances in copper mining, milling, and smelting, millions of tons of copper-bearing material have become ore, whereas they once were waste rock with some copper in it. Because of advances in the design and operation of steam power plants, the great increase in the production of electric power in the United States between 1920 and 1970 was about two-thirds "paid for" by increased efficiency; only one-third of the new electricity generated in that period needed more fossil fuel than was used in 1920 to generate electricity. This is an example of extending the product by better use of the raw material. Similarly, a great deal more useful heat is obtained today from a unit of energy than was obtained fifty years ago.

Another kind of technology, that might be called *provident technology,* creates a resource from material, an energy potential, or flux that did not previously constitute a resource, or provides a device or system for the conversion or use of the resource that in one swoop creates a vast demand for a resource previously of comparatively little use. Examples of provident technology are the wheel, the sail, the stirrup, the geared water mill, the steam engine, the electric motor, the automobile, and the atomic pile. It was provident technology that created from an iron-bearing rock called taconite a great new resource, the present basis of more than half the iron ore produced in the United States. It is provident technology that, we hope, will make fusion power a reality.

LIMITS TO WORK PROFIT
FROM ENERGY RESOURCES

The work profit or natural subsidy to be gained from renewable energy resources such as plant food and running water theoretically has no limit in total quantity or duration of availability. The limit is one of rate, not quantity, and the sustainable rate of profit depends on the rate at which the natural energy flows are tapped and on the work cost of tapping them.

The work profit to be gained from nonrenewable energy resources theoretically is limited only by the total quantity of the resource that can be recovered and used economically, although the capacity of an environment to absorb waste and the willingness of a community to endure pollution may put local limits on the rate. There are several ways by which a limit to work profit from a nonrenewable energy resource may be reached.

The resource may become physically exhausted. Although physical consumption of all the crude oil, coal, natural gas, or uranium in the world is highly improbable, under exceptional circumstances local deposits can be exhausted almost completely. Cerro Azul Number 4, in Mexico, one of the world's greatest oil wells, after producing 60 million barrels of oil suddenly produced nothing but salt water. Although some oil may not have been flushed out, that pool was exhausted and there was no hope of ever producing more oil from it no matter how high the price might go or how efficient new extraction technology might become. Probably most natural gas accumulations are physically exhausted if the extractive engineering has been carried out properly. It is likewise possible to mine out completely a lenticular coal bed by surface methods. High-grade uranium deposits usually are removed completely.

Much more common, however, than physical exhaustion of a resource is economic exhaustion. For every barrel of oil so far produced in the United States, more than two have been left in the ground. It is difficult to flush oil from a "tight" formation, that is, one with low permeability. Although it might be porous, the tininess of the pores or their lack of interconnection keeps much of the oil in the reservoir rock. In some cases, only 10 percent of the oil contained has been recovered. It has been found that the cost of recovering oil after primary recovery (which includes pumping) by any of various secondary-recovery techniques increases exponentially with the percentage of additional recovery. In other words, even a strong rise in the price of crude oil cannot pay for much more recovery effort.

Something rather similar happens in underground coal mines. In the first place, there is an economic limit to the thinness of a minable coal seam: minimum headroom for miners and equipment is about thirty inches, and to mine seams thinner than that entails removing waste material along with the coal, to be separated later. The costs of mining and waste separation rise exponentially as the thickness of the coal bed decreases. In the second place, substantial amounts of coal must be left in position to support mine openings; replacement of the coal pillars by mechanical supports or by waste material mined elsewhere would be prohibitively expensive. For these two reasons, the average recovery efficiency in underground coal mining is now about 57 percent, much better than the recovery efficiency for crude oil, but still indicating that an enormous amount of energy is left in the ground.

Economic exhaustion of an oil or gas field, a uranium or a coal mine can come about because of decreasing size of the individual accumulations (oil pools, ore bodies, or coal beds), because of increasing depth, or both. The cost of making holes in the ground increases exponentially with increasing depth. Increasing depth also entails greater difficulty in maintaining openings against higher hydrostatic and formation pressures—formation pressures being those encountered at depth that exceed the pressures to be expected from the weight of the overlying rocks and fluids. In general, the permeability of rock decreases with depth, the formations become tighter, and recovery of oil and gas from them is more difficult.

The work required to find and produce energy materials increases with the increasing hostility of the environment. An offshore oil or gas well in North America costs twice as much to drill and equip as does an onshore well of the same depth. The average Arctic well probably also costs about twice as much as a well drilled on land to the same depth elsewhere in the United States.

Although great advances have been made in the efficiency of transporting energy, distance from the centers of use plays a large part in the physical economy of energy systems. It costs about three times as much to haul coal from Montana to Indiana by unit train as it does to mine the coal and, although these costs are affected by labor practices and interest and amortization schedules, the three-to-one ratio is an indication of the relative work costs involved. Likewise, the costs of the intercontinental transport of natural gas in liquid form are about three times the cost of the gas at the wellhead.

Energy resources have ecological and environmental limits. The surface of some land areas is too valuable to permit mining, in spite of the profits that would be foregone. For instance, where the surface is valued for its scenery, ecology, or history, an implicit public assignation of a social value may outweigh the value of the energetic profit that could be derived from mining. In such places, the underlying material is not a resource, although it may later become so should the surface be devalued. With present technology, the use of some energy materials, such as high-sulfur coal, may be barred because the costs of using it without degrading the environment are considered to be too high. The remaining deposits of high-sulfur coal no longer are resources, although they may become reserves again if some new technology can, economically, solve the pollution problem. There may even be a social hazard. The hazard of blackmail with the threat of a bomb made from stolen plutonium, added to chances of accident and risks to health, may be evaluated as a social cost that outweighs the work profit to be gained from using an energy resource such as uranium in fission power plants. If such an evaluation were made, uranium would still be an energy resource, but only for bombs and other explosive devices.

It is even possible to visualize a worldwide decision to limit the disposal of wastes that would halt the use of fossil fuels while large amounts of recoverable coal remained yet unmined. The consequences of the continuing accumulation of carbon dioxide in the world's atmosphere may someday be adjudged too costly to warrant the continued burning of the fossil fuels for energy, and remaining reserves would then be devoted to other uses.

THE EXPONENTIAL IMPERATIVE

At least three kinds of costs follow exponential curves: the costs of speed, the costs of scarcity, and the costs of perfection. To a large degree, the economic efficiency of production has been the result of speed. Industrial historians tell us that American industry caught up with and surpassed the maternal British industry by running cheaper machines faster. As long as speed was cheap in human terms, it was an attractive way to increase the material produce of the natural subsidy. But inevitably we have come to the steep part of that curve: the cost per seat-mile in the Concorde is more than *three times* that of a seat-mile in the 747, and as the cost of fuel increases, the gap widens.

It was also inevitable that we should come to the steep part of the scarcity-cost curve. Nonrenewable energy resources are found in discrete concentrations that are tens to thousands of times greater than the average crustal abundance of the desired material. Not only do they have sharp boundaries but also they are small, occupying much less than one percent of the accessible volume of the earth's crust. Furthermore, they were formed near the earth's surface, say within forty thousand feet.

The more accessible and richer deposits are found and exhausted first. Increasing depth, leanness, and distance from centers of use impose rising costs as consumption progresses. The cost of drilling oil and gas wells, for example, doubles with each thirty-six hundred feet of depth, which means that a well drilled to thirty thousand feet, only six times as deep as a five thousand foot well, costs 130 times as much. To see the practical significance of this drilling cost curve, let us look at natural gas in Texas. All the natural gas so far discovered in Texas has been found at an average depth of about eight thousand feet. For

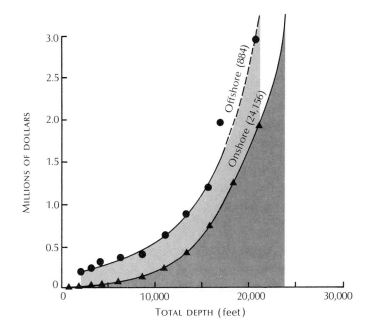

TOTAL DEPTH (feet)

FIGURE 5.4
The cost of drilling oil and gas wells increases steeply with increasing depth. This diagram is based on data for wells drilled in the U.S. in 1971, given in the Joint Association Survey for that year published by the American Petroleum Institute, Washington. The curves reflect the reported costs of drilling 25,040 wells, including dry holes.

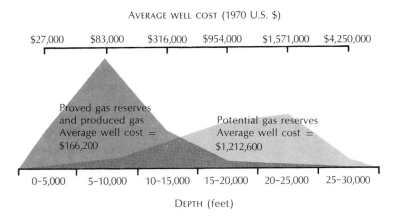

DEPTH (feet)

FIGURE 5.5
Potential natural gas reserves in Texas, if they exist where geologists think they may, will be much more expensive to produce—because of greater depth—than will the proved gas reserves.

geological reasons, substantial undiscovered or potential gas reserves are believed to lie at an average depth of somewhat more than twenty thousand feet. The cost of drilling and completing wells to that depth is more than seven times the cost of drilling and completing wells to the shallower depth. Thus several major gas-producing and pipeline companies are spending millions of dollars to import liquefied natural gas (LNG) and to build plants to make substitute natural gas (SNG) from coal; both LNG and SNG of pipeline quality (one thousand Btu per cubic foot) will cost five to six times the average wellhead price in 1976 for domestic natural gas (in interstate commerce), but these corporate decisions demonstrate a conviction that it will be even more costly to find and produce large new supplies of natural gas in the United States.

The cost of offshore drilling and production platforms increases with the depth of water in which they are placed. Oil from the North Sea will be expensive because it is found under deeper water than were older gas fields.

The exponential imperative makes it increasingly costly to get the last sulfurous whiffs out of effluents from an oil refinery. One of the principal gaseous pollutants in refinery effluents is sulfur dioxide (SO_2). As refineries in the United States process more and more high-sulfur imported crude oil, the containment of sulfur within the refinery will be a growing problem. The easiest way to get rid of sulfur has always been to oxidize it, discharge the oxide through a tall stack, and let nature's winds dilute and distribute the fumes. Rainfall causes sulfuric acid fallout and, recently, Sweden has complained about sulfuric acid fallout believed to have originated in British plants, an indication of how far harmful concentrations of SO_2 may be carried. It is not technically impossible to prevent the escape of SO_2 from a refinery, but the cost of controlling it rises with the amount to be contained. Likewise, the total cost of pollution control devices, both in the production of automobiles and in the waste of fuel they engender, rises with the degree of pollution abated.

The exponential-cost curve, whether based on energy or monetary units, can be useful for analysis and decision, if we can determine at

FIGURE 5.6
Drilling platform costs in the North Sea have been increasing sharply, partly because of inflation but more because of the increasing depth of water in which they are being installed.
(This figure is based on information given in The Oil and Gas Journal, *Nov. 25, 1974, p. 56.)*

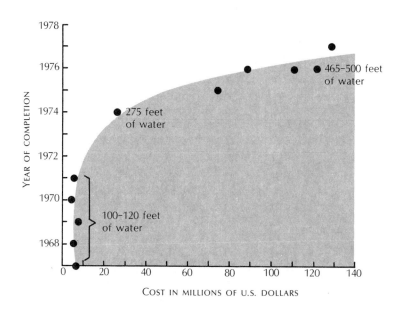

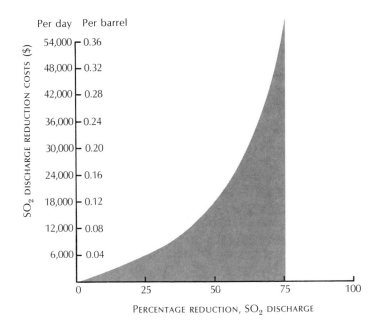

FIGURE 5.7
An example of the cost of approaching perfection. As the degree of reduction of sulfur dioxide discharged from an oil refinery approaches 75 percent, the reduction of another percentage point costs much more than did a one percent reduction earlier.
(Figure is redrawn from one in the 1972 annual report of Resources for the Future, Inc., p. 42.)

what point on the curve (assuming that all costs are represented) hazards to health and accidents have been dealt with adequately, at what point the hazards to land use, ecology and property have been taken care of satisfactorily, and at what point the major esthetic problems of smell, unsightliness, and noise will not arouse a political furore. We may also be able to plot the points at which substitutions or imported products or systems become economical.

Let us look briefly at this matter of substitution. The only way to eliminate pollution from automobiles would be to do away with the inefficient internal-combustion engine that powers them. The federal Clean Air Act as amended in 1970 requires that by 1976 the three most harmful automobile emissions, hydrocarbons, carbon monoxide, and oxides of nitrogen be reduced by an average of 95 percent from 1970 levels. The 1973 models achieved a 66 percent reduction in emission levels. The Mobil Oil Corporation in 1973 calculated that the direct cost of meeting the 1976 standards and maintaining them for ten years would be at least $100 billion, about $1,500 per family. An 83 percent reduction, which is what California requires, would cost each family only $500 over ten years and by the same standards the 66 percent reduction over a ten-year period would cost less than $200 per family. These figures do not include the cost of extra gasoline required because the engines are less efficient. For the estimated difference in direct costs between reductions of 83 and 95 percent, approximately $66 billion, we might be able to subsidize electrically energized or hydrogen-powered cars, at least for metropolitan areas, and achieve complete abatement more cheaply. These are the kinds of calculations and considerations that exponential-cost curves should lead to.

The exponential imperative also limits man's attempts to harness large but relatively diffuse or slow flows of energy. The lower Mississippi River, for example, potentially represents enormous power because of the great volume of moving water, but not enough to justify the work required to dam that great river. With existing technology, the work cost of a solar energy collection, storage, and conversion system would likewise be more than could be recovered. This may well turn out to be true also for "dry" geothermal energy.

The present stage of the industrial revolution is essentially a battle between economies of scale and the exponential imperative.

LONG-DISTANCE TRANSPORT OF NATURAL GAS

A great deal of natural gas, an irreplaceable and now highly valued fuel, has been wasted in the world since its production first began. Produced of necessity along with crude oil in which it was dissolved, associated natural gas was separated from the oil in the oil fields and allowed to escape to the atmosphere. As a precaution against untoward explosions, the wasted gas was usually ignited and "flared" into the air. This process goes on today on a large scale in the oil fields of the Middle East. The problem was, and in remote fields still is, one of economical transport of the gas to places of use.

To transport the energy at an equivalent rate, a pipeline for natural gas must be both larger and stronger than one for crude oil or petroleum products. Even compressed under considerable pressure, natural gas contains less energy than the same volume of crude oil, so that the investment in pipelines and pumping stations is economic only where there are enough users to justify large pipelines and relatively high flow rates. This condition obtains in much of the United States and Europe today, but does not in most of the countries now producing associated gas in large quantities.

There are at least two ways other than by a pipeline by which the energy of natural gas may be conserved for use rather than wasted by flaring. One is to liquefy it, the other to convert it into methyl alcohol (methanol). Both are expensive when compared with the construction and operating costs of a large pipeline, even one a thousand miles long. But a pipeline from Saudi Arabia or Indonesia to western Europe or North America is not economically feasible. If the gas is to be used, it must be transported as a liquid rather than as a gas. The liquefaction of natural gas requires a considerable amount of energy, the process being only about 70 percent efficient. Transport by cryogenic tanker is relatively expensive because the specially built vessels are expensive and are considerably smaller than the supertankers used for crude oil. Regasification is the only part of the process that costs little energy; if the "cold" of the liquid could be used for refrigeration as the liquid turns to gas, some of the energy invested in liquefaction could be used and regasification then would turn an energy profit.

Making methanol from natural gas is about as expensive as reducing it to liquid by refrigeration, but methanol can be transported in ordinary tankers built for crude oil and oil products. Additional advantages are the relatively low capital cost for a methanol plant, which means that there can be many of them widely scattered throughout the producing world, the fact that impure natural gas can be used as feedstock, the impurities being readily separated during the chemical process, and the fact that methanol itself can be burned directly as a fuel or can be reconverted to gas.

Another way to use the natural gas of the less-developed producing countries would be to build industrial plants such as aluminum reduction works, carbon-black and fertilizer manufacturing plants. Apart from national trade barriers, the main economic question would be whether it would cost less energy to liquefy the natural gas and transport it to the consuming nations than to manufacture goods with the energy from the gas at sites near the producing fields and to transport the goods rather than the liquefied gas.

FUEL-TO-FUEL CONVERSIONS

As crude oil and natural gas for most of the industrialized nations become constrained, in rate of supply, in cost, or in both, interest grows in fuel-to-fuel conversions in those countries that have access to larger reserves of solid fuels than they have to fluid fuels—as the United States does. Fuel-to-fuel conversions are projected for the purpose of maintaining on fluid fuels those energy systems that cannot readily be converted to solid fuels. The transportation system of most high energy societies, for example, is based on the use of liquid fuels— gasoline, diesel oil, and jet fuel. When the crude oil from which these fuels comes grows scarce, conversion to a solid-fuel base can be made either by electrifying the transport system, with the electricity being produced from coal-fired or nuclear power plants, or by establishing an industry to produce substitute liquid fuels from coal, oil shale, and tar sands. The choice of a nation will be influenced strongly by the fuel available, by the relative efficiencies of the various systems, and by relative effects on the environment. Here we shall consider only the

Table 5.1 Efficiencies of Converting Coal to Fluid Fuels

Gas Product	Btu/Cubic Foot	Conversion Efficiency* (from coal)
Natural gas (for reference)	1,027	——
Synthetic, pipeline quality	1,000	60%
Coal gas, low-temperature	780	80%
Coke oven gas	540	95%
Water gas	300	95%
Producer gas	130	70%

Liquid Fuel	Btu/Barrel	Conversion Efficiency
Substitute crude oil	5,600,000	60%
Methanol	2,720,000	65%

*The percentage of the chemical or latent-heat energy in the coal that can be obtained by burning the product of conversion.

second influence; fuel availability will be discussed in chapter 9 and the environmental impact in chapter 10.

Several different kinds of gas can be made from coal. The first such gas, a byproduct of coke making, is called coke oven gas. As this gas came to be used more and more for heating and lighting, an industry grew up that carbonized coal to produce gas, so that coke became the byproduct. It was discovered that all the coke could be gasified by passing hot steam over the glowing coke; with suitable controls, all the coke (carbon) would react with the water vapor to produce elemental hydrogen and carbon monoxide, a poisonous but combustible gas; the resulting mixture is called water gas. Another method of distillation involves no water; a restricted amount of air admitted to the coal is just sufficient to burn it to carbon monoxide. The resulting gas, which contains nitrogen, is called producer gas. Although producer gas has a much lower energy content than water gas, much more of it can be produced from a given mass of coal; where water is not plentiful, such as the regions of the Rocky Mountain coal mines, producer gas might be preferable to water gas. Coal gas is made by the distillation at low temperatures of the volatile hydrocarbons naturally present in most coal; by itself the production of coal gas is inefficient, because it leaves the elemental carbon in the coal unused; the latter may be burnt as fuel or used to make producer gas.

The underground gasification of coal has been the subject of extensive experiments in Europe and North America for some forty years. The obvious advantage to be gained is the elimination of the costs of cutting or breaking the coal, of lifting it to the surface, and of carbonizing it in ovens, a process that involves several handlings of the material and is both labor- and capital-intensive. Underground gasification yields a form of producer gas from the partial combustion of coal at the bottom of drill holes down which air is forced. The coal is ignited electrically and the gas is withdrawn through other drill holes, without the necessity for shafts and underground workings. Because of its low heating value, the gas could be used economically only in plants at the gasification site. An abundance of crude oil and natural gas, of much higher heating value, available at low cost to industrialized society and affording much greater flexibility in the siting of plants to use the fuels has so far limited commercial interest in underground gasification. Increasing costs of alternative fuels, rising costs of underground coal mining, and the growing reluctance to allow strip mining will bring renewed attention to the possibilities of getting energy from coal without mining it.

Making liquid fuel from coal is more complex and costly than making gaseous fuel from it. Although a process has been technologically feasible for half a century, only under rather special conditions has it been economically practicable. The energy problem of making liquid fuel from coal is the same as that of making gas from coal: how to produce a fluid with a high heating value. Elemental carbon, the principal constituent of coal, is a solid. It can be transformed to a gas, carbon monoxide (CO), by partial combustion, but at the expense of some of its initial heating value. Most gaseous and liquid fuels are hydrocarbons. To make a hydrocarbon from coal requires an abundance of hydrogen, the source of which can be water or coal itself. Hydrogen can be obtained from water by thermal dissociation, as in the production of water gas, or by electrolysis. Water gas has a low heating value because the elemental carbon is oxidized to carbon monoxide in the process. In energy terms, this is a heavy price to pay for the liberation of hydrogen, but in monetary terms, it may be cheaper than obtaining free hydrogen through the electrolysis of water. Except for anthracite, coal contains an appreciable number of volatile hydrocarbons in addition to elemental carbon. These hydrocarbons can provide the hydrogen needed to make either a gaseous or liquid synthetic fuel from coal. To use coal, however, as the hydrogen source may involve a great waste, since a large quantity of carbon will be left over. That carbon, however, could be used to make water gas or producer gas, so that two products would come from a substitute-fuel plant based on coal: a high-value substitute natural gas (SNG) or substitute crude oil ("syncrude") and a low-value gas. The high-value product could be shipped long distances, the low-value product would need to be used near the synthetic fuel plant. It is also possible to make methanol directly from coal and in certain situations this single product might prove the most economical, in both energy and work-cost (monetary) terms.

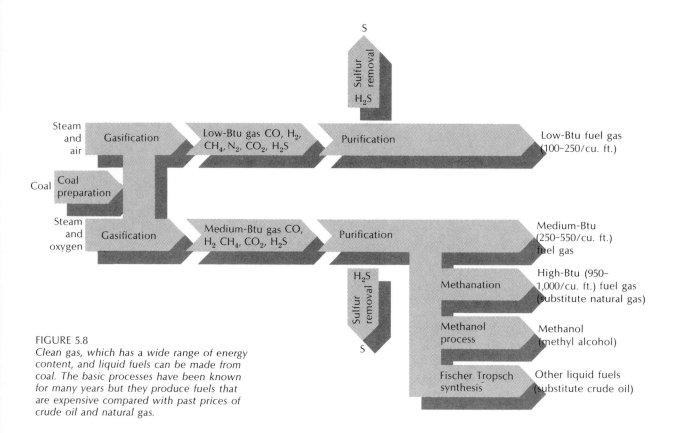

FIGURE 5.8
Clean gas, which has a wide range of energy content, and liquid fuels can be made from coal. The basic processes have been known for many years but they produce fuels that are expensive compared with past prices of crude oil and natural gas.

THE HYDROGEN ECONOMY

The advantages of hydrogen as a fuel have received considerable attention. They are:

1. It could be used directly as a fuel without any further refining or processing.
2. It could be transported cheaply by pipeline.
3. It burns cleanly, yielding only water as a byproduct; hence, neither water nor air pollution could result from its use and the engines in which it is burned will not be corroded or fouled.
4. The adjustments required so that hydrogen could be used in present applications of the internal-combustion engine and in furnaces would be relatively inexpensive.
5. If hydrogen were made from seawater, the fuel reserves would be almost infinite.
6. If widely used, hydrogen would greatly relax the demand for crude oil and coal, which then might be used to produce high-value petrochemical products.

The major disadvantage of a hydrogen economy is the thermodynamic inefficiency of systems that include hydrogen as an energy "carrier". To produce hydrogen from a fossil fuel, as is done in the manufacture of ammonia, is thermodynamically inefficient. To use electricity to produce hydrogen by the dissociation of water likewise is inefficient, because of the conversion losses in the power plants. If the hydrogen is to be used to produce work, its efficiency of conversion to work is unlikely to be more than 25 percent—unless fuel cells using hydrogen can be improved to function at an efficiency of 50 percent—whereas electricity can be converted to work with only a small loss.

For engines, such as those in automobiles, that cannot use electricity directly, gasoline or methanol from coal or oil shale has a higher system efficiency than hydrogen from water. The methanol system, assuming an engine efficiency of 25 percent, would convert the latent energy in coal into work with an efficiency of about 14 percent. That may seem low, but a system using hydrogen from seawater could hardly exceed 8 percent; if a nuclear power plant were used, less than 4 percent of the energy in the nuclear fuel would be converted to work by the automobile engine.

THE CHANGING ROLE OF TECHNOLOGY

Although the efficiencies of the energy systems in high energy society appear to be nearing asymptotic limits, there are still many ways of improving the efficiency of energy use. Electric power generation seems to be closer to an absolute economic limit than any other large sector of the energy economy; the present 40 percent efficiency for a large coal-fired power plant will probably be matched or slightly exceeded by an efficient breeder power plant. The possibilities of getting above 50 percent appear to require some provident technology in fusion power, magnetothermodynamics, or fuel cells.

New pressures on technology will include these demands:

1. To increase the efficiency of use where possible: transport, process heating, fuel-to-fuel conversions, comfort heating.
2. To develop renewable or quasi-renewable substitutes for nonrenewable energy resources; by making significant advances in breeder, fusion, solar, and geothermal power systems.
3. To decrease the adverse environmental impact of energy systems, at minimal cost in energy and work.
4. To develop materials-recycling systems to conserve energy as well as materials (the energy saved in remelting scrap aluminum rather than electrolyzing aluminum oxide, for example, could be much more important than the aluminum ore reserves conserved).
5. To push economies of scale to their limits in energy transport (supertankers, ultra high-voltage and cryogenic transmission of

electricity, unit trains and coal-slurry pipelines) and conversion (large power plants, solar collecting devices).

6. To standardize the components of energy-conserving systems (solar heating and cooling) and to design metallic objects for greater economy of recycling.

7. To minimize the adverse environmental impact of energy systems while retaining as much efficiency as possible in those systems.

ECONOMIES OF SCALE

At the same time there will be new pressures on social management systems to control the potentially adverse effects of the blind pursuit of engineering goals. The need for such control already is apparent in systems designed to achieve economies of scale.

The justification for growth in production units and therefore the promotion of increased consumption has always been that the economies of large scale mining, manufacturing, and power-producing plants would reduce the unit cost of energy, raw materials, goods, and even services, so that everybody could have more. Aside from the question of limits to the abundant life of the individual imposed by constraints of time, congestion, noise, and the duty to manage the trivia of a complex society, there is the pressing question of the limits to increases in economies of scale, or decreases in unit costs.

Neither the consumer nor the environment is a bottomless sink for products and waste materials. Pursuit of economies of scale ultimately requires infinite consumption and rates of waste disposal. The ultimate strength of real materials limits the physical size of objects constructed from them. Galileo long ago pointed out that enormous structures would collapse under their own weight. The exponential imperative of work cost limits the scale of man's activities. There probably is a limit to the scale of an effective social structure for managing production and consumption. It appears that the advanced industrial society is reaching these limits, and continued emphasis on increase of productivity through increases in scale may be having a boomerang effect, actually decreasing efficiency and raising unit costs, instead of continuing to raise efficiency and lower costs.

Economies of scale have frequently required or depended upon increases in waste, especially when the costs to the environment and other resources have not been charged to production. Huge electric shovels for stripping coal may decrease unit costs only when all the costs of maintaining and restoring the quality of the environment and the usefulness of the mined area are not in the balance sheet. Economies of scale in agriculture have not been recalculated with all the costs of decreased productivity per acre, depletion of soil nutrients, erosion, environmental impact of agricultural chemicals, and accelerated exhaustion of fossil fuel reserves added in. Economies of scale in the production and use of the automobile appear to require shocking inefficiencies in the use of nonrenewable resources, the dehumanization of the worker, a high toll in life and limb, and incalculable costs in air pollution and increased noise and congestion.

Good engineering has produced some great advances in efficiency of power plants at comparatively small cost, but not only are the limits of efficiency being approached with present technology, but also new technologies (nuclear, solar, geothermal), held by some to promise further decreases in unit costs of power, have questionable benefit-cost ratios when all costs are assessed. Efficiencies in other parts of the world energy system, however, can be improved substantially.

The evidence strongly suggests that further increases in productivity where possible and achievable can be made only at the expense of social and environmental effects that may be unacceptable, in other words, only if we continue to use incomplete and inadequate analyses of benefits and costs. The point can hardly be made too strongly that the efficiency of production and even of an energy system do not necessarily accord with the social efficiency of a production system.

SUPPLEMENTAL READING

Berry, R. Stephen, and Hiro Makino. 1974. "Energy Thrift in Packaging and Marketing." *Technology Review,* v. 76, n. 4, p. 33–43.

    The energy costs of various forms of retail packaging are detailed and diagrammed; savings obtained by re-use and recycling are shown.

Cottrell, Fred. 1955. *Energy and Society.* New York: McGraw Hill.

    An outstanding treatment of the relations among energy, social change, and economic development. The physical economy of energy use underlies the entire book, but is brought out in some detail in chapters 2, 5, and 7.

Leach, Gerald. 1972. *The Motor Car and Natural Resources.* Paris: Organization for Economic Cooperation and Development.

    Calculates the pressure the motor car places on natural resources and points out that the drain is restricted mainly to the industrial nations, particularly the United States.

Office of Emergency Preparedness. 1972. *The Potential for Energy Conservation.* Washington, D.C.: United States Government Printing Office.

    An interagency staff study that deals with energy conservation in the short (1972–75), mid (1976–80), and long (beyond 1980) term.

Price, John. 1974. *Dynamic Energy Analysis and Nuclear Power.* London: Earth Resources Research.

    A fine example of energetic analysis; shows how energy inputs and outputs are calculated and, perhaps more importantly, what social, economic, and political questions are evoked and illuminated by energetic analysis.

Soddy, Frederick. 1933. *Wealth, virtual wealth and debt,* 2d ed. New York: Dutton.

    A lucid redefinition of wealth and debt in energy terms; outlines a system of economic analysis based on energy.

Synthetic Fuels Panel, Federal Council on Science and Technology. 1972. *Hydrogen and Other Synthetic Fuels.* Washington, D.C.: United States Government Printing Office.

    Review of the potentials, for production and use as fuels, of hydrogen, ammonia, methanol, and other substitutes for conventional hydrocarbon fuels.

Technology Review. 1972. *Energy Technology to the year 2000: A Special symposium.* Cambridge, Mass.

    Informative articles on energy, economy, and the environment, electric power from nuclear fission, and energy systems and future transportation, energy, and pollution.

William Tenney

Chapter 6

THE CONSERVATION OF FREE ENERGY

Electricity seems destined to play a most important part in the arts and industries. The question of its economical application to some purposes is still unsettled, but experiment has already proved that it will propel a street car better than a gas jet and give more light than a horse.

—Ambrose Bierce, 1906

CONSERVATION AND EFFICIENCY

Where for several generations inanimate energy has been used with careless waste because of its actual cheapness and its imagined illimitability, people may find it hard to grasp the physical principles of energy use and to practice conservation when energy becomes scarce instead of abundant. Even language gets in the way. They speak of conserving energy who do not understand the laws of thermodynamics. Energy will be conserved, willy-nilly. We do not need to worry about that. What needs conserving is *free* energy, defined as the ability to do work. Heat energy at the temperature of its surroundings can do no work, no matter how much there is of it. So-called energy conservation is an attempt to minimize the increase of entropy while obtaining useful heat and work from available sources of energy. By this route we come to the concept of efficiency in the use of energy. The most efficient energy system is that which delivers useful heat or work with the smallest increase in entropy, with the minimum production of waste or unusable energy. Of course, there is another way to conserve free energy and that is not to use it.

The comparative analysis of the efficiency of energy systems is a powerful tool for making judgments on the conservation of energy resources. It can be used by itself, however, only when a single energy resource (coal, natural gas, or other) is involved. A renewable energy resource such as fuel wood, even used inefficiently, may conserve other fuels that are not renewable. The choice between two nonrenewable resources may depend more upon their relative abundance or availability than upon a comparison of efficiencies.

THE MEANING AND MEASURES OF EFFICIENCY

There are several ways to define efficiency. Take, for example, the automobile. The *thermal efficiency* of its engine is the ratio of the amount of work produced on the pistons by the expanding gas in the cylinders to the potential chemical energy of combustion in the gasoline used as fuel; this efficiency may be as high as 33 percent. But the efficiency of the automobile as a machine is nowhere near this high, because of energy losses in the power train, the tires, and the exhaust system, and because it takes energy to run the accessories such as lights, windshield wipers, power steering and power brakes, radio, and air conditioner. The *machine efficiency* of the automobile frequently is given as the ratio of work delivered to the wheels to the energy in the gas. For American cars, that is about 15 percent. But the true machine efficiency is the ratio of the work done in *moving the car* to the energy in the fuel and that is only about 6 percent because of so much of the fuel energy is used up in deforming (and heating) the tires for the sake of comfort. Finally we can calculate a *system efficiency* for the use of gasoline in motor cars; this is the ratio of the work done in propelling the car to the energy content of the crude oil from which the gasoline was made as it existed in the ground; that is about 5 percent. Then there is the *propulsion efficiency* of the machine, measured, not in percentages, but in vehicle-miles, ton-miles, or passenger-miles obtained for each gallon of fuel or unit of energy (such as 1 million Btu), or by the inverse of these ratios, which gives a measure of the energy consumed for each unit of useful accomplishment.

The efficiency of energy systems depends to a degree on the end to which the resource or its products are put, but mainly on whether or not the system requires conversion of heat to work, an inevitably costly step in terms of energy. The thermal efficiency of coal, fuel oil, or natural gas used for space or water heating in well-insulated structures of good thermal design can be high as 75 percent; but the national average probably is close to 50 percent, and the efficiency of the systems using fossil fuels for space and water heating probably averages about 45 percent. The conversion efficiency of the same fuel burned in a steam power plant is between 30 and 40 percent and is further reduced by energy losses in transmission lines, transformers, and appliances, so that the average efficiency of a system that heats by electricity that comes from a fossil-fueled plant is only about 28 percent contrasted to the 45 percent for direct use without conversion to electricity.

Advances from one state of society to the next are marked, not only by large increases in rate of energy use, but also by changes in the aggregate efficiency of its energy systems. The decreasing rate of increase in aggregate efficiency is as clear in table 6.1 as it is in the

Table 6.1 Social Systems and their Use of Energy

Type of Society	Approximate Aggregate Efficiency	Per Capita Daily Gross Energy Input (kcal)	Per Capita Daily Useful Energy Output (kcal)
Subsistance agricultural	10%	5,000	500
Advanced agricultural	15	20,000	3,000
Emerging industrial	25	60,000	15,000
Advanced industrial	35	120,000	42,000
Industrial-technological	36	225,000	81,000

record of energy efficiency in the United States, which rose from about 8 percent in 1850, when a great deal of fuel wood was being burned very inefficiently, to a high of about 38 percent in 1969.

EFFICIENCY OF ENERGY USE IN TRANSPORTATION

Efficiency of energy use rarely is the sole basis for choosing one mode of transport over another. It may not even be the main criterion. Power and speed are often given more weight, and comfort and dependability are valued in high energy societies. For example, the bicycle is the most efficient means of transporting passengers, requiring only about 13 percent of the energy input per passenger-mile that the best mass transit systems require (see figure 6.1). But the bicycle lacks power, speed, and comfort.

The early locomotive was not as efficient as the horse, but its inefficiency did not seem to matter. A locomotive using coal or wood at less than 2 percent efficiency substantially underperformed the horse, which used its "fuel" at 20 percent efficiency; but locomotives were more powerful and faster than horses and they produced an energy surplus because they burned wood, which grew where grain and hay would not, or coal, which came from underground and likewise did not compete with crops. The locomotive exploited new sources of energy; although thermally inefficient as an engine, its fuels were produced cheaply and represented a large net work profit. A coal miner, who consumed about one-fifth as much food as a horse, could deliver *through the steam engine* between four and twenty times the work of an average horse.

The sailing ship represented a larger proportionate return on man's work investment than the steam locomotive ever did. The two became partners, but not competitors; railroads, branching inland from ports in the middle half of the nineteenth century, carried passengers and

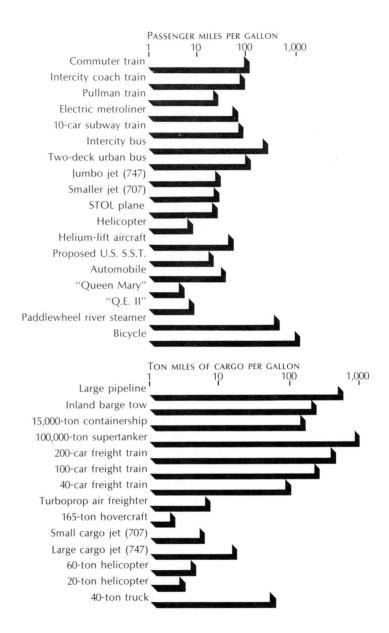

PASSENGER MILES PER GALLON

Commuter train
Intercity coach train
Pullman train
Electric metroliner
10-car subway train
Intercity bus
Two-deck urban bus
Jumbo jet (747)
Smaller jet (707)
STOL plane
Helicopter
Helium-lift aircraft
Proposed U.S. S.S.T.
Automobile
"Queen Mary"
"Q.E. II"
Paddlewheel river steamer
Bicycle

TON MILES OF CARGO PER GALLON

Large pipeline
Inland barge tow
15,000-ton containership
100,000-ton supertanker
200-car freight train
100-car freight train
40-car freight train
Turboprop air freighter
165-ton hovercraft
Small cargo jet (707)
Large cargo jet (747)
60-ton helicopter
20-ton helicopter
40-ton truck

FIGURE 6.1
*The propulsion efficiency of various transport
systems. The bicycle is shown on an
equivalent-energy basis. Note logarithmic
scales.*
(From Rice, 1972.)

freight to and from the sailing vessels. As the locomotive was to the
horse, so the steamship was to the sailing ship: markedly less efficient,
but faster, more powerful, and more dependable, all qualities that in
an industrial society have justified substantial premiums in energy
costs.

The automobile is an even better illustration of how unimportant
efficiency can be where energy is cheap. Even before the demise of the

steam locomotive, the automobile had taken over as the least efficient of all man's common machines. In considering the several meanings of efficiency, we have seen that the car has a machine efficiency and a system efficiency that appear rather low. Let us look into the question a bit further. Ayres and Scarlott described the automobile energy system thus:

> Something about the possession and operation of a motor-car provides effective anesthesia for any awareness of economy. . . . A few people for technical reasons keep an account of miles per gallon of fuel, depreciation per mile, and other costs of motoring, but the result is usually so appalling that the accounts are hurriedly discontinued and forgotten.
>
> The highly competitive sellers of motor-cars and motor-fuels search earnestly for features in their products to advertise. The advertising specialists, conscious of the public pulse, do not waste much space and money talking about economy. They talk instead, of performance, comfort, style, and reliability. Nearly everything said about a new car means lower fuel efficiency—for example, large body, longer wheelbase, greater weight, softer tires, more horsepower, more rapid acceleration, higher speed, automatic drive, improved flexibility of control. [1955, p. 130–131]

They cited (p. 134) calculations made by R. J. S. Piggott, past president of the Society of Automotive Engineers, as the basis for the figures shown in table 6.2 and viewed the situation with ill-concealed disgust and some alarm.

> The reason such huge amounts of energy are being wasted in the form of liquid fuel (and the reason the idea of wasting still more is being seriously entertained for the future) is that we insist upon operating heavy cars with dangerous and useless potentialities of speed and acceleration. This inclination, which is growing all the time, will do more to advance the end of the fossil-fuel era than any other factor . . . a car designed for a maximum speed of 100 miles per hour must have 3.5 times as much installed horse power per ton as a car designed

Table 6.2 The Energy-System Efficiency of the Automobile

Step	Step Efficiency	Cumulative Efficiency
Production	96%	96%
Refining of crude oil	87	83.5
Transportation of gasoline	97	81
Thermal efficiency of engine	29	23.5
Mechanical efficiency of engine	71	16.7
Rolling efficiency	30	5

for a maximum speed of 50 miles per hour. Other things being equal, the fuel consumption is about proportional to installed horsepower.

Histories written a few centuries hence may describe the United States as a nation of such extraordinary technologic virility that we succeeded in finding ways of dissipating our national wealth far more rapidly than any other nation. [Ibid., p. 136-137]

The energy-system efficiency of the automobile has continued to decline. Cars have continued to grow in weight; the average American car is now twice as heavy as the average car in the rest of the world and it uses twice as much gasoline. Auxiliaries designed for speed, comfort, and pollution control add to the weight and require energy; power steering, power brakes, automatic transmissions, and air conditioners likewise require energy. In 1940 the full-sized Ford or Chevrolet got more than twenty miles to the gallon of gas; today's average for the same makes is less than ten miles per gallon. A return to the fuel efficiencies of thirty years ago would greatly reduce the present need for petroleum imports and would help stretch out remaining reserves of crude oil in North America.

In addition to its inefficiency as a machine, there is a utilization inefficiency in that the automobile is commonly used to carry fewer passengers than its rated capacity and it is often used for short trips (under three miles) during which the efficiency of a cold engine is half or less than the efficiency of a warm one.

Figure 6.1 illustrates some differences in *net propulsion efficiency* (NPE), measured in number of passenger-miles obtained for each gallon of fuel used, of the main alternatives for passenger transport. The principal factors affecting propulsion efficiency are the per-passenger weight of the system and its load, the average speed of the system, and the work efficiency of the system.

The fuel consumption of the ground transport systems is approximately proportional to the weight of the system; consequently, for a given vehicle at a given speed, propulsion efficiency varies almost inversely with the number of passengers. A two-ton automobile carrying two 200-pound passengers (a total weight of 4,400 pounds) at an average speed of forty-five miles per hour may have an NPE of 36; the same automobile at the same speed, but carrying four passengers (a total weight of 4,800 pounds), will have an NPE of 66. In the first case, the car carries the two passengers eighteen miles on a gallon of gasoline; since fuel consumption at constant speed is proportional to weight, a gallon of gasoline will carry four passengers only

$$\frac{4,400}{4,800} \times 18 = 16.5$$

miles; but the NPE is 16.5×4 or 66, a substantial gain. By a similar calculation, it can be shown that the NPE for this car carrying one passenger is only 19.

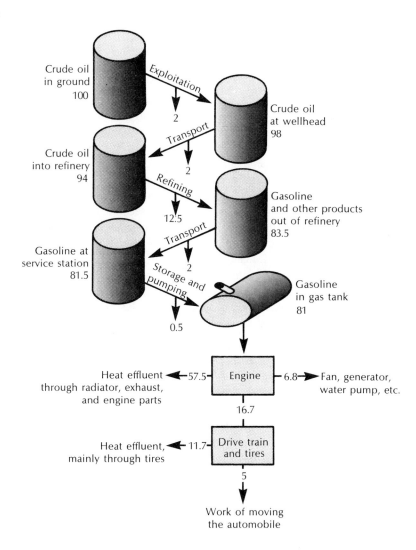

FIGURE 6.2
Energy-system efficiency of the American automobile.

In general, speed costs more than the proportionate amount of energy. Within the range for which a given transport system is designed, however, there will generally be an optimum speed, so that the efficiency of the system declines if the speed is lowered as well as if it is higher than the optimum. The optimum speed for highway vehicles lies between 30 and 50 miles an hour. The Volkswagen Beetle with two passengers and an urban bus with twenty may have the same vehicle and load weight per passenger, but the urban bus, operating at an average speed of 25 miles an hour or less, has a lower NPE (55) than that of the Volkswagen (60) operating within its optimum range at, say, 45 miles an hour. The cost of speed shows clearly in a comparison of the jumbo jet (Boeing 747) and the proposed United States supersonic transport (SST); for the 747 carrying 210 passengers (about half its capacity) the NPE is 27; for the SST, carrying 150 passengers, the NPE would be 10. At full load for both, the 747 is more than four times as efficient a mover of people.

Finally, different transport systems have different work efficiencies. The bicycle and the car driven in urban traffic may serve to illustrate this point. The bicycle and rider weigh about 200 pounds, the car and driver about 4,180. If we assume that the average speeds of the bicycle and car are not different enough to affect the comparison, the difference in NPE between the bicycle and the car should be the result of the combined effect of the difference in weight and the difference in work efficiencies. The weight difference will account for a 21-fold difference in NPE. What about the work efficiencies? The thermal efficiency of the human body in good physical condition is about 30 percent; if we take the mechanical efficiency of the bicycle as 90 percent, the work efficiency of the system is 27 percent. The work efficiency of the modern automobile in urban traffic can hardly exceed 4 percent. The difference in NPE, then, between the bicycle and rider and car and driver should be about

$$21 \times \frac{27}{4} \text{ or } 142\text{-fold},$$

figures similar to Professor Rice's estimates (1972, p. 37) of an NPE of 1,000 for the bicycle and 7 for the car.

The net propulsion efficiency for freight transport is measured in cargo-ton-miles per gallon of fuel. The most efficient movers of freight are supertankers and oil pipelines, both of which move only crude oil and its liquid products. The most efficient movers of packaged freight are long freight trains. Towed barges and containerships both fall behind the 100-car railroad train in net propulsion efficiency, while trucks and aircraft are far behind.

The efficiency of the automobile is a matter of particular importance to North America, Western Europe and Japan, where most of the world's automobiles are used, where the automobile population is increasing faster than the human population, and where cities and transport routes are increasingly designed for the automobile. But these regions also are net importers of crude oil, on which the automotive engine depends for fuel. Increases in automotive fuel consumption raise questions of unfavorable trade balances and economic security for the motorized nations as they raise questions of cost and availability to the individual consumer. The automobile is an almost living symbol of the individual's freedom of choice as well as an expression of personal style and power; for these reasons, it is cherished far beyond its economic utility. Historically, the automobile has been a luxury for which needs were created so that it might become a necessity.

Whether symbol, luxury, or necessity, the automobile is an energy wastrel. This is not the fault of automotive engineers and fuel chemists whose efforts to increase the efficiency of the internal-combustion engine have been overwhelmed by demands for comfort, power, speed, bulk, style, ease of operation, quietness, and (lately) inoffensive exhaust emissions. Air conditioning the interior of a car can decrease its propulsion efficiency as much as 20 percent. The fact that grossly inefficient design criteria have been accommodated by manufacturers of automobiles is a measure of the affluence of modern industrial society, which in turn is a measure of the cheapness of energy resources in terms of the human work required to make them available. One of the early effects of the realization that crude oil and its products are running out has been a pervasive fear that the automobile may be doomed as a means of mass transport.

The work or propulsion efficiency of the automobile is a product of many things, the weight of the vehicle and the softness of the tires being among the most important. The vehicles used in efficiency contests in which new records for vehicle-miles per gallon of fuel are sought are stripped of every ounce of superfluous weight and roll on tires inflated to rock hardness. They also have the engine mounted over the powered axle, to which it is linked by a drive chain, a design that greatly reduces the substantial resistance of the transmission and differential gearing in a stock car.

THE ENERGY COSTS OF THE AUTOMOBILE

To compute the total energy costs of the automobile in the United States is an instructive exercise in physical economy that not only reveals some hidden energy costs, but also shows how much of the nation's energy is devoted or diverted to the automobile. The energy required to manufacture the equipment used in drilling oil wells, to drill the wells, to extract the oil from the ground, to transport it to a refinery, and to refine it has been calculated to average 0.206 kilocalories for each kilocalorie of refined petroleum product. One may then calculate the amount of energy used to produce and refine the crude oil that yielded the gasoline used in cars in the United States during any year by multiplying the total gasoline consumption for that year, in gallons, by its unit kilocalorie content, and then by 0.206 (see table 6.3).

The energy used in the manufacture of automobiles has been calculated at 5,850 Btu (1,463 kcal) for each dollar of the value of the cars when shipped from the factory, in 1968 constant dollars. However, the energy required, on the same basis, in iron mining and transport,

Table 6.3 Energy Requirements for
Automobiles in the United States

	Energy Used (10^{15} kcal)		
	1960	1968	1970
Petroleum production and refining	0.29	0.41	0.46
Iron mining, steel making, etc.	0.17	0.23	0.16
Automobile manufacturing	0.02	0.03	0.02
Selling cars	0.19	0.25	0.20
Gasoline consumption	1.40	1.99	2.24
Repairing and maintaining: oil and replacement parts; parking, insurance, highway construction, etc.	0.76	0.99	1.11
TOTALS	2.83	3.90	4.19
Percentage of national energy consumption devoted to automobiles	25.2	25.0	24.4

Source: Eric Hirst, 1972, p. 21

in steel making, rubber manufacturing, and so on, is more than eight times the direct energy component:

Energy required in mining, steel making, etc.	12,105 kcal/$
Energy required in automobile manufacture	1,463
TOTAL	13,568 kcal/$

In calculating the energy used in selling cars, a conversion factor of 35,200 Btu (8,400 kcal) per dollar of sales in 1968 constant dollars was used. Hirst computed the energy requirement for repairs, oil, insurance, highway construction, and other automobile-related activities by multiplying three available figures, all for 1968: The average cost of an automobile, after deducting gasoline and depreciation costs, 6.71 cents a mile; the energy consumed per dollar of GNP, 18,065 kcal a dollar; and the total automobile mileage for the year, 814×10^9 miles. The product of these figures, for 1968, is 0.99×10^{15} kcal. The numbers given in table 6.3 for gasoline and repair and maintenance probably are inaccurate, as Hirst points out in his analysis, but they do indicate the magnitude of the indirect energy costs of the automobile, which are approximately equal to the direct energy cost in gasoline consumed.

Even a brief study of the physical economy of energy use in transportation suggests that railroads would be better than trucks and aircraft for hauling freight, trains and buses would be better than the automobile for carrying passengers, large subsonic jet planes better than the SST, and low-powered light-weight vehicles such as the bicycle, the motorcycle, and the compact car preferable for the transport of one or two passengers.

THE TRANSPORT OF ENERGY

The transport of energy itself is an important subject. The most efficient means yet devised is the supertanker, now exceeding four hundred thousand tons in size, and the second is the crude oil pipeline. The supertanker and the oil pipeline, however, can transport only fuels that are liquid at ordinary temperatures; these are crude oil, some of its products, and alcohol. Natural gas can be transported most economically by a high-pressure pipeline designed specifically for gas and not for liquids. Gas pipelines transport energy at three times the cost of oil pipelines. If natural gas is to be transported as a liquid, it must be refrigerated to a very low temperature and carried in cryogenic tankers or pipelines, both of which require a large capital investment.

Transporting solid fuel is a less efficient process than a large-scale transportation of fluids but can be more efficient than the small-scale transportation of fluids, as, for example, the haulage of gasoline by truck. The efficiency limits of hauling coal by unit trains have not been reached, and hauls of more than one thousand miles (two thousand miles round-trip) are now planned. Where construction and maintenance energy costs of a railroad must be charged largely or mainly to coal haulage, the coal slurry pipeline is a more efficient transportation system, comparable to the natural gas pipeline. Both types of pipeline transport energy at between 60 and 70 percent of the price of unit-train haulage, although in energetic terms all three are comparable.

Considerable economies of scale have been achieved in the past few decades in the transportation of electricity. Transmission losses from the power plant to the distribution transformers once dictated that a power plant be situated within the city it was to serve. In recent years, however, stimulated by the construction of hydroelectric plants far from load centers and by the obvious advantages of interlinked regional distribution systems, ultra high-voltage transmission has been developed to a point where it may well save energy to situate a power plant for Chicago in Montana, because the energy cost of transmitting electric power from Montana to Chicago, including the energy equivalent of amortized construction costs, may be less than the energy cost of hauling coal from Montana to Chicago by rail, when allowance is made for the fact that about three times as much coal energy needs to reach Chicago in order to produce the desired amount of electrical energy. This problem in energy economics will become more complex with the development of large plants to produce substitute natural gas from coal and others to produce substitute crude oil from tar sands and oil shales. Then there will be an array of alternative methods for moving fossil energy from the Rocky Mountains to other parts of the continent, and comparisons of the alternatives will be based on different energy and environmental costs.

EFFICIENCY OF ENERGY USE IN INDUSTRY

Energy is used in industry for process heating, for producing mechanical work, for comfort heating, and for lighting. In this section, we will discuss the first two uses, leaving comfort heating and lighting for the next section.

The making of steel, the reduction of aluminum, and the manufacture of glass require large amounts of heat. Most steel is made in coke-burning blast furnaces where iron ore and scrap steel are melted. Aluminum is produced from alumina (Al_2O_3, a product of aluminum ore processing) by electrolysis at high temperature. Glass is made by melting a solid charge of glass sand and soda ash in a reverberatory furnace, the heat being supplied by natural gas or fuel oil. Estimates of how efficiently heat is used in these three processes vary considerably. The thermal efficiency of existing blast furnaces probably does not exceed 15 percent on the average; most of the heat created by the combustion of the coke escapes up the stack. The thermal efficiency of aluminum vats has doubled in the past forty years and now is about 50 percent. The thermal efficiency of a well-constructed glass furnace is about 35 percent, with the average falling somewhat below 30 percent.

In both steel and glass making, electric furnaces are being introduced. The electric furnace used in making steel has much greater thermal efficiency than the blast furnace, approaching 50 percent, and thus compensates for the inefficiency of converting fossil fuel to electricity and for the losses in transmission (see table 6.4). In glass manufacture, however, where the change from gas-fired to nuclear-powered electric furnaces is being made because of the shortage of natural gas, the system efficiency with electric furnaces is lower (see table 6.5).

The most efficient way of using energy in process heating (where heat is used in changing the state of materials or the form of objects) is in a device that uses the heat of the combustion of a fuel directly and that has a thermal efficiency of 20 percent or better. In some process heating, such as high-temperature electrolysis, it is not possible to use direct heat and, in other cases, such as that of the blast furnace, the mechanics of getting heat diffused uniformly at a high temperature

Table 6.4 Comparative Energy-System Efficiencies of Steel-Making Furnaces

Blast Furnace			Electric Furnace		
Step	Step Efficiency	Cumulative Efficiency	Step	Step Efficiency	Cumulative Efficiency
Coal mining	96%	96%	Coal mining	96%	96%
Coal transport	96	92	Coal transport	96	92
Coke manufacture	90	83	Conversion to electricity	40	37
Thermal efficiency of furnace	15	12	Transmission of electricity	92	34
			Thermal efficiency of furnace	50	17

Table 6.5 Comparative Energy-System Efficiencies of Glass-Making Furnaces

Gas-Fired Furnace			Electric Furnace (Nuclear Power)		
Step	Step Efficiency	Cumulative Efficiency	Step	Step Efficiency	Cumulative Efficiency
Natural gas production	97%	97%	Uranium mining	95%	95%
Natural Gas transport	93	90	Uranium milling	90	85
Furnace efficiency	33	30	Uranium transport	99	85
			Fuel enrichment	70	59
			Reactor thermal efficiency	26	15
			Transmission of electricity	92	14
			Furnace efficiency	50	7

NOTE: Should electricity from a coal-fired power plant be used, the energy system efficiency would be 17, the same as for the electric steel furnace, but still much less than for the gas-fired furnace.

through an initially solid charge almost precludes any substantial thermal efficiency in the furnace, although there would be no technological barrier to trapping and recovering for use much of the heat that now escapes from a blast furnace. To conserve energy, the more efficient of two alternatives should be chosen, but the comparison of the two glass furnaces shows that a less efficient energy system is no longer available—which points up the need for weighting efficiency comparisons in favor of those energy sources that are more available or renewable.

Process heating itself suffers no limitation because of the second law of thermodynamics. In other words, there is nothing in the laws of physics that says that the thermal efficiency of process heating need be low. Consequently, it comes as a surprise to find the average thermal efficiency of process heating in the major industries of the high energy world hovering at about the same level as the average efficiency of thermal electric power plants, which *do* labor under the cross of the second law of thermodynamics. The main reason appears to lie in the cheapness of energy; it has generally been more econom-

BOX 6.1   The Uranium Fuel Cycle

The uranium fuel cycle consists of seven main steps:

1. Exploration for, and mining of, uranium ore, which commonly contains less than 1 percent of uranium, bound with other elements in various mineral forms.

2. Milling of the ore to produce a concentrate of about 85 percent uranium oxide (U_3O_8), called yellow cake. In milling, most of the weight and volume of the ore is discarded as waste or tailings, which however may contain significant amounts of radioactive substances, such as radium.

3. Further purification and conversion to uranium hexafluoride (UF_6), which is a gas.

4. Isotopic enrichment of uranium 235 (U-235), which comprises only 0.7 percent of natural uranium, to about 2.0 percent. Separation depends on the minute difference in density between U-235 and uranium 238, the main constituent of natural uranium. Enrichment, carried out by gaseous diffusion through several thousand stages, is an energy-intensive process necessary to obtain uranium suitable for bombs and other explosive devices and useful in the fuel cycle to improve the power and performance of the boiling water reactors that in the United States are the first generation of commercial power reactors.

ical to waste thermal energy than to expend mechanical energy in constructing more efficient heating devices or auxiliary systems to use some of the wasted heat. Hybrid heating and power systems, sometimes called "total-energy" systems, have barely been developed, because the savings, although large in kilocalories, have not been large in dollars.

Mechanical work in industry, for more than a century after James Watt's inventive years, was done mainly by steam engines, at first very inefficiently, and later, particularly after the development of the steam turbine, much more efficiently (see table 6.6). After 1910, steam power in industry started to give way to electric power, which was much more adaptable. Electric motors were built in a wide range of sizes and all but the largest were easily moved. The large steam turbines that were developed for generating plants in the first third of the present century were more efficient than the reciprocating steam engine and they increased the energy-system efficiency when steam-electric conversion replaced direct steam drive (see table 6.7).

5. Chemical and mechanical conversion of uranium hexa-fluoride (UF_6) into uranium dioxide (UO_2) pellets, which are loaded into metal tubes, which in turn are bundled together to form fuel elements.

6. Fuel elements are placed in the reactor core, where some of the U-235 is fissioned, releasing kinetic energy that is quickly transformed into heat, which in turn is transmitted to water, causing it to turn into steam, which is then used to drive the turbines of an electricity generating plant.

7. During the operation of the reactor, only part of the U-235 is fissioned or burnt up. The spent elements, containing residual uranium, plutonium formed during irradiation, and a variety of other radioactive products of fission, are taken from the reactor core and, after being cooled at the plant, are shipped to a fuel repro-cessing plant.

8. At the fuel reprocessing plant, the spent elements are dissolved in a strong acid solution from which the residual uranium and plutonium are recovered to be used again and the other products of fission are reduced in volume by partial evaporation and then stored until they solidify and can be shipped to a permanent repository.

Industry so fortunately situated as to have hydroelectricity available benefited from the most efficient power-delivery system of them all (see table 6.8).

The overall efficiencies of power-delivery or mechanical work systems in industry today, except in such countries as Sweden, Switzerland, and Canada, that produce a large part of their electric power from hydroelectric plants, probably is about 25 percent.

Table 6.6 Comparative Energy-Systems Efficiency of Steam Powered Mills

| Step | 1850 | | 1910 | |
	Step Efficiency	Cumulative Efficiency	Step Efficiency	Cumulative Efficiency
Coal mining	99%	99%	98%	98%
Coal transport	85	84	93	91
Reciprocal steam engine	05	04	10	09
Driveline efficiency	80	03	85	08

Table 6.7 Comparative Energy-System Efficiencies of Direct Steam Drive and Electric Drive Engines in 1930

Direct steam drive			Electric drive		
Step	Step Efficiency	Cumulative Efficiency	Step	Step Efficiency	Cumulative Efficiency
Coal mining	97%	97%	Coal mining	97%	97%
Coal transport	96	93	Coal transport	96	93
Reciprocal steam engine	12	11	Conversion to		
Drive efficiency	85	10	electricity	15	14
			Motor efficiency	90	13

Table 6.8 Comparative Energy-System Efficiencies of Hydroelectric Power Plants

	1930		1970	
Step	Step Efficiency	Cumulative Efficiency	Step Efficiency	Cumulative Efficiency
Water falling through turbine	90%	90%	95%	95%
Generator efficiency	90	81	95	90
Transmission efficiency	80	65	92	83
Motor efficiency	90	58	90	75

EFFICIENCY OF ENERGY USE IN HOME AND COMMERCE

Palmer Putnam (1953, p. 106) estimated that the aggregate energy efficiency in the Soviet Union in the 1950s had fallen from a high of about 75 percent achieved in the early years of the present century and that, when Imperial Russia's aggregate energy efficiency was 75 percent, that of the United States was only about 15 or 20 percent. Here it is necessary to modify the statement made earlier in this chapter that the aggregate efficiency of the energy systems of a country is a measure of its industrial development. This is only strictly true if we eliminate the space heating system from the calculation. In the United States, the efficiency of space heating improved as the efficiency of process heating, transportation, and other work improved. But in Russia and most of Central Europe space heating was a highly efficient process—mainly because of the high value of firewood—long before industrialization had made more than a primitive beginning in those countries. Space heating comprised the main part of Putnam's calculation (he ignored the efficiency of work performed in Russia by men and animals that, at about 20 percent, would have lowered his estimate greatly) so he presented a rather misleading, or at least not very useful, picture of a feudal agricultural society that somehow knew how to use energy twice as efficiently as we in a highly industrialized society do today. There are several indices of the progress of industrialization, each one of which by itself may not be comprehensive. Perhaps the best single index is the rate of production of work and useful heat per capita. The gross energy consumption has been used, but this is a crude index that places East Germany ahead of West Germany. There is also a tendency to equate high energy consumption with average standard of living. Not only does this fail in the case of the two Germanys, but even the work-and-useful heat rate would likewise fail, the first mainly because of the lower efficiency of energy use in East Germany (and in any country where much coal is used instead of petroleum products and natural gas and where there is little or no hydroelectric power), the second because much of the product of the work of East Germany has been exported on terms that reduced the provision of domestic goods and services below that which was realized from equivalent work in West Germany.

In a high energy society, less of the energy consumed is used for space heating than in a low energy society, but still the amount is 30 percent or more. In cool climates (no high energy society has developed in the tropics), 50 percent or more of the inanimate energy used in the home will be devoted to warming rooms and heating water.

In the past hundred years, the economic use of firewood in the homes in high energy societies has all but disappeared. Coal is used almost only in homes in coal-producing areas, fuel oil and natural gas are the most used direct sources of heat, and electricity is widely used to heat water, to power cooking stoves and ovens, to provide power for refrigeration and air conditioning, for lighting and a multitude of appliances, and even for space heating.

Although heating systems using natural gas or fuel oil may be 75 percent or more efficient when in good condition and run at full load, many receive inadequate maintenance and most are seldom run at full load. Moreover, most American buildings are not designed for efficient heating and cooling. The fact that the builder of an American home or office building, more often than not, will not be bearing the costs of heating and cooling the structure, encourages corner cutting on insulation and design that is motivated by appearance rather than thermal efficiency. While energy is cheap and systems run automatically, there is a tendency to ignore efficiency and to neglect maintenance. The average of the efficiency of the space heating in residential and commercial buildings in the United States probably is less than 60 percent. A study by Westinghouse (Dunning, Geary, and Trumbower, 1974) indicated an average efficiency for gas heating systems of about 50 percent; fuel-oil heating was estimated to be in the range of 35 to 45 percent. The national energy-system efficiency for space heating is

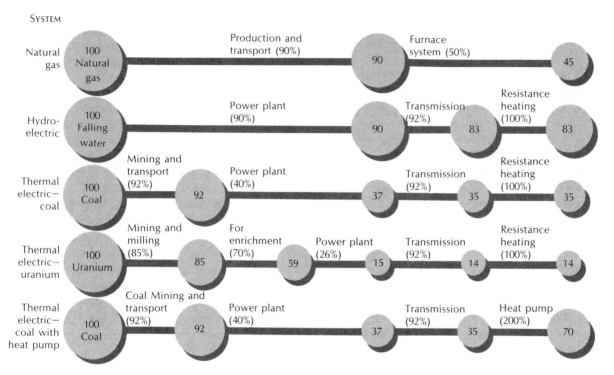

SYSTEM

Natural gas — 100 Natural gas — Production and transport (90%) — 90 — Furnace system (50%) — 45

Hydro-electric — 100 Falling water — Power plant (90%) — 90 — Transmission (92%) — 83 — Resistance heating (100%) — 83

Thermal electric—coal — 100 Coal — Mining and transport (92%) — 92 — Power plant (40%) — 37 — Transmission (92%) — 35 — Resistance heating (100%) — 35

Thermal electric—uranium — 100 Uranium — Mining and milling (85%) — 85 — For enrichment (70%) — 59 — Power plant (26%) — 15 — Transmission (92%) — 14 — Resistance heating (100%) — 14

Thermal electric—coal with heat pump — 100 Coal — Coal Mining and transport (92%) — 92 — Power plant (40%) — 37 — Transmission (92%) — 35 — Heat pump (200%) — 70

FIGURE 6.3
Comparison of heating system efficiencies.

close to 40 percent, ranging from about 18 percent for electric heating when the electricity comes from a nuclear power plant to about 83 percent for electric heating when the electricity comes from a hydroelectric power plant (see figure 6.3). Because only a fraction of the nation's heating systems will ever be supplied with hydroelectricity, the most efficient alternative is the heat pump running on electricity from a fossil-fueled power plant.

Heat pumps could be used also for water heating. Charles Berg (1974, p. 17) has calculated that the same amount of water heating might be obtained using a heat-engine heat-pump system for as little as 8 percent of the fuel consumption required for direct water heating. The next most efficient system is direct heating. Although an electric water heater has a thermal efficiency of 92 percent compared to the 64 percent of a gas heater, it falls behind the gas heater in a system efficiency (see figure 6.4). With the best thermal power plant an electric system delivers only (100) (0.90) (0.40) (0.92) (0.92) = 30.5 percent of the energy in coal in the ground to water in the tank, whereas a gas system delivers (100) (0.90) (0.64) = 57.6 percent of the energy of the natural gas in the ground. If available supplies of coal and natural gas were approximately equal in energy content, water heating by gas would be much more conservative of resources than electric heating.

Stanford Research Institute (1972, p. 18) has calculated the cooking efficiency of gas as 37 percent and of electricity as 75 percent. The

The Conservation of Free Energy

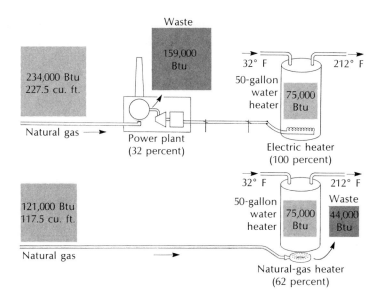

FIGURE 6.4
Efficiencies of heating water with natural gas indirectly by generating electricity for use in resistance heating (top) and directly (bottom) are contrasted. In each case the end result is enough heat to warm fifty gallons of water from 32° Fahrenheit to 212°. The electrical method requires substantially more gas even though the efficiency of an electric heater is nearly 100 percent.

system efficiencies are reversed: gas = (100) (0.90) (0.37) = 33 percent; electricity = (100) (0.90) (0.40) (0.92) (0.75) = 17 percent.

The reason for the marked improvement in the efficiency of space heating in the United States during the past 150 years is indicated in table 6.9.

Any conversion of energy to light is inefficient. The energy-system efficiency of the most efficient lamp, the fluorescent, is only 6.8 percent and that of the common home lamp, the incandescent, is only half that (see table 6.10).

In the United States, rapidly increasing air conditioning of homes and business structures has accelerated the use of electricity in the residential and commercial sectors (where it is increasing faster than in industry) and decreased efficiency of the use of energy there. Anyone paying summertime electric bills for a home in the southern United States equipped with central air conditioning knows what a large proportion of the annual energy input to the home is represented by the electricity needed to power the air-conditioning system.

Table 6.9 Fuel Efficiencies

Open fireplaces	10%
Wood-burning heater	25
Hand-fired coal furnace	40
Gas furnace in average domestic use	50
Automatic coal furnace	60
Industrial oil or coal furnace	80

Table 6.10 Comparative Energy-System Efficiencies of Fluorescent and Incandescent Lighting

Step	Step Efficiency	Cumulative Efficiency
Production of coal	96%	96%
Transport of coal	96	92
Generation of power	40	37
Transmission of power	92	34
Light from fluorescent	20	6.8
Light from incandescent	10	3.4

151

In low energy societies the principal uses of energy in the home are for space heating and cooking, and the fuels range from dung through fuel wood to coal. Commerce in low energy societies is apt to use considerably more electricity and petroleum products than is the residential sector.

EFFICIENCY OF ENERGY CONVERTERS

As we have just seen by working through several examples, the efficiency of energy conversion is the key factor in determination of the amount of useful heat and work that a given society obtains from the energy resources available to it. The size of the natural energy subsidy depends on the overall efficiency of the energy systems in use as well as the availability of energy resources. The efficiency of engines and power plants, those machines that convert other forms of energy into mechanical work or electric power, was discussed in chapter 2, and the efficiency of other conversions was mentioned in chapters 3 and 4. A recapitulation shows the extremely wide range of efficiencies among the common converters, from about 0.3 percent for the conversion of radiant (solar) energy to chemical by the corn plant (*Zea mays*) to 100 percent for the conversion of electrical to thermal energy by the electrical-resistance heater (see figure 6.5). Again, it will be noted that the "uphill" conversions, from thermal to electrical or mechanical energy, from radiant to chemical or electrical, and from electrical to radiant, are those with the lowest efficiencies and the greatest losses; no converter in these categories achieves 50 percent efficiency. The "downhill" conversions, those from chemical to thermal, mechanical or chemical to electrical, and electrical to mechanical or chemical, fall below 50 percent only when coupled with one of the "uphill" conversions.

EFFICIENCY OF ENERGY USE IN FOOD PRODUCTION

The process of photosynthesis on which all life of the earth and the seas is dependent is not very efficient from man's point of view. Corn, for example, under the best conditions, yields 0.3 percent of the radiant solar energy falling upon the land on which it is grown; and this only if the cobs, stalks, and leaves are burned to produce useful heat in addition to the kernels' being eaten as feed or food. It is estimated that only 0.1 percent of the solar energy reaching the earth is converted by photosynthesis to vegetable matter.

Man likes to eat meat. Not only does it provide him with essential protein, but also it increases his ability to survive in hostile environments since it can be stored in a variety of ways (including on the hoof)

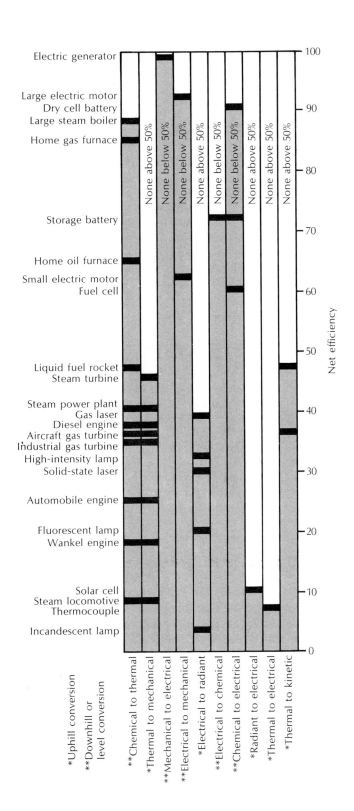

FIGURE 6.5
The efficiency of energy converters runs from less than 5 percent for the ordinary incandescent lamp to 99 percent for large electric generators. The efficiences shown are approximately the best values attainable with present technology. The figure of 47 percent indicated for the liquid-fuel rocket is computed for the liquid-hydrogen engines used in the Saturn moon vehicle. The efficiencies for fluorescent and incandescent lamps assume that the maximum attainable efficiency for an acceptable white light is about 400 lumens per watt rather than the theoretical value of 220 lumens per watt for a perfectly "flat" white light.
(From Claude M. Summers, "The Conversion of Energy." Copyright © 1971 by Scientific American, Inc. All rights reserved.)

and because meat and milk animals can digest food that man cannot and can graze on land unfit for food crops. When domesticated animals are used in this manner, they contribute to the physical economy of man's use of available resources; when, however, as in most industrial societies, meat animals are fed on grain raised in competition with food crops, their low efficiency as food-energy converters works against the physical economy of the food system. Since man gains through eating animals and their products only a fraction of the energy contained in the plants eaten by those animals, he may waste food energy by eating meat instead of plants. But this depends on the kind of plant feed his meat animals have had.

Three things keep meat, eggs, and milk from becoming prohibitively expensive in an industrial society: the large fossil fuel subsidy to agriculture, the eating of grass and the cheaper grains by the animals, and the use by these animals of otherwise unusable or only marginally useful land. The relation of meat production to population was outlined incisively by Fred Cottrell.

> One of the early evidences of population pressure is the reduction in the number of food animals; followed by the reduction in the number of draft-and-food animals, such as cows and horses, in favor of those draft animals, such as the water buffalo, which can survive on the plant products of land which will not yield nearly as much energy in the form of humanly edible food. Thus many areas which once supported draft animals and food animals now make use of almost none. [1955, p. 22]

Buck (1937, p. 6) found in the 1930s nearly 90 percent of the potential farm land in China in crops, only 1 percent in pasture. In the United States 42 percent is in crops and 47 percent in pasture. Moreover, much of the United States crop land is used to raise animal feed rather than human food. The United States could support a much greater population, perhaps one billion, if food animals were eliminated and all crop land devoted to raising plant food for humans. Because of the large amount of meat in his diet, the average American's daily food supply of 3,290 kcal represents more than 16,000 kcal of plant food, about five times what he would need to stay alive and healthy on an adequate vegetarian diet.

The energy costs of mechanized food raising are higher than those of hand cultivation because the work is done faster, the tools and equipment are larger and take more energy for manufacture, operation, and maintenance, the farm machinery is idle most of the year, raising the amortization cost per unit product, production units are larger and further from the places of consumption, requiring more energy for transportation, and yield per unit of energy input and per acre of land are decreased because machine cultivation works against selective treatment of areas in which productivity differs from the average of the tilled unit.

EFFICIENCY OF AGRICULTURE

The most efficient agriculture is not the mechanized agriculture of the industrial world, but that which minimizes energy input and maximizes energy output, without depleting the resources of the soil. In many environments this goal is best achieved by the heavy use of human labor in the fields and by employing draft animals or tractors only where short growing seasons or shortages of human labor prevent the full utilization of the available arable land through human power alone.

In the United States we have crop surpluses and a large amount of food is produced by each agricultural worker, but agriculture is not efficient. Fuel energy for tractors and self-propelled farm machines approximately equals the total food energy consumed by all the inhabitants. To fuel for farm engines, on the energy-input side of the balance, must be added the energy required to produce fertilizers, pesticides, and farm machinery, and to transport these items to the farms, as well as the electricity and human labor used in agriculture. Mechanized agriculture does not always run at an energy loss. Pimentel and others (1973, p. 445) at Cornell have shown that, in the United States corn (maize) production, for example, returns about three calories for each calorie invested. If some of the plant other than the grain is used, as in silage, the return can be doubled. Other crops also yield positive returns under mechanized agriculture, ranging from 1.4 for rice and peanuts to 5.3 for sorghum, according to a study by Heichel (1973), which shows clearly the strong energy-efficiency advantage of human labor as opposed to mechanical power; hand cultivation of rice in the Philippines is 12.3 times more efficient than mechanical cultivation of rice in Louisiana (see figure 6.6).

When transport and processing energy costs are added, a farmer's energy profit can disappear. Production of sugar from sugarbeet in California returns only 0.8 calories for each calorie invested in growing and processing the beets—and the costs of packaging, transport, storage, and selling are not yet added in. Foods that must be decorticated, canned, and cooked require even more energy. The processing, transportation, and storage of frozen foods raise energy costs greatly. Indeed, in a high energy society there are probably no foods that reach the consumer at an energy profit.

But if most crops do yield more food energy at the farm gate than has been invested in their culture, how can the 1:5 input-output ratio mentioned earlier for agriculture in the United States be anything but wildly inaccurate? The answer to that lies in the efficiencies of conversion of *plant* energy to *animal* food products: meat, milk, and eggs, and in the fact that a great deal of the crops produced by mechanized agriculture goes to feed animals. Mechanized agriculture in no way *requires* the diversion of crops to meat, milk, and egg production; but

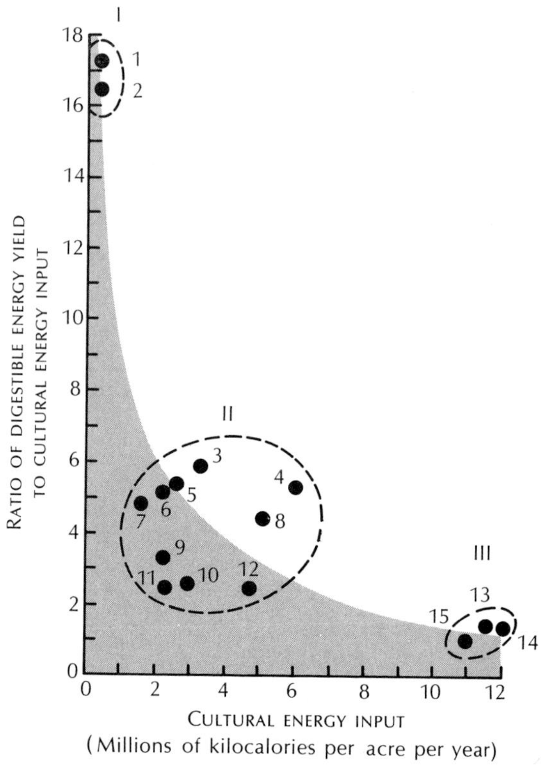

FIGURE 6.6
This diagram shows that, in general, the more cultural energy (labor, fuel, electricity) one invests in agriculture, the lower his ratio of energy "profit." What does not show in the diagram is the productivity factor. An investment of human muscle power (Cluster I) may yield a return seventeen times greater than the energy input, but the output per acre may feed fewer people than the output per acre of mechanized agriculture (Clusters II and III).
(Based on data in Heichel, 1973, 1974.)

the nations with well-developed high energy agriculture so far have been able to afford this pleasant diversion of plant energy.

The efficiency with which an animal converts the energy in its feed into food energy varies with the kind of the animal (cattle, hog, sheep, chicken), with the nature of the food product (meat, milk, eggs), with consumers' definitions of what is edible and what is not (some peoples consume the blood, some do not), and with feeding practice, health care, and physical regimen (scientific feeding, antiobiotic injections, and a short and slothful life may lead to higher overall conversion efficiencies, if not better meat). Efficiencies can be calculated from experiments with test animals or from feed-disappearance statistics and meat, milk, and egg production records. (Feed disappearance refers to feed raised that is not accounted for by exports and storage.) Although feeding experiments yield what is called the "total efficiency," the calculation of efficiencies from records of feed consumption and food production gives a better picture of the actual energy flow within a country.

The calculation of the conversion efficiency for beef in the United States is as follows. The estimated input for each pound of liveweight is ten corn-equivalent feed units. A corn-equivalent feed unit (CFU), which relates all feed to the energy content of a pound of corn, contains

1600 kcal.[1] Consequently, to produce a pound (liveweight) of beef for the American market requires 16,000 kcal on the average. This figure, derived from actual records, represents both cow and calf, and accounts for losses due to disease, storm, drought, and accident; the resulting efficiency is a *system* efficiency. In the United States each liveweight pound produces only 0.60 pound of carcass; hide, bones, blood, and inedible offal may be recovered for use but not as food. The food energy in the edible portion of the average beef carcass is about 1,240 kcal per pound. Therefore, 0.60 pound of beef carcass (the food output from 16,000 kcal of feed input) contains 744 kcal. The conversion efficiency is

$$(100) \; \frac{744}{16,000} \; \text{or 4.65 percent.}$$

It takes 21.5 calories of plant energy as feed to produce a calorie of food energy as beef. However, in "breaking" a beef carcass into retail meat cuts, 25 percent of the carcass is removed (about 20 percent as fat or tallow and 5 percent as bone). If some portion of this 25 percent does not subsequently reenter the human food system, the conversion-efficiency calculation must be based on the retail cut, which contains an average of 1,020 kcal in each pound (because it is only 19 percent fat, whereas the carcass was 34 percent fat) and represents only 45 percent of the liveweight. The new conversion efficiency is

$$(100) \; \frac{(1,020) \; (0.45)}{(16,000)} \; \text{or 2.87 percent,}$$

and the meat multiplier, the number of plant calories required for each calorie of human food, is 35. The probable upper limit of beef-conversion efficiency, based on Byerly's estimate (1967, p. 893) of 7.5 CFU per liveweight pound is 6 percent, in terms of food energy in the carcass, not in retail cuts.

Similar calculations for other animal food products (table 6.11) indicate that the pig is by far the most efficient converter of plant-energy to meat energy, the sheep is the least efficient, cattle are about four times more efficient as milk producers than as beef producers, and that the chicken is more efficient as an egg-layer than as a fryer or roaster.

Now we can begin to see the effect of a large number of animals on the efficiency of agriculture. If we assume, for illustration, that all feed eaten by beef cattle in the United States is in the form of sorghum (table 6.12), the energy profit (4.3 units) in raising sorghum is more than dissipated in beef production; in fact, we lose 75 percent of all the

---

[1]Animals differ in their ability to use energy from a particular feed, and this is a major factor in their differing conversion efficiencies. Sixteen hundred kilocalories is the energy of combustion in a pound of corn and is used here as the basis for comparison of the direct plant-food system with various feed-food systems.

Table 6.11 Summary of Plant-to-Animal Food Conversion Efficiencies (United States practice only)

Product	Conversion Efficiency (percentage)	Multiplier (as calculated herein)	Multiplier (as calculated from Jennings, 1958)
Beef	4.7 (carcass)	21.5	23.2
	2.9 (retail)	35	
	5.5 (NRC test)*	18	
Pork	14.0 (carcass)	7.1	5.2
	22.2 (NRC test)*	4.5	
Lamb	2.7 (carcass)	37	37.9
	1.7 (retail)	60	
Chicken	5.6 (broilers)	18	12.5
	3.7 (farm-raised)	27	19.4
Turkey	6.2	16.2	12.0
Milk	18.4	5.4	5.7
Eggs	8.0 (1941)	12.5	11.2
	9.2 (1966)	10.9	

SOURCE: Byerly (1966, 1967), Jennings (1958), Merrill and Watt (1973), Watt and Merrill (1963).
*From tests carried out on animals by the National Research Council.

Table 6.12 Input-Output Energy Relations in Plant-to-Animal Food Conversions

Input		Output	Profit (+) or Loss (−)
1.0	Corn (Illinois, 1969)	4.4	
4.4	Pork	0.6	− 40%
1.0	Sorghum (Kansas, 1970)	5.3	
5.3	Beef	0.25	− 75%
1.0	Hay (Missouri, 1970)	5.1	
5.1	Lamb	0.1	− 90%
1.0	Hay (Missouri, 1970)	5.1	
5.1	Milk	1.0	Break even
1.0	Corn (Illinois, 1970)	4.4	
4.4	Egg	0.4	− 60%
1.0	Vegetables (New Guinea, 1962)	16.4	
16.4	Pork	2.3	+ 130%

SOURCES: Heichel (1973), Pimentel and others (1973).

energy invested in beef production. There is no energy profit in animal-food production if one uses crops raised and harvested mechanically. If, however, one puts a pig next to a New Guinea garden and feeds it on vegetables from that garden, the resulting energy profit is 130 percent. Although this profit is much lower than the 1,640 percent obtained from a vegetarian diet in the same locale, the pig does provide protein in which the vegetables are deficient and, raised and slaughtered in limited numbers, is not a severe drain on the energy surplus of the community.

A distinction must be made between these conversion efficiencies and what may be called *feed-utilization efficiencies*. Despite the rapid growth of beef feedlots in the United States recently, the average beef animal (and his mother) still gets more than half his energy input from pasture forage, and the average sheep far more than half. Much of the pasture land used for cattle and sheep grazing is not suitable for growing grain or vegetables to be consumed directly by humans. Consequently, although the overall efficiency of American agriculture, if sheep and cattle raising were eliminated, would be *increased* in proportion to the meat multipliers (37, 21.5) of those animals applied to that portion of the total crop production representing their feed that could be diverted to direct human consumption, it would be *reduced* by an amount approximately proportional to the pasture and woodland forage input. Milk production in the United States, although breaking even on the direct-conversion basis, makes a profit on the resource-utilization basis, because a fair portion of the feed comes from land not convertible to food-crop production. Another way of looking at the system is to assess the energy in uncultivated grass, like the energy in sunlight, as free energy in terms of benefit-cost analysis, although it must be accounted for in an energy input-output analysis.

THE EFFECTIVENESS OF RECYCLING

About half the energy used in industry goes into mining and processing raw materials taken from the earth because very few resources come out of the ground ready for use. Most of the energy used is devoted to breaking and grinding ores and smelting concentrates. These two energy-devouring steps are not necessary if one can simply remelt used aluminum, copper, steel, and lead objects. The energy required to melt a metal is much less than the energy required to smelt its concentrated ore and to refine the impure product of the smelter.

Since in most of their uses metals are neither "consumed" nor widely dispersed (lead in gasoline and titanium in paint are exceptions), the potential energy savings makes recycling attractive. Recycling also lowers the pressure on diminishing reserves of nonrenewable resources. The main problem in recycling is the cost, whether measured in kilocalories, man-hours, or dollars, of separating and collecting the used objects. As the costs of primary metals (those obtained from ores) rise, more attention will be paid to economic ways of increasing the amount of secondary metal (that recovered from scrap) going to the melting furnace. Metal recycling could be increased in several ways: objects, machines, and structures could be designed so that the metallic elements can be easily separated from the rest; the dispersive use of metals could be reduced as much as possible, for instance, their use in nonreturnable containers; incentives could be established for returning disused metallic objects such as junked automobiles; separation and recovery plants could be developed at municipal garbage dumps.

Substances other than metals can be recycled. Glass and paper are well-known examples, but the energy costs of collecting them tend to be higher than for metals, so that glass and paper recycling on any significant scale requires the effort of many volunteer collectors, subsidy, coercion, or some combination of these.

An energetic comparison between the recycling of paper and the manufacturing of paper from wood pulp shows that, although recycling requires less energy, it may still be more wasteful. The energy cost of making a ton of paper out of wood pulp is much more than the energy cost of making a ton of paper out of waste paper, but, because wood is a renewable resource and waste wood and liquors provide part of the fuel in the manufacturing process, recycling, if powered by fossil or nuclear fuel, is more of a drain on such nonrenewable sources of energy. In this example, a choice of the more conservative paper-making process requires not only a comparative energy analysis, but also knowledge of the sources of the energy used in the mill and the rate (if they are nuclear or fossil fuels) at which they are being depleted or (if they are wood-pulp forests) are renewable.

WAYS TO CONSERVE FREE ENERGY

Free energy can be used only once. So, if we are to conserve reserves of free energy in nonrenewable stocks or flows of free energy in renewable stocks, we must pay attention to the efficiency of our use of energy.

The efficiency of some of the energy systems in a modern high energy society can be improved considerably. By far the greatest inefficiencies are at the points of energy use. It has been shown (Berg, 1974, p. 16) that tea kettles may transmit as little as 15 percent of the heat in the flame or electric heating element beneath them to the water within them and that this transfer could be increased to 60 percent if the tea kettle were redesigned. The American automobile's machine efficiency could almost surely be doubled without any decrease in its transport capacity, and then its consumption of fuel could be cut in half again if its weight were reduced, so that the same number of vehicles could run on one-quarter the amount of gasoline they now use. Many heat-transfer processes in industry are inefficient; major savings can be effected either by redesigning the equipment, such as steel furnaces, smelter roasters, and cement kilns, or by recapturing wasted heat and using it in or near the plant.

The use of reject heat from power plants, refrigerators, and air conditioners would improve efficiency and conserve free energy but, in most cases, would require a new system design rather than a simple modification of the heat-producing machine. For example, the reject heat from a modern power plant is not hot enough to be of much use in most places (but still hot enough to cause unwanted ecological changes). Plants can be designed to produce somewhat less electricity and heat of a temperature high enough to be distributed and used to heat buildings although systems might be difficult to install in cities that have no system of pipes for distributing steam—and where the heating fuels and the electricity are now supplied by competing firms. Domestic refrigerators help with winter heating but hinder summer cooling; a vent system to the outside that could be open in summer and closed in winter could conserve energy in an air-conditioned house.

The use of heat pumps—as heaters in winter and coolers in the summer—could effect substantial savings in free energy and at the same time substitute solar energy for fossil energy. The heat pump is an available, well-tested solar-energy device that is underused because of its relatively high initial cost and the low efficiencies and poor maintenance records of some early installations. As the cost of fossil energy rises, the heat pump becomes more attractive.

Space heating and cooling are inefficient when the spaces to be heated or cooled are not well insulated. Insufficient insulation in the walls and beneath the roof of a house or an apartment building can waste a great deal of energy, as can large glass windows. The ideal place for efficient space heating and cooling is underground, and some offices and business establishments in Kansas City and in a few other places are proving that free energy may be conserved and the cities revitalized if one is prepared to go underground.

According to Charles Berg (1974a, p. 19), the largest losses of thermodynamic availability (free energy) occur in combustion, heat transfer, and consumers' equipment and practices, which last leads us to the matter of the social efficiency of energy use. Social efficiency differs from technical efficiency. The technical or machine efficiency of the automobile changes very little when five people are riding instead of one, but its social efficiency changes a great deal. The thermal efficiency of electric lights being used by students in a college dormitory may be precisely the same as that of electric lights being used by gamblers in a Las Vegas casino, but one might argue that the social efficiency differs. Using an automobile for two trips to the supermarket where one trip would have sufficed is socially inefficient because it wastes a nonrenewable fuel, it exposes the occupants of the vehicle and the vehicle itself to unnecessary risk (the home is still safer than the street or highway), and the individuals may have spent their time more usefully.

Much of the social inefficiency of energy use in the industrialized world is the result of careless habit, hardened by the availability of cheap energy. Rising costs may change those habits, as well as the wasteful practices in industry and commerce. The contrast between Sweden and the United States of America in the economic efficiency of energy use reflects a difference in social efficiency. Sweden, with a per capita income of not much less than that of the United States, uses only a little more than half as much energy per person—despite a climate that requires more energy for heating and lowers the operating efficiency of automobiles and trucks. The greater efficiency of the Swedes is due to an organization of society and an attitude toward thrift and waste that promote the efficient use of energy and materials. Small cars, well-developed public transport, compact towns and cities, and an awareness of conservation result in a much lower energy consumption than in the United States where heavy cars, inadequate mass transport, expensive use of air carriers, sprawling cities, and a throwaway economy have inflated the consumption of energy.

Some of the energy waste in the United States is enforced by the laws of the land. Laws, for example, restricting the kinds of goods that may be carried by some trucks and specifying the trucks that may carry them result in the increased mileage of trucks running empty or partly loaded. The laws that prevent loaded trucks from using the shortest distance between two points encourage a waste of time and energy. The federal government is committed to encouraging and protecting competition even when that competition causes scarce resources to be wasted, as is the case when two large airplanes fly the same route, perhaps only minutes apart, each with fewer than half the passengers it could carry. Each time such things happen, a heavier burden is laid upon those who will not have the bounty of cheap fossil energy.

SUPPLEMENTAL READING

Ayres, Eugene, and C. A. Scarlott. 1952. *Energy Sources—the Wealth of the World.* New York: McGraw-Hill.
Excellent historically oriented treatment of energy resources and use. Includes a scathing commentary on the inefficiency of the automobile.

Berg, Charles A. 1973. "Energy Conservation through Effective Utilization." *Science,* v. 181, p. 128–138.
The chief engineer of the Federal Power Commission highlights places in the energy system in the United States where efficiency can be improved.

Berg, Charles A. 1974a. "A Technical Basis for Energy Conservation." *Technology Review,* v. 76, n. 4, p. 15–24.
A good outline of the opportunities for energy conservation in industry in the United States; points out advantages of using heat pumps.

Berg, Charles A. 1974b. "Conservation in Industry." *Science,* v. 184, p. 264–270.
Information on the efficiencies of various kinds of industrial furnaces.

Berry, R. Stephen, and Margaret F. Fels. 1973. "The Energy Cost of Automobiles." *Bulletin of the Atomic Scientists,* v. 29 (December), p. 11–17, 58–60.
Suggests that the life of an automobile could be tripled by an expenditure of only 15 percent more energy than is now required to make a car.

Byerly, T. C. 1967. "Efficiency of Feed Conversion." *Science,* v. 157, p. 890–895.
A concise presentation of the basic facts of the conversion in the United States of feed for animals to food for humans.

Committee on Agricultural Production Efficiency. 1975. *Acricultural Production Efficiency.* Washington, D. C.: National Academy of Sciences.

Agricultural productivity in the United States has increased dramatically in the last thirty years, but curves of production efficiency appear to be leveling off, and the reservoir of reserve land has been depleted. Does not recommend changes in American eating habits.

Heichel, G. H. 1973. *Comparative Efficiency of Energy Use in Crop Production.* Bulletin 739. Connecticut Agricultural Experiment Station (New Haven).

Energy input-output analysis of fifteen plant-food systems ranging from vegetables in New Guinea to oats in Minnesota.

Hirst, Eric. 1972. *Energy Consumption for Transportation in the U. S.* Oak Ridge National Laboratory (ORNL-NSF-EP-15).

Shows that the automobile accounts for about 25 percent of the energy consumed in the United States.

————. 1973. *Energy Intensiveness of Passenger and Freight Transport Modes, 1950–1970.* Oak Ridge National Laboratory (ORNL-NSF-EP-44).

Shows relations between shifts in energy-use patterns in transportation and changes in energy intensiveness and efficiency.

————. 1974*a. Direct and Indirect Energy Requirements for Automobiles.* Oak Ridge National Laboratory (ORNL-NSF-EP-64).

Packed with information, this brief report can be used as a handbook for calculating the efficiency and the energy costs of automobile transport.

————. 1974*b. Energy Use for Bicycling.* Oak Ridge National Laboratory (ORNL-NSF-EP-65).

Uses the method of his report on the automobile to calculate total energy inputs to bicycling, the energy and money costs of bicycle trips of different lengths, and the savings to be made by substituting the bicycle for automobile transport.

————, and J. C. Moyers. 1973. "Efficiency of Energy Use in the United States." *Science,* v. 179, p. 1299–1304.

Overview of efficiencies in major sectors of energy use.

Jennings, R. D. 1958. *Consumption of Feed by Livestock, 1909–56.* Production Research Report 21, Agricultural Research Service, United States Department of Agriculture.

A basic reference for information about the relation of feed for livestock to food from animal products.

Kummer, Joseph T. 1975. "The Automobile as an Energy Converter." *Technology Review,* v. 77, n. 4, p. 26–37.

Relates design for acceleration ability and comfort and the driving mode in America to the low machine efficiency of the American automobile.

Pierce, John R. 1975. "The Fuel Consumption of Automobiles." *Scientific American,* v. 232, p. 34–44.

Shows how propulsion efficiency of the American car could be sharply increased without any drastic changes in fuel or motor.

Pimentel, David, and others. 1973. "Food Production and the Energy Crisis." *Science,* v. 182, p. 443–449.

An input-output analysis that comes up with ratios similar to those of Heichel and that stresses ways to make agriculture more energy-efficient.

Putnam, P. C. 1953. *Energy in the Future.* New York: Van Nostrand.

A thorough systems analysis of the efficiency of the energy system of the United States, some information on the efficiencies of foreign energy systems, and a look into the future from mid-century America.

Summers, C. M. 1971. "The Conversion of Energy." *Scientific American,* v. 224, p. 148–160. (Available as *Scientific American* Offprint 668.)

Estimates of energy-conversion efficiencies in all sectors of the economy of the United States.

Tansil, John. 1973. *Residential Consumption of Electricity, 1950–1970.* Oak Ridge National Laboratory (ORNL-NSF-EP-51).

Contains interesting observations on the efficiency with which electricity is used in households; the inefficiency of many air conditioners and the heat pump's potential for energy saving are stressed.

Chapter 7

SOCIAL EVOLUTION AND ENERGY

If . . . you were taken on board a ship
which went . . . with a strong tide
before a following wind, you would
undoubtedly be much impressed with
the power of that ship. You would be
wrong; and yet in a way you would
also be right, for the power of the
waters and the winds might be said
rightly to belong to the ship, since
she had managed, alone amongst all
vessels, to ally herself with them.

—Isak Dinesen, 1934

THE NEED FOR HUMAN SOCIETY

As a solitary individual, man is basically helpless but with society and tools he has become the master of his environment. Since Darwin, scientists have thought that the reduction of the canine teeth, first noted in *Ramapithecus,* who lived 14 million years ago, might be related to the use of tools, for man with a club (or a group of men with clubs) would not need his teeth for offense or defense. In this view, the use of tools or weapons preceded physiological changes, including the enlargement of man's brain.

The old idea that man's big brain and erect posture came first and allowed or caused him to use tools has died hard, but the evidence now appears conclusive that, as man started to use weapons and other tools, he was free to walk and run more on his hind feet, allowing his forelegs to become the tool wielders, the arms. In a double sense, *arms made the man.*

Without fire, man's vegetable diet would be limited to fruits, nuts, and similar products. But man can digest all kinds of meat without needing to cook it. Thus it appears that man became a hunter before he learned to master fire. And indeed, this is what the fossil record tells us. Crude implements or weapons have been found even with *Ramapithecus.* The evidence of weapons, of reduced canines, and the crushed bones of other animals all converge with the deduction from man's digestive abilities to show he has been a hunter for a long, long time. Hunting requires organization, especially when one is weaker than the prey. Hunting requires a society, a system for communication, not only of commands but also of ideas, and learning. Hunting, in other words, begets culture, a complex pattern of learned responses to the environment and to other humans. Hunting requires that one learn the habits and behavior of prey, the cycles and impact of weather and climate, the ways to hunt, capture, and kill, the ways to lead and command in a society.

But many species hunt and only one hunting family has produced man. If indeed man's culture grew out of what he did, there must have been some physical characteristics of man's progenitors that gave them the unique opportunity. We have seen already what these were: the binocular, stereoscopic vision of the primitive arboreal primates, and their prehensile claws. But these only made man's emergence possible. What caused it was his ability to respond and adapt by means of culture to the wide fluctuations of climatic change that started in the late Miocene and characterized the Pleistocene epochs. These changes may have given some evolutionary advantage to the more mobile creatures and to the less specialized animals, but they gave a huge advantage to the primate genus which chose to adapt through cultural development rather than to wait out the uncertain and time-dependent hazards of mutagenesis. Man has developed greatly the ability to pass on his experiences from one generation to another and has found that cultural evolution is very much faster than genetic evolution, but requires strong social organization if it is to be effective. Cultural evolution has gone so far, or worked so well, for man that he is now the only animal that shows a strong predisposition for learning, in the sense of accepting knowledge from others or from his own experience in preference to following impulse or instinct.

Throughout his history there has been a fairly close correlation between the forms and amount of energy available to man, the forms of his society, and the size of his communities (see figure 7.1). In his most primitive stage, the capacity of man to use energy was limited to the food that he ate, about two thousand kilocalories a day a person. With the capture of fire, the rate at which he could use energy may have doubled. In primitive agricultural society, the rate rose to between ten and twelve thousand kilocalories a person each day. At the height of the Industrial Revolution (1850–70), the per capita daily energy consumption reached seventy thousand kilocalories in England, the United States, and Germany. After a temporary plateau, the rise of the consumption curve was resumed with the proliferation of the automobile and the central electric power station. Today it averages about one hundred and twenty thousand kilocalories in the developed countries. There is now among human societies an enormous range of energy consumption, from four thousand to two hundred and twenty thousand kilocalories per person per day.

The classification of nations as developed, less developed, or underdeveloped is also a classification that reflects energy use, a classification that accords with the cultural development of man, in stages that we

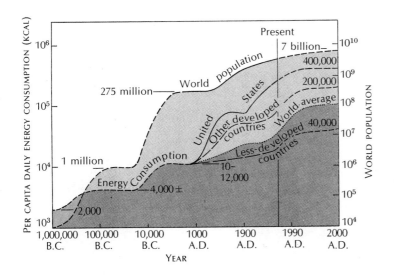

FIGURE 7.1
Human population and energy use have grown in surges representing revolutions in man's control of energy, from the discovery of fire through reliance on hunting, agriculture, industry, and technology. Note the logarithmic scales.

label for convenience primitive agricultural, advanced agricultural, industrial, and technological; the latter two constitute what may be called high energy societies. The industrial regions, with 30 percent of the world's people, consume 80 percent of the world's energy.

ENERGY IN THE HUNTING SOCIETY

A hunting and gathering society has been the frame of man's activities for well over 99 percent of his earthly tenure. In a hunting society energy flows in from food and fire and flows out as bodily energy spent in hunting, gathering, sleep, and leisure activities. In order to minimize the energy cost, settlements are mobile and follow the game. Knowledge of the habits and habitats of game and the monitoring of its numbers and location are vitally important to the hunting society. Hunting groups and settlements are small, because the maximum number of persons in a viable community is determined by the density of game and the energy cost of harvesting it.

Hunger is a constant threat to a hunting society. Food is shared; after a successful hunt, there may be a village feast. There are no set mealtimes; people eat when they are hungry or when the successful hunters return. The average daily calorie intake, depending mainly on climate and the density of game, ranges from two thousand to three thousand and the diet is high in protein and animal fats.

Dependence on the ecosystem is recognized. Animism is common, as is a belief that the animals on which the group depends for food

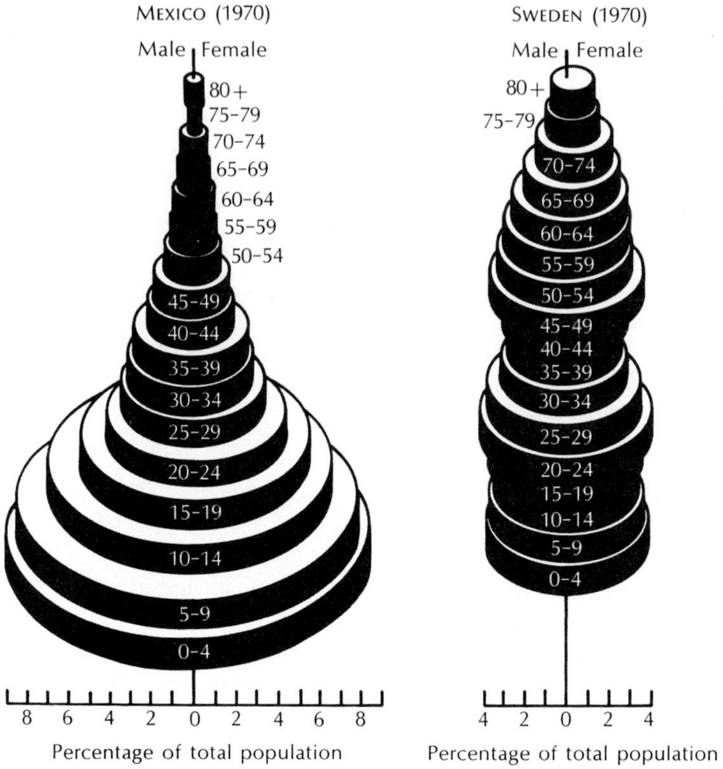
MEXICO (1970)
Male Female
80+
75-79
70-74
65-69
60-64
55-59
50-54
45-49
40-44
35-39
30-34
25-29
20-24
15-19
10-14
5-9
0-4
8 6 4 2 0 2 4 6 8
Percentage of total population
SWEDEN (1970)
Male Female
80+
75-79
70-74
65-69
60-64
55-59
50-54
45-49
40-44
35-39
30-34
25-29
20-24
15-19
10-14
5-9
0-4
4 2 0 2 4
Percentage of total population

FIGURE 7.2
Population distribution by age group in a dominantly agricultural country, such as Mexico, differs from that in a mature industrial society, such as that of Sweden. The population pyramid for a hunting society would be similar to that of Mexico, but more squat.
(From Ronald Freedman and Bernard Berelson, "The Human Population." Copyright © 1974 by Scientific American, Inc. All rights reserved.)

have souls that must be dealt with in another world. Because the relation between survival, resources available, and numbers of people to be fed is clearly recognized, there is usually some form of population control: infanticide, long lactation periods, a tabu against intercourse before weaning, restricted breeding opportunities, and geriatricide have all been noted in hunting societies.

The population structures of hunting, agricultural, and industrial societies show characteristic differences (see figure 7.2). The hunting society has the shortest average life span, the industrial society has the longest, and the agricultural society is intermediate; since all three have about the same percentage of persons above seventy years in age, the differences show most strikingly in the middle-aged group which tends to be small in the hunting society, large in the industrial society, and middling in the agricultural society.

ENERGY IN UNSUBSIDIZED AGRICULTURE

An unsubsidized agricultural society is one that derives all or most of its energy from crops and animals that it raises; one that is not subsidized by fossil energy. Until recently, the agriculture of China, India,

Indonesia, and much of Africa and Latin America was of this type. The power sources are human muscle and draft animals (if available). The primary energy sources are plant food and meat (if available). Fire is used, sparingly if wood is scarce, and fuel must be made from animal droppings, chaff, and straw. The tools used are generally simple.

Such agriculture is labor intensive, has a high productivity per unit of land, and is much more efficient in terms of energy return for energy input than is mechanized or subsidized agriculture. It is not necessarily true that unsubsidized agriculture is subsistence agriculture, with the farmers always on the verge of starvation. A crop energy return of almost twenty to one was noted by Rappaport (1971, p. 127) in a study of swidden farming in New Guinea (see figure 7.3). Although much of the plant harvest was fed to pigs whose keeping was not economic in an energy sense but was crucial as a protein reserve, there was still surplus to be spent by the farmers as leisure time. Such surpluses exist and are harvested in the same way—as leisure time—in hunting societies as well as in those unsubsidized

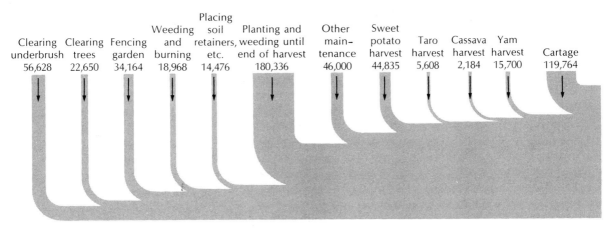

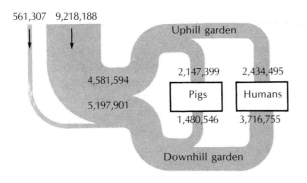

FIGURE 7.3
Energy flow diagrams of a primitive garden culture in New Guinea. The top diagram shows the twelve major inputs of energy (in kilocalories per acre) required to prepare and harvest a pair of gardens. The lower diagram shows the complete energy flow system; much of the harvest is used as pig feed. Crop yield gives more than a sixteen-to-one return on the human energy investment; the surplus is taken in pork and leisure.
(From Roy A. Rappaport, "The Flow of Energy in an Agricultural Society." Copyright © 1971 by Scientific American, Inc. All rights reserved.)

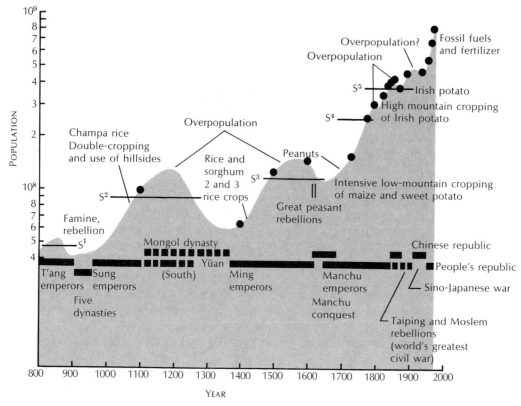

FIGURE 7.4

Oscillations in the population of China appear to result from introduction of new food plants, expansion of cropped land, increase in food supply, and subsequent overpopulation, soil erosion, famine, and warfare. Note the logarithmic vertical scale. The horizontal lines marked S^1, S^2, S^3, S^4, and S^5 represent estimated population levels that would have been stable with contemporary food supply. Present vigorous efforts to reduce China's population growth rate may emerge from an estimate of S^6, not shown here.

(Basic data from Ping-Ti Ho, 1959.)

agricultural societies that manage to keep their population within the range of available surplus. Those societies that do not control their population must let nature do it for them and spend most of their days working to exist, without amassing leisure credit.

Chinese society for twelve hundred years was a striking example of an unsubsidized agricultural society without internal population control (except sporadic infanticide, usually of females) that suffered repeated famines and political as well as physical catastrophes as the inevitable substitutes for self-restraint. For many centuries in China periods of rapid population growth would be followed by famine and civil unrest (see figure 7.4). Until very recently, the overwhelmingly dominant source of energy for the Chinese society was the chemical energy in food plants, a renewable resource with a limit on the rate at which it could be made available. This rate limit shifted upward with the introduction of each new food crop that could be grown on previously uncultivated land or that would give two or three crops each year where only one was harvested before. Periods of agricultural plenty did not change the surplus into industrial capital, they merely afforded high living for some and a rapid increase in the number of mouths to be fed, and in each cycle these mouths ultimately turned surplus to deficit and encouraged overplanting, soil erosion, flooding, infanticide, famine, and the government to collapse.

The development, in about the year 1000, of fast-maturing varieties of rice and of some requiring less water, allowed both double cropping and planting above the easily irrigated floodplains. The introduction of sorghum as a major food crop in the fourteenth century, of maize and the sweet potato in the sixteenth century, and of the Irish potato at the beginning of the nineteenth century improved the availability of food. Each of these new food plants could be grown on previously uncultivated land. In the late eighteenth century a great deal of forest land on mountain sides was cleared for maize and sweet potatoes; with the introduction of the Irish potato, even the mountain tops were cleared, as the population kept pressing on the food supply. The resulting soil erosion, depletion of fertility, flooding, and famine were inevitable.

Higher population densities may be maintained in an unsubsidized agricultural society than exist in many industrialized countries. Energy is used very efficiently, even night soil being returned to the fields. Small differences in soil characteristics can be dealt with efficiently and mixed crops can be planted for high photosynthetic efficiency and to take maximum advantage of the volume of soil available. If such a society can generate an energy surplus, it is usually consumed internally, either directly as feed for animals (an effective way of using up a surplus of plant energy), or indirectly as leisure time. If such a society runs an energy deficit, it loses both its leisure and its meat, becoming a society of work-ridden vegetarians in which the elderly starve if they do not have children to take care of them and in which, ironically, all go hungry because they took pains to have enough children so that they would not starve when they got old.

The daily calorie intake in an unsubsidized agricultural society ranges from eighteen hundred (insufficient)[1] to twenty-four hundred (enough to provide some combination of meat and leisure). Where population presses on production, the diet tends to be seriously deficient in protein, because meat cannot be raised without some humans starving to pay for it. Droughts, floods, and pests are constant threats to an unsubsidized agricultural society, the more serious as the economy nears the margin of subsistence. Such an economy cannot marshal the large infusions of capital necessary to fight droughts with

---

[1]Recent studies by Alexander Leaf (1973) of three isolated agricultural communities (Vilcabamba, Ecuador; Hunza, Pakistan; Abkhasia, Soviet Caucasus) throw some doubt on this generalization. Unusually old people in these communities appear to maintain good health and physical vigor on daily diets that supply between twelve hundred and eighteen hundred kilocalories.

pumps and irrigation, floods with levees, pests with chemicals sprayed from aircraft, and the depletion of the soil with fertilizer.

In an underdeveloped country, energy for space and water heating and for cooking comes from dung, chaff, and wood rather than from the fossil fuels. Electricity production and use is much lower than in a developed country. Animal and human labor in the fields may exceed that of tractors. The industrial sector may take a minor amount of the total energy consumed despite the desire of the country to develop industry. Malnutrition is always present and mass starvation a recurrent peril.

Only if the population can be reduced, or if it possesses hydrocarbon or mineral resources that can be exchanged for industrial capital, can an underdeveloped country hope to increase the amount of energy available to its industrial sector and thus raise its standard of living. The environmental hazards of an underdeveloped country are the threats to subsistence because the soil is being depleted and eroded under the pressure of a hungry population and because there is no way to store and distribute food reserves that might carry them over drought, flood, and storm destruction of food crops.

ENERGY IN SUBSIDIZED AGRICULTURE

A subsidized agriculture is one that does not pay for itself in energy terms. Energy flows into the society from sources other than the harvest. There is a strong tendency to grow a single crop for the economies of scale and to use chemical fertilizers and pesticides intensively to maintain the fertility of the soil and safeguard the crops.

Mechanized agriculture has a high yield per agricultural worker, but may have a lower yield per acre than unsubsidized agriculture as well as a lower efficiency of energy use. Such agriculture exists only by subsidization on a massive scale by cheap energy from fossil fuels or hydropower. No unsubsidized agricultural society can make the transition to mechanized, capital-intensive agriculture without such an energy subsidy. David Potter saw that "an infinite supply of free land would never, of itself, have raised our standard of living very far, for it would never have freed us from the condition in which more than 70 percent of our labor force was required to produce food for our population (1954, p. 162)." The parallel rise in the amount of work derived from abundant fuel wood and cheap coal and the productivity of the agricultural worker in the United States is striking.

Mechanized agriculture produces large energy deficits, in terms of energy input-output analysis. The surplus it produces is in energy credits: the food coming out of the straight-rowed fields will enable the miners and oilmen to keep producing cheap energy to plow back into those fields. As long as fossil energy is available at a low work cost in relation to the useful energy that may be obtained from it, this kind of agriculture will be profitable.

ENERGY FLOW IN AN INDUSTRIAL SOCIETY

Present industrial societies are characterized by per capita energy consumption rates of ten to twenty times those of unsubsidized agricultural societies and by the fact that this high energy consumption is at the expense of fund rather than income resource. In other words, industrial energy comes mainly from coal, petroleum, and natural gas rather than from solar radiation, food energy, or water, wind, and muscle power. The advanced industrial societies are further characterized by their increasing use of electricity, a trend that has direct effects on gross energy consumption and indirect effects on environmental quality.

Since 1945 coal's share of the energy in the economies of all industrialized nations has declined sharply while petroleum and natural gas have increased their shares. The reasons are not hard to find. The railroads' use of diesel engines represented a major substitution of petroleum for coal. Great new discoveries of oil and gas in Africa and Asia and the development of very large tankers and pipeline networks increased the supply and reduced the cost of petroleum and natural gas at a time when the direct and indirect costs of producing coal were continuing to rise. Finally, the air pollution caused by burning coal, especially coal with a high sulfur content, in urban areas has accelerated the trend toward substitution of the cleaner burning fluid fuels.

The modern industrial society, at least in the unplanned economies, is strongly oriented toward the consumer. Not only does much of the energy used in industry go to produce consumer goods, but also much of the energy used in transport is consumed by private vehicles. The household and commercial sectors of energy use are large. In the United States, electricity used in homes may surpass that sold to industry within the present decade.

The greatest choice in human history to the average citizen is today available in the high energy societies of the developed countries. The availability of energy is used to exploit resources both at home and abroad and to develop military might, national pride, and prestige. Energy is used for research and development of technologies designed to increase the efficiency of energy use and to enlarge the availability of other resources. The desired form of energy can be chosen by many in the society, regardless of the social or technical efficiency involved;

thus, one may choose to heat a house with electricity and to eat beef-steaks, careless of more efficient forms of heating and protein production.

The preservation of choice in a high energy society is, as Cottrell saw (1955, p. 255), brought about neither by the free market nor by the government but by the growth of unions, trade associations, professional societies, environmental-action groups, and other associations of citizens whose role it is "to prevent the operation both of the free market and the total state."

The use of energy in a developed country incurs serious environmental and social hazards and costs. The environmental hazards include the health and ecological hazards inherent in air and water pollution, the accident hazards of nuclear power and of energized vehicles and missiles, and the biological perils of monoculture and the extinction of other species. Social hazards include the danger of blackmail by a manipulation of the sources of energy, of social unrest stemming from an unequal distribution of the benefits of energy use, of a malaise engendered by the unexamined use of leisure time and disposable income, and of a concentration of political power in the hands of those who control the flow of energy in the society.

IMPACT OF HIGH ENERGY ON LOW ENERGY SOCIETIES

High energy, industrialized society has consistently upset the energy balances of low energy societies where the two have come into contact with each other, and that is now almost everywhere they could. The four most common disturbances have been disease, medical technology, humanitarianism, and consumer status symbols. The first brought catastrophic declines in native populations, the second increased survival and population growth rates, the third frowned on infanticide and provided food to delay starvation, and the fourth introduced the market economy into places where it had not been needed.

Shortly after the Spanish conquistadores came to what is now Latin America, the native populations declined precipitously, mainly because of the lethal effects of diseases unknown to them to which they had no resistance, and also because of "wars" they were ill-equipped to wage on anything like even terms (see figure 7.5). The same thing happened to the Indians of North America, under the onslaught of two sets of new diseases—one brought from Europe by the early settlers, the other from Africa by the imported slaves—and under the guns of the Manifest Destiny of westward expansion of the white population, which killed them as well as the game they depended upon.

Low energy societies that survived or avoided the holocaust of

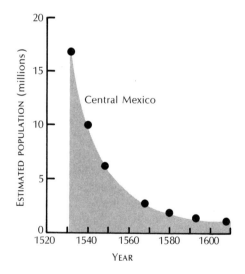

FIGURE 7.5
During the period represented in this figure, 1532-1608, most of the population of central Mexico consisted of natives. The almost incredible decline reflects the deadly impact of Europeans on the native population.
(From Cook and Borah, 1971. Copyright © by the Regents of the University of California.)

disease and pacification later were subjected to quite a different set of hazards, stemming largely from successful attempts to bring the benefits of civilization to them. Infant mortality was reduced strikingly, but the surviving children have inadequate diets. Industrialization has come to some low energy societies but its material benefits tend to be reserved for a small managing class. New wants and machines have been introduced that are both socially destructive and physically wasteful (for example, snowmobiles in Eskimo communities). Often the old way of life and the old balance between resources and population has been destroyed quickly. One of many examples is that of the Eskimos of Ungava whose population is reported to be rapidly outstripping their resources for hunting and fishing and who are drifting into unskilled labor and dependence on welfare payments.

Mexico, which has made valiant efforts to change from an agricultural nation to an industrial nation and which was well on the road because of its petroleum and mineral resources, which could be traded for industrial capital and used to subsidize agriculture, is running into trouble because its population is increasing at the rate of 3.4 percent per year, one of the highest rates in the world, its arable land is overused, its food production per capita is declining, and its industry is not creating enough jobs to take care of the growing labor force. As in many other countries, the attempted transition has created a highly visible small group of affluent people, but has not benefited much the great mass of the people, in whom it has, however, engendered great expectations and desires.

SLAVES AND SERFS
IN THE INDUSTRIALIZING SOCIETY

Our present industrial society is out of agriculture by way of serfdom, monasticism, and piracy. There are two ways in this world to rise above the common level of material welfare. One is to subjugate your fellow man and force him to produce an energy surplus that you appropriate; the other is to harness the energy resources of nature in such a way that a surplus is produced for your discretionary enjoyment. Better than either is to combine the two, as did Great Britain with her rubber, jute, and sugar workers in the colonies and her coal miners at home.

It is doubtful that industrial society could have arisen in a society that did not practice slavery or serfdom, in which the primary energy flow was controlled so that a surplus was assured for the managers. A surplus of energy was assured by keeping the number of slaves or serfs to the level needed for the sustained production of agricultural and mineral commodities and the provision of desired services. The human energy converters were discouraged from breeding to the point where their food needs would burn up the energy surplus otherwise available. Moral sanctions *against* infanticide were relaxed for the workers while the value of the individual was becoming sacred for the masters. Until the Middle Ages, societies that had used large numbers of humans as controlled energy converters or combustion engines had or had perceived little opportunity to use their energy surpluses in any way other than for the comfort and aggrandizement of the ruling classes and for warfare. Then a fortuitous conjunction of the social and physical environments sparked a new adventure for mankind, an adventure that was to lead to unimagined energy surpluses and to a marvelous range of choice in the use of those surpluses. For the first time, the emphasis of management shifted from stern control of the total energy demand of the serfs and servants toward a means of increasing their productivity. As this new emphasis began to pay off, some of the surplus was allowed to trickle down to the human converters; as their life improved in northwest Europe, hope began to fashion its own chains to reinforce the political and economic bondage of the serf to the baron, of the peasant to the burgher.

Ubbelohde suggests that

> the decisive reason why the steam engine was not introduced 1700 years earlier, with incalculable consequences for world history, appears to lie in the acceptance of the practice of slavery . . . . The economic incentive for developing inanimate power was neutralized by [the] facile harnessing of animate energy in the ancient world . . . [moreover] the thinkers and leaders of Greece and Rome were turned away by the influence of [their] philosophies from any intellectual interest in the development of technology. [1955, p. 58]

The strong Greek prejudice against the crafts and technology was reflected in Plato's statement that no philosopher should allow his daughter to marry the son of an engineer. Slavery and serfdom allowed population to be limited while a surplus still was possible and concentrated control over the surplus so that it could be used in trade. The ratio of the rate of population growth to the rate of the growth of energy from sources other than humans, as Cottrell emphasized (1955, p. 149–150), is a critical factor in the course of industrialization, for there is always a choice available "between using energy to make converters which will increase the capacity to use cheap surplus energy and using energy to increase the production of food . . . . If population can be limited, the increased numbers of high-energy converters can be used to increase the supply of surplus energy and of additional high-energy converters at a rate which results in a mounting per capita output of goods . . . produced by machine." If the rate of nonhuman energy input increases faster than population, the society is likely to move in the direction of high energy, but if population increases faster, the society stays or becomes a low energy society.

Now, why was it possible, on the plains and in the forests of Germany and France, to develop a new social order, a new kind of political economy, and a new relation of man to resources? As in most questions of historical interpretation, the answer has to be an informed guess. The area was rich in potential agricultural and forest resources and the human population in the middle of the first Christian millennium was small. There was, in other words, a frontier rich in resources capable of creating an energy surplus if adequately exploited. The plains and broad valleys of Germany and northern France allowed the development of productive agriculture that did not rob the soil of its nutrients and expose it to destructive erosion.

Another factor of incalculable importance in the emergence of industrial society was that the solar plexus of European trade routes, some established well before the Roman conquest, was in southern Germany and northern France. Along these routes flowed metals obtainable from only one or two sources in Europe: tin and zinc from Cornwall, gold and copper from Spain, iron from Sweden and Britain, silver and lead from Austria and eastern Germany. Exotic goods from Asia and Africa, such as spices, fragrances, and precious stones, were carried along these routes. Itinerant smiths and traders bore news and knowledge from place to place, exposing craftsmen, burghers, and farmers to new ideas and new desires. Later, as Christianity moved northward, some of these ancient trade routes became pilgrim pathways and the pilgrims, too, returned with new and potentially useful ideas from other places.

Roman control of southern Germany, the Low Countries, and northern France had resulted in great improvement in roads and waterways for transport and communication and in the establishment of the city as a center of government and trade, to which the countryside or hinterland was tributary. This cultural imprint and the lack of strong centralized government allowed the later emergence of the *hansa,* city-states governed by burghers. The Romans also introduced new crops and new agricultural methods into their Atlantic provinces; agricultural progress was stimulated most near the frontiers, because of the needs of the soldiers stationed there for food, clothing, and weapons.

By the fourth century industry was well established in the Rhineland and northern Gaul; cloth, clothing, arms, glassware, jewelry, earthenware, as well as bronze, brass, and pewter objects were being made; there were mines and mills, the latter powered mainly by water. The Germans were already using the heavy wheeled plow, which had been used in the Tirol as early as the second century.

About the seventh century two factors of great significance in the creation of an energy surplus first appeared: the horse collar, the forerunner of the shoulder-bearing harness, and the three-field rotation system which allowed an extra crop every other year (three crops in two years with winter planting).

From about the eighth century agricultural surpluses existed and from these surpluses specialists in harness-making and horse-shoeing could be paid, exchanges for barter could be set up, and towns could be more than clusters of farmers' houses. In addition, the social structure was basically tribal; there was no strong centralized government to tax agricultural production as much as it would bear, pull men away from productive labor to fight wars for the ruling family, still the peasants' hope of material improvement, and stifle social and technological innovation. On the contrary, innovation appears to have been encouraged: the stirrup, the hydraulic saw, the heavy sod-turning plow, and the shoulder-bearing harness are major examples of technological innovation that appeared in what is now Germany and northern France in the Middle Ages. The plow and the social organization required for its use may have initiated democratic capitalism in the world. The thick, heavy sod of the German and French fields could not be broken with the scratch plow that was developed for the thin soils of the lands round the Mediterranean. Consequently, a plow with a cutting edge and a turning face was imported; this plow could not only cut through the sod and deeply into the rich soil beneath, but also would turn the cut sod in furrows, allowing it to be broken further by hoe or harrow. The plow, however, had to be heavy to be useful. Oxen were the draft animals of the time and region, and it took teams of between four and eight oxen to pull a heavy plow. No peasant owned that many oxen; indeed, few probably owned even two. Therefore, it was necessary to pool animal power and to establish a system for sharing the ultimate proceeds of the power pool. It was decided that each should own a share in the field according to his contribution of power, the peasant supplying two oxen receiving twice as much area and its produce as the peasant supplying one ox. Since the new plow allowed the production of considerably more feed and food than required by the peasants and their families, and since in general they still had a surplus after taxes for defense paid in kind or labor to the local baron (ex-tribal chief), they wanted something to do with the surplus. The need and desire to barter for goods and services encouraged the enlargement of towns, the division of labor, and the establishment of a credit economy. The invention of the shoulder-bearing harness also encouraged the coalescence of villages into towns, because now horses replaced oxen and the farmer could go farther from home to field and back in the same time with a horse than with an ox.

Of the period 550–700, so critical in the development of a high energy society, Lewis (1958, p. 178) mentions instability, political formlessness, and an inability of kings to rule effectively. "This age," he notes, "was politically and administratively anarchical, inchoate and independent. It had vigor and growth, but it lacked discipline."

THE ROLE OF MONASTICISM IN INDUSTRIALIZING SOCIETY

Cultural historians, notably Max Weber and Lynn White, have argued that a particular religious ethic, called Protestant, Puritan, or Judeo-Christian, was a causal or motive force in the orientation of Western society toward production and growth. There appears room to doubt that an ethic, whether derived from religion, philosophy, or a particular perception of human needs, is ever much of a motive force. It seems more likely than an ethic acts as a moral template that tells man, of all the things he can do, those that he should not do, and of the things he may do, those that are better than others.

The agricultural and social revolution sparked in southern Germany by the introduction of the heavy plow appears already to have been under way when Christianity was imposed there by the ruthless and brutal Charles Martel. If the Judeo-Christian moral view, which separates man from nature and gives him dominion over her, played any substantial role in the early development of industrial society, it probably played that role subtly and by example. It was the Benedictines, founders of the world's greatest agricultural extension service, who set the example. The order founded in 529 by Saint Benedict at Monte Cassino first made work as virtuous as prayer in the Christian world. Not only did the Benedictines make work virtuous, but also they made it efficient and productive, regulating their work by schedule and directing it toward increasing agricultural productivity. Expanding rapidly in the Middle Ages, partly because of the conversion of the priest who became Pope Saint Gregory I to the rules of the order, the Benedictines grew to have some fifteen thousand monasteries, many of which were truly agricultural experiment stations, where new methods and new crops were tested. The Benedictines came to Anglo-Saxon England at the turn of the seventh century and moved into southern Germany early in the eighth century. Towns and occasionally cities grew up around the monasteries they had taken pains to situate in the wilderness. It seems likely that the Benedictine monks demonstrated to a peasant populace already experimenting with agricultural surpluses how to increase further those surpluses and, later, how to mill, and cure, and weave, to make superior "value-added" products from their raw agricultural produce. At the same time, the Christian campaign against pagan animism may have removed some lingering psychological barriers to the unrestrained exploitation of timber, soil, and falling water.

In the twelfth century, the Cistercians, picking up the moral baton they claimed the Benedictines had forfeited by soft living brought on by material success, started what amounted to an engineering as well as agricultural extension service. They reclaimed land from the sea, built roads, dug canals, built mills to manufacture textiles and furnaces to work metal. They operated twenty-five of the thirty hammer forges for iron working mentioned in twelfth-century French documents (Forbes, 1968, p. 17). The Cistercians were the first monastic order to make extensive use of lay brothers. They helped spread the water mill throughout France, Germany and Britain in the twelfth and thirteenth centuries. A great demonstration program mounted by two religious orders led western Europe toward the Industrial Revolution. Not the least important of the monks' innovations was the development of the mechanical clock to regulate work and prayer. Without the clock and the concomitant concept of linear time, industrial society would be impossible. It is appropriate that the oldest document known to have been printed on Gutenberg's press is a calendar, not a Bible (Mumford, 1934, p. 135).

THE ROLE OF PIRACY
IN INDUSTRIALIZING SOCIETY

The early stage of contractual capitalism represented by the division of agricultural surplus according to the contribution of power and capital and by the hanseatic towns and cities dominated by burghers did not lead to a market economy and a mercantile society. Trade was rigidly controlled by the burghers, who set prices and allowed or prohibited sales, kept the countryside under economic subjugation, set up alliances (the greatest of which was the Hanseatic League) with other cities, and let no free market develop.

A market economy might eventually have emerged from this system, but it seems unlikely. In any case, pirates made it unnecessary to wait for social evolution. Pirates are highwaymen of the sea. In the Middle Ages those peoples living near the shores of the North Sea had a hard and uncertain life if they chose to subsist entirely as hunters on the sea and gatherers on the littoral. As agricultural life improved in the hinterland, and as the *hansa* grew rich, the temptation to share by force the wealth of the interior became irresistible and some of the bolder Danes, Norsemen, Angles, Picts, Vandals, Celts, Normans, and Saxons turned from fishing to daring raids on the burghers' strongholds. In the third century both Saxon and Irish sea rovers began raiding the English coast and the continental side of the English channel. By the fifth century, the Irish had colonies in Wales and Cornwall and Saxon pirates were settling in France. Vandal pirates raided Rome in 455. The Vikings were latecomers in this freebooting parade but, from 787 to about 920, they were the masters of the northern seas, penetrating inland as far as Strasbourg on the Rhine, Paris on the Seine, and Chartres on the Loire.

Although the pirates preferred tribute in gold, silver, and gems to loot of a bulkier nature, such tribute required that they find a market in which to exchange it for food, drink, and finery. So the first stirrings of the market economy can be seen in the pirates' quest for buyers of their valuable acquisitions for which they themselves had little use. Considering the nature of the goods, one may assume the existence of middlemen or "fences."

When the pirates raided interior continental towns, they appear to have retired as soon as either defeat was certain or tribute was in hand. When they invaded England, Ireland, or Brittany, sometimes they retired, but sometimes they stayed. The descendants of these immigrant pirates joined forces toward the close of the Middle Ages to form what we call today a mercantile nation, powered by the wind, riding on the wave, thriving and growing by intelligent, adventuresome commercial aggression, overwhelming lower energy societies and leaving the burgher-dominated continent in their wake as they built the British Empire, while proclaiming the blessings of laissez faire and the virtues of the market economy.

The economy of the continent also benefitted from the activities of the Vikings. Lewis (1958, p. 298) notes with some astonishment that Germany's economic life appears to have been stimulated during the period of the Viking invasions and raids, despite the heavy tribute paid. Trading profits certainly helped pay as well as replace the ransom paid to the Vikings. During her steady economic advance in the ninth century, southern Germany "was in contact with both the world of the Mediterranean and that of the Northern Seas." In 955 Otto I by walloping the Hungarians at Lechfeld succeeded in uniting most of Germany and part of central Europe; there was "an immediate upsurge of economic life" as Otto moved on to found the Holy Roman Empire.

By 920 the Vikings were turning from hit-and-run piracy to aggressive commerce. They had received Normandy as a fiefdom in 911. They had settled along the shores of Ireland and had made Dublin, Limerick, and other cities rich before the year 1000; in the interior, the Norsemen produced, or had produced for them by the native Irish (who regarded them as gluttons and commercial travelers), surpluses of wheat and honey. The high-water mark of pure Viking energy was the short-lived maritime empire of Canute, made up of Anglo-Saxon England, Norway, and Denmark, which came apart in 1043.

Lewis remarks that "free merchant individualism" was Scandinavian whereas the burgher-dominated guild economy was Flemish and German.

THE ENERGY OF THE HIGH MIDDLE AGES

For whatever combination of causes, our modern industrial society did have its beginnings in what are now Germany and northern France, where an agricultural revolution fostered by a favorable combination of natural and social environments produced an early version of contractual capitalism and led to the domination of cities by men of commercial power rather than men of hereditary ruling families. In the late or High Middle Ages, possibly because of the youthful vigor of the emerging mercantile state, but certainly related to rapidly growing energy surpluses in western Europe, there was a remarkable flowering of intellectual interest and mechanical invention.

Water mills proliferated along the streams of western Europe and were put to many uses. The invention of shoulder-bearing harness made the horse the premier draft animal of the world. Mechanical clocks, already known in the Arab world, were installed in monasteries and city clock towers to regulate work, religious activity, and commerce. The windmill, a Persian development, became a major factor in land reclamation in the Low Countries. Canals and roads were improved and extended. Mining technology and geology made great advances. The compass was reinvented, and marine technology blossomed.

In the High Middle Ages a tremendous interest in the physical and biological world developed in western Europe. The Moslem world was cultivated for its knowledge at the same time as it was being fought for territory and religious glory. Constantine the African, a Benedictine at Monte Cassino in the eleventh century, translated a large number of Moslem scientific and philosophic works from Arabic into Latin. Later at Toledo in Spain a veritable translation institute flourished under the resident Christian bishop. Man was turning his attention from trying to get through this hellish world into a promised heaven toward trying to make the earth more heavenly, or at least more comfortable and rewarding. He now knew it was possible to enjoy life, because he had accumulated an energy surplus and was using it for everything from creature comforts to Crusades. There was a new world ahead.

The thirteenth century in western Europe epitomized the changes being brought by new energy systems. The aberration of the Children's Crusade, the signing of the Magna Carta, and the founding of some of Europe's major universities marked the first quarter of the century. Also in the early part of the century flourished such diverse men as St. Francis of Assisi and Frederick II of Hohenstaufen. St. Francis preached a reverence for life and a community of man with animals, as well as the virtues of absolute poverty, humility, and devotion to mankind. Frederick II, supposedly a Christian ruler, maintained a harem, was a patron of the arts, founded the University of Naples, wrote the world's first scientific treatise (*The Art of Falconry*), was excommunicated by two popes, led the Fifth Crusade, and talked a Saracen commander into delivering a town to him. In the latter part of the century, Roger Bacon carried out some of the first truly scientific experiments, taught at Oxford and Paris, and had a prophetic vision of the future of applied science, writing in the year 1260:

> Machines may be made by which the largest ships, with only one steering them, will be moved faster than if they were filled with rowers; wagons may be built which will move with incredible speed and without the aid of beasts; flying machines can be constructed . . . . [Quoted in Forbes, 1968, p. 19]

Marco Polo went to China, the world's first representative parliament was established in England, windmills reached the Low Countries, and coal smoke became so bad in London that the gentry complained. All in all, quite a century.

FROM ROGER BACON TO JAMES WATT

In the almost five hundred years from the death of Roger Bacon (1294) to the perfection of the double-acting steam engine by James Watt (1782), water, wind, and horses were the sources of power for industrializing society. The energy surplus was spent in war, in exploration, in the development of foreign trade, in art, music, and literature, in the adornment of palaces and cathedrals, and in support of a population that even two devastations of plague, numerous wars, and overwork could not keep from growing rapidly. Weapons were greatly improved during this period. The peasant girl–soldier Joan of Arc was burnt at the stake. Notable inventions included watches and movable type. Painting flourished, and many great works of art were created. The Protestant Reformation shook the foundations of Christendom. Galileo and Copernicus turned the established order on its head by challenging the official doctrine that the sun went round the earth. Machiavelli analyzed the politics of his age with a dispassionate eye. The British monarchs Henry VIII and Queen Elizabeth left indelible signatures on Western history.

From 1488 when Bartholomeu Diaz sailed around the Cape of Good Hope until 1588 when the Spanish Armada was defeated by the English fleet under Sir Francis Drake, there was a Golden Age of discovery and conquest (and plunder and piracy) led by Portuguese and Spanish adventurers, followed by the British, who took over and started their long, successful campaign of seeking the world's raw materials for appropriation. One not inconsequential energy event of these five centuries was the introduction of the humble potato into England by Sir Francis Drake in 1585—an edible by-product of the quest for gold.

Scientific inquiry and discovery quickened in the seventeenth century with the publication of Kepler's laws of planetary motion (1626) and Newton's *Principia* (1687). Then, in the early part of the eighteenth century, engineers (like Newcomen, Darby, and Watt) for the first time came into view.

SOCIAL CHANGE AND
THE INDUSTRIAL REVOLUTION

The Industrial Revolution was built not only on invention nurtured by opportunity but also on the remarkable expansion of British overseas trade in the seventeeth and eighteenth centuries. Great Britain had command of the seas, had developed overseas sources of cheap raw materials as well as overseas markets, had accumulated the capital required for industrial expansion, had maintained political and economic stability at home, and had understood the role of applied science in industrial development.

As food and other farm products began to be imported in large quantities, displaced farm workers streamed into the new factory towns, went into the mines, worked upon the railroad, or took to the sea. The agriculture of nineteenth century Britain was not mechanized or otherwise subsidized directly by the fossil fuels, it simply had its burden lightened as some of the energy surplus was used to buy the labor of farmers overseas and of the crews of ships bringing food to Britain.

Conditions in the early factories, mines, and mills were almost unbelievable by modern standards. It was not unusual for men, women, and children to work for fifteen hours a day under hazardous and unhealthy conditions for a subsistence wage; their living quarters were dark, cramped, and often infested by vermin. Many of these workers, or their forebears, were farmers who were forced to sell their farms when the commons, which they used for pasture, were enclosed for sheep raising. The only alternatives of the unemployed was the workhouse or crime—both were degrading, and the latter risked monstrous penalties for small crimes.

Gradually, the owners of the factories, mines, and mills recognized that healthy workers were more productive than sick ones; the so-called sanitary revolution began. Cheap soap, better nutrition, and the establishment of municipal sanitation services improved the health and increased the longevity of the worker and his family, while laws governing safety practices, working hours, and eventually requiring health and accident insurance gradually improved the lot of industrial labor.

Characteristic of the Industrial Revolution and the subsequent Technological Revolution was the speed with which a technological innovation could change life-styles. When the Liverpool and Manchester Railroad was first opened in 1830, the stage-coach capacity between the two cities was 688 passengers a day and the fare was ten shillings inside, five outside. The fare by the new railroad was five shillings inside, and three shillings and sixpence outside (the original second class coaches were open to the elements). The journey by stage had taken four hours; by rail it was now one and three-quarter hours. By 1833 the railroad had averaged 1,100 passengers a day since it was opened. (More persons were travelling and some were going more often.) Whenever demand has been increased by technologic advance, life-styles have been changed. In its great hundred years, the railroad changed the lives of millions of people. The automobile was to do much the same thing in the following century.

The reins of government and the power of the professions, the church, and the military were in the hands of a managerial elite made up partly of the nobility and partly of the sons of the mine owners and manufacturers. One of the remarkable characteristics of this latter group was the manner in which it spawned scientists and supported intellectual pursuits. James Prescott Joule was the son of a brewer and became one himself. Josiah Wedgwood, the china manufacturer, was the father-in-law and firm supporter of Charles Darwin.

The great growth in the use of energy and the accumulation of surplus energy in Great Britain was helped by the Puritan ethic, which held that idleness was sinful and that selling was good but spending bad. Consequently, energy surplus tended to be reinvested in still greater energy production, as well as in the military strength of the Empire.

Although there was a good deal more social mobility in industrial Britain than there had been, it was not until the Industrial Revolution was transplanted to North America and established there without nobility, squirearchy, or established church, that a meritocratic society was created, in which, presumably, each man could rise according to his ability. It was never a classless society but, compared with the European industrial nations, America has always been a place in which no one seems to know his station in life although everyone has a pretty good idea of his current position.

As production units became larger, the basic consumption unit, the family, became smaller. Increasingly, the family was no longer a production unit, and its social importance ebbed. Loyalty to family and community were partially replaced by loyalty to factory owner, corporation, trade union or association, professional society, and the state.

SOCIAL CHANGE AND DOMESTIC POWER

The Industrial Revolution involved a concentration of machines and energy devices in mills and factories, in steamships, and on railroads.

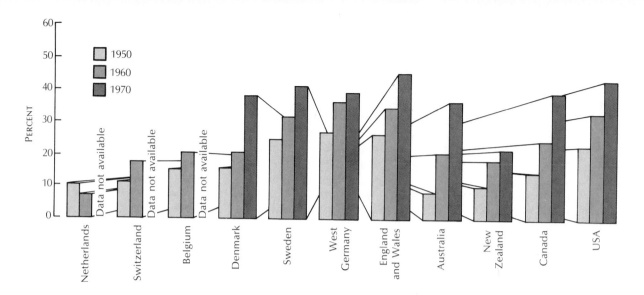

FIGURE 7.6
A recent significant cultural change in the high energy countries lies in the increase of women in the working force. The speed of this change is shown by steep rises, in only twenty years, in the number of married women who work shown as a percentage of all married women aged 15 to 64. Note Denmark, Australia, and Canada in particular. (From Judith Blake, "The Changing Status of Women in Developed Countries."

These were not power plants or power devices for the benefit of the individual, they produced goods and services in gross, not particular. With the building of the first central electric power station in 1881, another great change started to take place. Now power technology was moving into the home and becoming available to the family directly, not simply through mass-produced goods and mass transit. Home lighting and powered appliances became possible. The family automobile, motorcycle, power boat, snowmobile, and airplane gave the average individual in industrial society the feet of Mercury and the wings of Pegasus. Cities began to be places to escape from, not to live in. The growth of suburbia, explosive in recent decades, started in the seventeenth century when the gentry of England escaped from the polluted air of London. In today's suburbs, the automobile has become as important as the home and has brought further stress to the family.

Women's role in society changed radically. As men's occupations became so specialized and separated from the social dynamics of the community that it became difficult for mother to explain to the children "what Daddy does," the wife's social role as a supporter and sharer of her husband's status dwindled. As corner groceries and shoe-repair shops coalesced into shopping centers, as residential areas receded from the central city and sprawled away from schools and post offices, the housewife began to drive the family car more than the husband did and her daily schedule became more like the time-table of a milk-run train than the routine of a household manager and maintainer of social position. Women started to enter the work force in large numbers (see figure 7.6); old distinctions between the roles of men and women in society were eroded further, and the traditional symbolic distinctions of role based on dress, adornment, and scent became much less sharp. Perhaps in no other way are the differing social gradations of low energy and high energy societies so clearly calibrated as in the status and appearance of women.

Social Evolution and Energy

SUPPLEMENTAL READING

Albion, Robert G. 1926. *Forests and Sea Power.* Cambridge, Mass.: Harvard University Press.

The maritime power and prowess of England for many years depended on English oak, of which the supply diminished because of clearing and charcoal making.

Blake, Judith. 1974. "The Changing Status of Women in Developed Countries." *Scientific American,* v. 231 (September), p. 137–147.

Women who want primarily to be wives and mothers are becoming in relatively short supply.

Bloch, Marc. 1967. *Land and Work in Mediaeval Europe.* Berkeley: University of California Press.

Essay on the triumph of the water mill is particularly enlightening.

Cook, S. F., and Woodrow Borah. 1971. *Essays in Population History: Mexico and the Caribbean.* Berkeley: University of California Press.

Conclusive demonstrations of the disastrous effect on native populations of the Spanish conquest.

Cottrell, Fred. 1955. *Energy and Society.* New York: McGraw-Hill.

The concluding chapter, "Not One World but Many" is especially relevant to the subject of social evolution and energy.

Forbes, R. J. 1968. *The Conquest of Nature.* New York: Praeger.

Contains much of interest on the relations of energy use to social evolution.

Landes, David S. 1969. *The Unbound Prometheus.* Cambridge: Cambridge University Press.

Well-written explanatory economic history of industrial development in western Europe from 1750 to 1968.

Lewis, A. R. 1958. *The Northern Seas.* Princeton, N. J.: Princeton University Press.

An exciting story of piracy, technological innovation, trade, and social change in northwest Europe during the Middle Ages.

Mantoux, Paul. 1961. *The Industrial Revolution in the Eighteenth Century.* New York: Macmillan.

First published in 1928, this book traces the institutional evolution of capitalistic enterprise in England against a changing backdrop of social and working environments.

Mumford, Lewis. 1934. *Technics and Civilization.* New York: Harcourt, Brace, and World.

The evolving relations of man and machine.

Nichol, Hugh. 1967. *The Limits of Man.* London: Constable.

Concludes that fossil fuel is the main source of food for modern man, who "can never rise again from the ashes" of his predecessors' use of it.

Rees, William. 1968. *Industry before the Industrial Revolution.* 2 vols. Cardiff: University of Wales Press.

Valuable source of information on early coal mining, the charcoal-supply problem, and the relations of fuel supply to industrial location and growth.

Scientific American editors. 1971. *Energy and power.* San Francisco: W. H. Freeman and Company.

Eleven articles; see especially chapters 5, 6, and 7 by Kemp, Rappaport, and Cook, respectively.

Ubbelohde, A. R. 1955. *Man and Energy.* New York: Braziller.

Some interesting passages on the role of slavery in the advent of high energy society.

Vanek, Joann. 1974. "Time Spent in Housework." *Scientific American,* v. 231 (November), p. 116–120.

Working women spend less time in housework than did their mothers and grandmothers; changes in the distribution of time spent are shown for period 1926–1968.

White, Lynn. 1962. *Medieval Technology and Social Change.* Oxford: Clarendon Press, 194p.

The social impact of the development of the stirrup, the horseshoe, the heavy plow, shoulder-bearing harness, and other medieval inventions.

Workman, Herbert B. 1913. *The Evolution of the Monastic Ideal.* London: Epworth.

See especially chapters 3 and 5, on the Benedictines and the development of monasticism from Saint Benedict to Saint Francis.

Chapter 8

LIFE-STYLES, GOVERNMENT, AND ENERGY

Democracy is clearly most appropriate for countries which enjoy an economic surplus and least appropriate for countries where there is an economic insufficiency.

—David Potter, 1954

IMPORTANCE OF ENERGY SURPLUS

Life-styles in a society depend to a great extent on whether or not the society produces an energy surplus and, if so, on how the surplus is achieved and maintained. The nature of government in a society relates also to the question of energy surplus and has a direct bearing on the life-style or styles of the society.

Energy surplus in a society may be obtained from hunting or agriculture, from inanimate income resources such as wind power or water power, or from fund resources such as the fossil and nuclear fuels in nature. By modern standards, no society has ever managed to achieve much energy surplus through agriculture alone. Furthermore, the persistent maintenance of an agricultural-energy surplus in a society appears to require cultural homogeneity and control, a requirement that seems to set an upper limit on the size of the population that can enjoy such continued surplus. An agricultural population without new land to spill into must not become so large that internal dissension can breach the rigid structure of social control necessary to keep the population and the total consumption of agricultural products always within the limits of the surplus. Especially where there is danger of recurrent drought, flooding, or pestilence, agricultural surplus must be sufficient to maintain the human population over several years of poor harvests. The pressures against the required social self-control may be formidable. If there is no institutionalized and dependable state welfare system for the aged, the tendency is to inculcate filial devotion in numerous offspring; parents thus attempt to procreate their own social security systems, without regard to the cumulative impact on the resources of the society. If there are no strong social mechanisms working against the preferential amassing of surplus by an individual, or by a comparatively small group of individuals, the tendency is to monopolize resources within families, with a resulting social stratification based on control of surplus. Much of the surplus may ultimately support a police establishment designed to maintain the uneven distribution of the surplus and to supplant the mores, tabus, and conventions that appear to work effectively in some cooperative societies, where social distinctions reflect position and status in the communal power structure, but not the actual possession for selfish consumption of unequal portions of the total energy surplus.

A society that imposes no social controls on the rate at which it consumes the resources available to it and that can manage to keep increasing its per capita energy supply becomes a *growth* society. Its consumption will expand to the limit imposed by available resources, which in turn will depend upon economic technology and on political control of the resources. If the rate of availability of resources reaches a plateau, as for example in an alpine canton with small climatic variation, subject to no disastrous storms, and without additional land to cultivate, the population will stabilize; persons above the limit supportable will be exported, allowed to starve, or will die of disease abetted by malnutrition; the society will have made a transition from a growth phase to a steady-state or equilibrium phase. If there is additional land on which food crops may be raised, or if agricultural science and technology can increase the productivity per acre, the population or the level (standard) of living, or both, will increase to new levels dictated by new limits of food availability. It is the product of population and living level that is limited by the availability of energy. A doubling of population requires a doubling of available energy. A doubling of living level requires approximately a doubling of available energy. Doubling of both requires that the available energy be quadrupled. The impact of an industrial growth society on energy resources is geometric, because population and living levels both grow.

Surpluses obtained from *renewable* resources, although historically much smaller than those now enjoyed from nonrenewable resources, at least in theory, can be perpetuated indefinitely. Moreover, the way in which such surpluses are achieved may not be strongly antithetic to the social measures required to stabilize consumption below the zone of marginal supply; in other words, a steady state may be achieved by cultural controls. Surpluses obtained from *nonrenewable* or fund resources are very different. In the long run there is no possibility whatsoever of perpetuating such a surplus; moreover, the time during which a fund-derived surplus may be available to a society diminishes with the growth of that society, no matter how much the surplus *appears* to be growing at the same time. The apparent growth of a fund surplus is illusory; what we see is growth in the rate at which the society is using up a finite resource. In this using up, there is always a certain amount of the resource being prepared for use and, as our consumption grows, so may these about-to-be-consumed stocks, which we euphemistically call "reserves," when they no more represent true reserves than do the cattle being fattened in a feedlot.

The importance of an energy surplus to the level of living of a society is almost self-evident. An energy surplus is required before there can be leisure, art, sport, entertainment, and comfort beyond mere existence. The nations with high levels of living, measured by per capita gross national product (GNP) or income are those that consume or have available large supplies of energy (see figure 8.1). But this standard of comparison does not allow for the physical economy of energy use in the society. What really counts is how much work and useful heat is available to the society, not the gross energy availability. Switzerland, for example, uses energy efficiently and therefore achieves almost the same per capita GNP level as Canada with a much smaller per capita energy input. Nor can it be held that GNP is an adequate measure of all the values inherent in the meaning of "standard of living." Perhaps we should distinguish between level of consumption and level of life, with the former defined by access to motor cars, television, and all-electric kitchens, and the latter by some measure of the average enjoyment of life on the yardsticks of health, education, and leisure.

It seems obvious that an energy surplus must exist for there to be much meaning to the phrase "level of life." But beyond that, the great question is, how much surplus, how much energy, must be available to each of us in order that we may experience a satisfying life?

It has been a facile assumption, indulged in by Rousseau and other philosophers, that a primitive society, in "harmony with nature," uncluttered by the artifacts and unhampered by the strictures of civilization, has the highest standard or level of life. Anthropologists, the great debunkers of such romanticism, have shown that many if not most primitive peoples live short lives of fear and discomfort that appear to be in harmony with nature only because their means for destroying nature are comparatively impotent. Some primitive societies have achieved an effective equilibrium with the resources available to them; they practice selective breeding and long weaning periods, as well as infanticide and geriatricide if normal disease and disaster are not sufficient to keep the population within the bounds of their resources.

Since the days of Adam Smith, many persons have opposed the concept of the noble savage, contending that continuous growth in the use of energy and materials is the best state for man, that his ingenuity can be unleashed only in the "growth state," that his material and even moral progress depend upon the existence of such a state. Industrial man's conquests of disease, his power over nature, his ability to

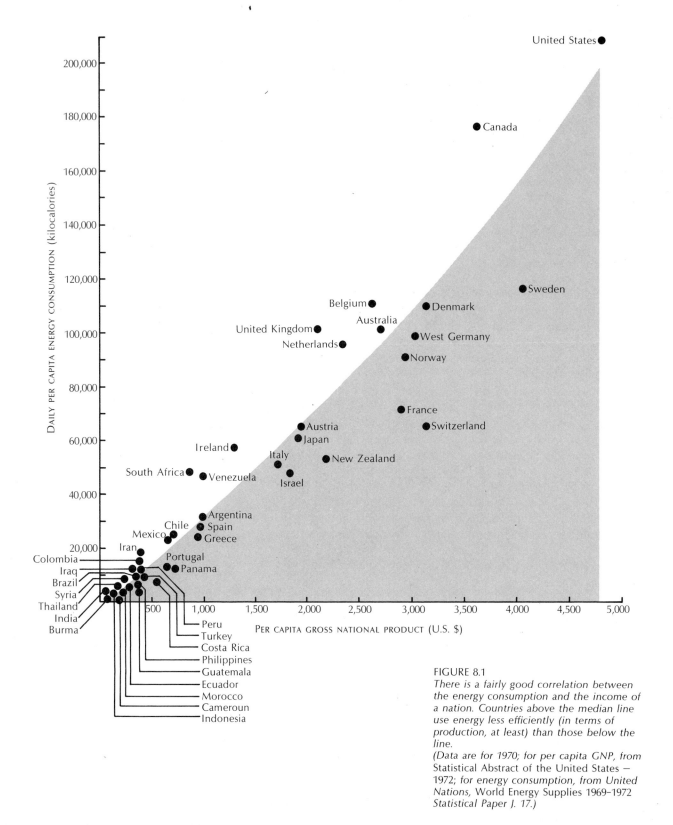

FIGURE 8.1

There is a fairly good correlation between the energy consumption and the income of a nation. Countries above the median line use energy less efficiently (in terms of production, at least) than those below the line.

(Data are for 1970; for per capita GNP, from Statistical Abstract of the United States — *1972; for energy consumption, from United Nations,* World Energy Supplies 1969–1972 Statistical Paper J. 17.)

go to the moon, the comforts with which he is surrounded, have all resulted from the reinvestment of energy surplus in still more energy-producing systems. That man's material welfare has been enhanced enormously as a consequence of the increasing exploitation of energy resources can hardly be denied. But not all the consequences of increased energy use have been beneficial. There have been some substantial costs, assessed as warfare, pollution, and economic slavery. Today the important questions are whether and for how long the growth state can be maintained. Will the continuance of the growth state be likely to result in a better or worse life for those now in high energy societies? And will it give those now in low energy societies a better or a worse chance of advancement toward a high energy society?

The importance of an energy surplus is that without it a society has no choice. With a small surplus, there is a small choice; with a large surplus, there is a large choice. The form of society appears to change to accommodate choice as the per capita energy surplus mounts. In other words, a society seems to become more democratic in structure as it becomes more able to tolerate individual choice in energy use. This change does not, however, insure that the society increases its ability to make good decisions as the energy it commands increases. On the contrary, an increase in individual choice probably is purchased at the expense of the society's ability to make decisions that will protect the integrity and future of the society. A society that has few alternatives will always find it easier to choose, and choose wisely, than will a society with many alternatives.

With a surplus of energy, high-energy man has, in an extremely short evolutionary time, come from a short life of pain and fear to one more than twice as long, in which he may enjoy more comforts than any king of earlier times, in which fears of disease and evil spirits have been pushed into the shadows of the mind. He has brought himself to a position where he alone, among all creatures past and present, can determine the fate of his species.

DISPOSING OF SURPLUS ENERGY

There are four ways to dispose of a surplus of energy in a society. The surplus may be taken in *leisure*. Primitive hunting and agricultural societies, which did not over harvest the food or degrade the land available to them, did not have to spend all their waking hours in work. When one reads a good contemporary account of life in a Sioux village before the white man had decimated the buffalo (advocated by General Philip Sheridan as a means of "solving the Indian problem"), or of life in Polynesia before missionaries and syphilis, or of life in an

Eskimo community before snowmobiles and the welfare state, one is looking through a window of the past into a low energy, steady-state society, in which a more or less continual energy surplus was spent in leisure.

In other, more complexly organized agricultural societies, control of the relatively small surplus by the upper part of the social pyramid has resulted in leisure for the few and toil for the many. In this situation, a second way to dispose of the energy surplus becomes evident: the provision of *unproductive goods and services,* whose enjoyment usually is limited to the same group that has the leisure. The manufacture of jewelry and other articles of adornment, the provision of servants, fine clothing, and comfort heating are examples of such unproductive goods and services. In ancient Mesopotamia, even bricks were luxuries, for they required so much work (and fossil energy) to produce that they were used only in palaces and temples. The pyramids of Egypt fall in this category, although it seems probable they were only incidentally tombs and served a larger purpose in uniting the Egyptians into the world's first state of modern outline. The immense amounts (by contemporary standards) of treasure returned to Spain from America by the conquistadores were lavished by the Spanish royalty on unproductive goods and services, leaving Spain fully as poor as before the Papal scission of the temporal world.

The third way of using an energy surplus is in the *construction of productive machines* that convert energy into work and into process heat. The modern concepts of a growth society and a growth economy are predicated on this way of using surplus energy. Neither of the other two ways, nor both combined, can produce the growth state. The history of the development of industrial society has been a history of plowing surplus energy back into more energy converters—more horses, more water mills, more windmills, and more sailing ships, then more steam engines and more blast furnaces, and then more electric motors, generating plants, and internal-combustion engines. In this process, some of the surplus usually has been taken in leisure and unproductive goods and services, but relatively not much. In the rapid development of industrial society, a true leisure class has found considerable difficulty in maintaining itself against a pervasive disapprobation of idleness and growing disinclination of the many to see the few in flagrant consumption.

The people of northwest Europe could have chosen, at any time during the past several centuries, to take a larger proportion of their surplus in leisure and luxuries but they chose instead to reinvest most of it in productive machines. The reasons are not hard to guess. First, there was a tradition of aggressive accumulation coming down from Viking days. Second, there was a social structure attuned to trade that generally allowed the accumulation of money by almost anyone and consequently facilitated upward mobility; where the social structure unduly restricted such mobility, as in late eighteenth century France, it became susceptible to violent attempts to restructure it. Third, the fact of social mobility brought many people to points of decision whose background of privation impelled them to pursue greater and greater financial (and contingent political) security.

WAR AND SURPLUS ENERGY

Still another way to dispose of surplus energy is to expend it in war and preparation for war. Without energy surplus, no nation can wage sustained war. China, for example, until recently seldom has had sufficient surplus to maintain a military force adequate for defense. Because human pressure on crop land discouraged the maintenance of horses, China has been helpless against invasion by mounted troops. In ancient Egypt the rich Nile floodplain for centuries provided an energy surplus to maintain both a priesthood and an army; slaves captured in warfare were an important part of the Egyptian economy. It appears from recently deciphered inscriptions that the Egyptians may have used some of their surplus energy in astounding transocean voyages of discovery (Carter, 1975; Fell, 1974).

For most of man's military history, armies had to live off the land they traversed and their routes were planned accordingly. Until the recent development of the rapid mass transport of supplies, an effective defense tactic has been to "scorch the earth" in front of the advancing enemy. Navies could not become effective at long range until methods of preserving food were fairly well developed. Even then, naval units were tied closely to coastal supply points for fuel. Only now does nuclear propulsion promise to free naval forces from dependence on fuel supply stations.

We have already noted that Roman troops stationed along the northern frontier of the Empire stimulated the frontier economy by their demands for food, clothing, weapons, and fortifications. The fact that the northern frontier was able to bear this burden indicates that it was producing an energy surplus before the Christian era. The Roman Empire persisted as long as it did mainly because it tapped energy surpluses outside the home peninsula. Later, the North Sea raiders, with a vastly different style, replaced the Romans as parasites on the producers of the energy surplus of the northern plains; then these raiders settled down to lead western Europe into the Crusades. Without the mobilization of a substantial energy surplus, reflected in horses, armor, and men, the Crusades could not have happened. The period of the Crusades lasted for more than a century, after which the religious confederations they represented broke up into armies representing powerful families who turned against one another for control of the emerging nations of western Europe.

Preindustrial war differed greatly from later, industrial, warfare. For instance, it was carried out almost without interrupting trade and communication between the warring parties. During the long struggle of the Christian West to push the Moslems out of Spain, it was common for Spanish noble families to send their sons to school in Moslem universities, and there was an almost uninterrupted flow of cultural, medical, and scientific information from the Moslem to the Christian regions of western Europe. Another way in which preindustrial war seems strange to us is that there was only partial social involvement. Although the energy used in war was drained from the living standard of the common man and although soldiers and sailors came mainly from the lower classes, the average person in times of preindustrial warfare probably did not feel the nearness of war unless an actual battleground involved his property or himself. Because battlegrounds were limited in extent and battles occupied only a very small part of any war, there was little sense of participation on the part of most persons. Total war was yet to come.

The French Revolution and the Napoleonic Wars represented the transition from preindustrial to industrial warfare, from limited warfare sustained by an energy surplus from renewable resources (agricultural, water, wind, and fuel wood) to unlimited warfare supported by the energy surplus obtainable from nonrenewable resources (coal, crude oil, and nuclear explosives). In the short period of eighty years (1785-1865), civil war between classes (as opposed to civil war between families) erupted in France, China, and the United States, leaving millions dead or socially dislocated. In the next eighty years (1865-1945), international wars on a scale previously unimaginable, carried on by the new high energy societies, devastated large areas of the industrialized world and brought about total mobilization, both physical and psychological, for the purpose of waging war.

In mass warfare as in individual fist fights, the good big man always beats the good little man. Big and little apply also to energy resources available to the combatants. In the United States Civil War, the North defeated the South by a campaign of attrition supported by coal mines, steel mills, and railroads. The South was clearly superior in military talent and social determination, but it could not prevail against the inexorably grinding superior power machine of the North. By 1870, the energy system of Germany had surpassed that of France, both in quantity and efficiency of organization; in 1871, Germany defeated France. Meanwhile, Great Britain had been waging war more shrewdly since her expensive participation in the Napoleonic Wars, avoiding confrontations with other industrial nations while expanding her worldwide empire over natural resources. For a century after Napoleon's defeat in 1815, she fought wars only with low energy societies for clearly perceived material benefits. The United States likewise fought no high energy society from 1815 to World War I. Both nations, however, were drawn into World War I and again into World War II. These two wars consumed energy in great quantities. After each much energy was expended in rebuilding, as well as in preparing for the next war.

The impact of war on the production of energy may not be obvious in statistics of gross energy consumption. For instance, neither world war is reflected in the petroleum-consumption curves for the United States or for the world. The reason seems to be that the additional energy devoted to building and propelling tanks, planes, warships, and missiles during wartime comes mainly from the cessation or slowing of consumer-oriented activities and not from new oil wells completed and old ones rejuvenated. Consumption or production patterns that do reflect the stimulus of war are those of iron, steel, and the strategic metals such as copper, tungsten, nickel, vanadium, and molybdenum. Mining, processing, and the fabrication of metals are energy-intensive processes. One of the main ways of reducing energy requirements in the metals industry is to recycle used metal. Recycling usually conserves two kinds of nonrenewable resources at the same time: metallic ores and fossil fuels. But recycling in warfare is difficult and generally not practicable. It has been estimated that 4 billion dollars worth of equipment was destroyed during the brief Arab–Israeli war of 1973; many times that amount was sacrificed in World War II.

The use of energy in modern warfare hastens the eventual exhaustion of vital nonrenewable resources, produces goods and services that are at best merely unproductive, and has proved strikingly ineffective in reducing gross populations, although it alters the demographic structure. Warfare can result, despite the wholesale deliberate destruction of productive energy-conversion machines, in an increased productive, or energy-conversion, capacity, not only for the winners, but also sometimes for the losers. The productive capacity marshalled by the North in the American Civil War was afterwards turned to civilian progress and environmental conquest. Aftersurges of civilian production and material prosperity followed both world wars in the countries that fought those wars. A curious feature of war is the strong encouragement it gives to the efficient use of energy. The 35 mph speed limit, the saving of fats and grease, and the regimentation of workers into effective production teams in the United States during World War II decreased the amount of energy required to produce a dollar of Gross National Product during those years. Immediately after the war there was a sharp increase in that ratio, representing a return to the normal inefficiencies of freedom.

The ability of a modern nation to wage war for any appreciable length of time depends critically on the energy surplus available to it. Those whose access to energy stocks and productive capacity has been diminished by depletion, interdiction, or both, will not be able to carry on full-scale warfare beyond the limits of standing stocks of energy materials, mainly fuel and explosives. Such nations will tend to rely on a "knockout punch," on a retaliatory threat, or on useful neutrality for protection.

THE ROLE OF ENVIRONMENT IN CULTURAL ENERGETICS

The idea that environment is the independent variable in human culture is old and persistent. Basically the argument between those who hold this concept and those who oppose it is similar to the argument over the role of religion in cultural evolution. In each case, the basic question is one of motor force: is human behavior determined by environment (or religion) or simply modified by it?

From the early Greeks to the modern northwest Europeans, savants have maintained that certain physical environments impart traits such as laziness, aggressiveness, or intelligence to their inhabitants. It was early noted that people differed in color, size, and shape more or less according to the parts of the world they inhabited. It was further remarked that some tended to move more slowly than others and to spend less of their waking hours in "productive" activities and, again, that there seemed to be a correlation between such traits and the physical nature of the environment. Tropical coastal plains, for instance, tended to be inhabited by slow-moving people who seemed to avoid manual labor and to be not much interested in either altering their environment or wondering about it.

Although simplistic theories of environmental determinism appear defeated in most scientific circles, it cannot be denied that environment sets some limits on what man is able to do. It is a striking fact that the high energy societies, until the ascent of Japan, were all based on an agriculture dominated by wheat and barley, despite the fact that probably more people in the world have always lived off rice than off wheat, and that many of the rice eaters have enjoyed more propitious climates and had have had the cultural advantage of abundant waterways for transport and communication. Gordon Childe, the British anthropologist, proposed (1951, p. 60) the following explanation.

In a rice culture, most of the year can be spent in growing and harvesting rice. Draft animals are not required, although they can be used. Soybeans or pigs will supply the necessary protein. Because much hand labor is required, children are an obvious advantage; therefore the population tends to grow to the limit of food availability and leisure becomes very limited. Little division of labor is required (men, women, and children often work at the same activity) and there is small incentive to try new ways of "earning a living." In a wheat or barley culture the growing season limits the time during the year when the principal group activity is farming. Pressure to sow and reap as much as possible in the short time available makes draft animals more useful than human hands. The need to augment the plant food available encourages hunting as well as the raising of food animals. The diversity of such income activities and the need to prepare food for storage and to protect it from spoilage and theft encourage a division of labor so that men, women, and children rarely work together but specialize. In a hunting society children are a liability and even in a mixed hunting and agricultural society they are not apt to be regarded as such an unmixed blessing as they are in a purely agricultural society. Consequently, increase in population may be controlled, and, under favorable circumstances, there may be considerable leisure available to such a society, especially during that part of the year when occasional hunting forays and the tending of livestock are the only income activities. During such leisure times, there is opportunity for planning new and more effective ways of hunting, of using domesticated animals, of raising crops, and of preserving food. Consequently, it is from such a culture that one would expect to see industrial, high energy society emerge, and not from the rice culture.

As we have seen, the extraordinarily variable world climate of the Pleistocene epoch enormously stimulated cultural evolution as man strove successfully to adapt to a changing world. In recent human history, we have abundant evidence of the role that social and technical innovation can play in advancing the material welfare and relative power of a society and strong suggestions that such innovation is stimulated, not by inherited traits of behavior, but by environmental opportunity and cultural success. Accordingly, Gordon Childe's hypothesis appears to stand the tests of historical analysis rather well. In other words, human beings are not inherently lazy or industrious, conservative or inventive, stupid or bright, because they live in a certain part of the world, but what they do and how they do it may be largely determined by their cultural adaptation to environmental stimuli and constraints. Cultural evolution involves both positive and negative feedback loops. A positive feedback loop is provided the wheat or barley culture by the enforced leisure of winter, which provides an opportunity for planned cultural advance. A negative feedback loop is imposed on the rice culture by the tendency for population to grow to the limit of food availability.

LIFE-STYLES AND GOVERNMENT
IN LOW ENERGY SOCIETY

In a low energy society, family and community are of great importance. Not only are many goods and services produced within the family, but also the family is a major instrument of social control. The code of conduct that governs relations between individuals and between groups is inculcated in the family. The family is the source of most of an individual's status in the society. As the family organization is hierarchical, so is that of the society. Patterns of behavior and group relations tend to be repetitive; nonconformity and innovation are discouraged. Social relations are regulated for the security of the society and conduct is guided into paths that have been found to promote social security. Social efficiency is given strong preference over individual choice. Skills are valued for their contribution to social welfare and wisdom may be valued more than strength; consequently, one's value and status tend to increase with age. Access to positions of leadership or power may be denied those of unfortunate parentage as it may be granted others of more fortunate lineage.

If we follow Northrup (1956), time in a low energy society tends to be cyclical instead of linear. Because most of the members of a low energy society are close to the sources of their sustenance and not walled away from the natural world, they feel time in terms of seasons of planting and reaping, of living and dying, rather than in hours and days. Their religion is apt to be animistic and their rites propitiatory. Sex and fertility are not separated in their philosophy. Reasoning about the world tends to be highly inductive, based on interpretations of what can be seen, felt, tasted, and smelled. There is usually a strong element of fatalism in their view of the world. Human virtues such as cunning and bravery may be viewed as being shared with animals, which may be held to have souls. Tabus, totemism, and fetishism are common. Art tends to be representational, and beauty is regarded as an integral part of an object rather than as "in the eye of the beholder."

A tabu of early Judeo–Christian religion (and still a feature of the Moslem religion) attempted to discourage moneylending by forbidding the acceptance of interest. In a low energy agricultural society the moneylender is a symbol of disaster; for centuries, one of the major defects of Chinese society, for example, was the leechlike virulence of those who lent money to farmers at high interest rates and thereby helped create a mass of hungry, landless, insecure people, not as well off as some of the slaves in western civilization. Since moneylending is essential to the creation of an industrial or growth society, this tabu eventually had to give way but, by the time it did, moneylending had become the peculiar province of the Jews both in Christendom and

Islam.[1] It was Thomas Aquinas, the great philosopher and theologian of the thirteenth century, who broke this bond by declaring that interest on loans for business purposes was not immoral, because the money was to be used to produce new wealth, and that it was sinful only to pay or receive interest on loans made for the purchase of consumer goods. Three centuries later, in the context of what became known as the Puritan work ethic, John Calvin withdrew all onus from interest and enlarged the eye of the heavenly needle so that the businessman if not the camel could pass through. A relic of Aquinas's view persisted, however, in the widespread belief in early industrial Britain (and in New England) that, while it was virtuous to sell consumer goods, it was sinful to buy them beyond necessity. In addition to raising business, and its mistress the military, in the scale of social virtue (something that never happened in China and occurred late in Japan), Calvinism made leisure a sin instead of an option.

What geographers call "the sense of place" is a feature of low energy life. Unless compelled to emigrate by the exigencies of life at home, a person tends to be a part of one community all his life and to occupy a well-defined position in the affairs of that community. There is thus a double sense of belonging and of continuity, of belonging to a place that is part of one's family and community history and to which one has an historical obligation (and the ancestors may be watching!), and of belonging to a social unit in which one's contribution, even though lowly, is valued by one's fellow citizens. Such mutual appreciation is abetted by the fact that labor is not very specialized and it is easy to understand what it is that everyone else does. Wives tend to share the status and responsibilities of their husbands; in European agricultural communities it was once common for women to be addressed as "Mrs. Dr. Doolittle," or "Frau Burgermeister Glockenspiel" and to be expected to carry out well-defined social responsibilities associated with those positions.

The energy of low energy societies comes mainly from plants and animals: food and wood, vegetable wastes and dung, animal and human

---

[1]Deuteronomy 23:20 contains this proscription: "Unto a stranger thou mayest lend upon usury, but unto thy brother thou shalt not lend upon usury." Usury in this context meant interest. Jews interpreted this passage as allowing them to accept interest from anyone except a fellow Jew; through the Middle Ages, Christians, who accepted the ideal of universal brotherhood, were constrained to forego moneylending entirely. Thus, for some centuries, Jews were the bankers of Europe, and even churches and monasteries borrowed money from Jewish lenders; in England, the building of nine Cistercian monasteries and the great abbey of Saint Albans were financed by the Jew Aaron of Lincoln (Durant, 1950, p. 377). Like Christians, Moslems were prohibited from taking or giving interest. Today, Moslems accept commissions for lending money, but not interest; Moslem courts still settle suits over debts by awarding the principal in dispute, but never interest.

FIGURE 8.2
Electricity can bring considerable changes in life-style. A light bulb hangs from the roof of the lean-to kitchen of this Mayan hut in Mexico.
(Photograph by Hilda Bijur for the World Bank.)

power (see figure 8.2). The power of flowing water for transport and to lift water for irrigation and of the wind for small sailing vessels may be important locally. These energy resources are all renewable, limited by rate of supply rather than by a total quantity available. In most of the so-called underdeveloped countries, such energy-supply limits were reached long ago; only the introduction of a new food plant that has greater range of habitat, greater productivity, or greater resistance to disease occasionally provides an increase in the energy supply, an increase quickly used up by new mouths to feed. The energy resources of low energy society are available locally and do not need to be imported. Most low energy societies that have been in more or less continuous contact with high energy societies in recent years have imported as much petroleum and petroleum products as they could pay for and are currently finding that imported fossil energy can be very expensive indeed when it must be paid for by the products of their labor.

The diet in a low energy agricultural society tends to be low in animal products and rich in starches. Even during the height of the Roman Empire, when resources were drawn from many lands, meat was for the rich. The Roman farmer's main dish was a thick soup made of emmer (a kind of wheat) or bean flour. Even the Roman soldier marched and fought on a diet that was mainly vegetarian, flavored by pork fat. Products of the cereal grasses, notably wheat and rice, are the mainstays of the modern world's low energy societies. Even societies in which the herding of goats and sheep is a major element in the agricultural economy cannot afford to eat much meat; Iran, Pakistan, and Saudi Arabia represent such societies.

The governments of low energy societies tend to be hierarchical and undemocratic. Where little energy is available, either the population or the resource use or both must be controlled and such control is effected by conventions, tradition, and religion on the one hand and by the police power of the state on the other. Together, social forces and government set limits on available life-styles. Most low energy societies today are agricultural; a few, such as that of the Eskimo, are still hunting societies. Probably 80 percent of the working members of an agricultural low energy society are engaged directly in agriculture. Both physical and social mobility are limited, as are educational and health services. Most of the agricultural products are consumed near where they are raised. In a primitive agricultural or hunting society where the energy supply is barely sufficient for the subsistence of its members, government in the sense of the ability of the society to make and implement decisions calculated to preserve the society is carried out largely through attitudes, mores, and conventions that have the force of law and that direct behavior along set lines. It is difficult for an individual or a group—even a new government—to effect change in such a society. There is no surplus energy in such a society to support the planners, innovators, and administrators required to develop a decision capability. The low energy society tends toward overspecialization in the evolutionary sense and cannot cope with adverse environmental conditions such as drought or conflict with a high energy society. An increase of energy supply to such an institutionally hardened society is more apt to result in an increase in population rather than in the creation of the energy surplus needed to raise its cultural level and enlarge its perceptual scale for decision.

FIGURE 8.3
Where electricity is relatively cheap, it can transform kitchen work from dimly lit drudgery to brightly light pushbutton pleasure.
(Photograph courtesy of General Electric Company.)

LIFE-STYLES AND GOVERNMENT IN HIGH ENERGY SOCIETY

Life in a high energy society is in sharp contrast to life in a low energy society. Family and community are subordinated to the state because most goods and services are produced outside the family and because the means of social control do not depend upon the family's and community's allocating status and inculcating behavior. All goods come from production units much larger than the family and for the most part outside the consuming community. Services are performed by specialists, only the more common of whom will be found within the immediate community, and by consumer-operated power appliances. The centralization of mass production and the proliferation of service centers require extensive transportation and communication networks; the speeds at which bodies and goods are moved, and words and pictures transmitted, become important indicators of the vitality of a high energy society. Claude Summers has defined (1971, p. 149) a high energy society as "a complex machine for degrading high-quality energy into waste heat while extracting the energy needed for creating an enormous catalogue of goods and services." The energy comes mainly from the fossil fuels, with lesser amounts from hydroelectric and nuclear power plants, and is widely available as electricity, gasoline, and fuel oil. In a high energy society movement of people, goods, and messages must be coordinated by a uniform system of linear time. Timetables and clocks are vital. Life may be inconvenienced by the seasons and made less productive by the inefficiencies of infancy and old age, but it is not governed by such cycles and does not seek to accommodate itself to them.

The values of a high energy society are oriented towards goods. Merit is a function of productivity and is given prime weight in choosing persons for positions in economic and political structures. To a great extent status depends on the acquisition of power. In high energy societies with unplanned economies and individual capitalism (the so-called free world), the power comes from personal wealth, high managerial status in industry, election to high office, or some combination of these three. In high energy societies with planned economies and state capitalism, power comes principally from high managerial status in government. In both types of high energy society status (but not power) can also be achieved by eminence in the performing arts, literature, and science. Capitalism, individual or state, runs on credit. Individual capitalism institutionalized the money lender as the most respected person in the production system, quite the reverse of his status in low energy society. In state capitalism, the moneylender is the state, a fact that may account in part for the high esteem in which the state appears to be held in the planned economies.

Advertising, whether in business or politics, is characteristic of high energy society. The cost advantage of mass production depends upon mass consumption. Beyond a certain natural demand, mass consumption of goods and services (including political) must be stimulated. The communications media, needed to carry out the affairs of a high energy society, provide a means for stimulating consumption by influencing belief in the potential rewards to be had from the purchase of a specified deodorant, motor car, or politician. Cottrell pointed out

Life-Styles, Government, and Energy

FIGURE 8.4
Where energy is cheap, choice is abundant, competition flourishes, and the esthetic as well as the natural environment may suffer. (Photograph courtesy of Atlantic Richfield Company.)

(1955, p. 250) that a high energy society, like the production units within it, may spend a good deal of its energy "creating those values by which its existence is justified."

In high energy societies, life-styles can be diverse because the energy-consumption requirements inherent in a particular life-style are not vital to the stability of the society. More choices are available; one may live in pastoral simplicity or in urban complexity. Choice and the availability of energy being almost synonymous, a high energy society tends to be free and wasteful, low energy society tends to be controlled and conservative. Social mobility is great in the high energy society mainly because of the emphasis on economic growth, industrial production, and individual merit. Less than 10 percent of the working force in a high energy society may be engaged in agriculture. In 1820, 72 percent of the work force in the United States was on the farm; in 1970, only 6 percent.

Although people in a high energy society tend to eat more and to need food less than those in a low energy society, the striking contrast is in what they eat and how much land it takes to produce it. The average diet in North America, for example, contains more than ten times the meat and twenty times the protein of the average diet in southeast Asia and takes eight times the area of cultivated land to produce. There are only a few exceptions to this general statement. Hunting societies such as that of the Eskimo eat a lot of meat and do not rely on cultivated land at all. Pastoral societies in various stages of industrial transition may eat abnormally high quantities of meat; Argentina is an example.

Life-Styles and Government in High Energy Society

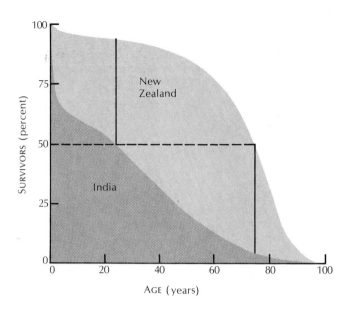

Better food for several generations has produced larger people in the high energy nations, has advanced the age of menarche (from seventeen to just over thirteen in Norway between 1843 and 1970), and has helped greatly to reduce the incidence and impact of disease, especially in infancy. That more food is not necessarily better food, however, is shown by statistical correlations of mortality with diets heavy in animal fats and with daily regimens lacking in exercise. Obesity is common in high energy society, even in childhood. Life expectancy for adults in the United States has not improved very much in the past hundred years, during which time there have been marvelous advances in public health practices and medical engineering. What these advances should have gained for the adult of the high energy society seems to have been nullified by the citizen's own eating habits and increasingly sedentary life-style.

The average age in a low energy society tends to be considerably lower than in a high energy society because of higher birth rates and shorter life expectancy (see figure 8.5). In recent years, the contrast in life expectancy has diminished because of the exportation to the low energy societies of some of the medical technology of high energy societies and the populations of the former have therefore grown more rapidly than they once did.

People in low energy societies try to keep warm by wearing more clothes and to keep cool by shedding them; people in high energy societies keep warm or cool by heating or cooling their buildings and conveyances, a far more expensive and wasteful method, as Potter pointed out twenty years ago (1954, p. 197).

Life-Styles, Government, and Energy

High energy society is urban society. The advantages of bringing people together as the division of labor grew ever finer, as human time grew more valued, and as the geography of energy grew more important were recognized early. The development of agricultural technology encouraged the growth of cities, as did the exploitation of power sites and the development of transportation. Mass production requires the concentration of materials and labor. Business and government are facilitated by the nearness of people, even with the development of rapid electronic communication. Consequently, the rise of the industrial technological society has been characterized by an accelerating urban implosion that depopulates the countryside and masses people into cities.

Bringing large numbers of people together creates problems of health, of social stability, and of the efficient use of human resources. Crises of urban sanity, in the broad sense of that word, have occurred since there were cities. Infectious plagues and insurrections have bred and spread in cities. But so have the effective countermeasures of public health, medicine, education, and social reform. The economies of scale inherent in the beehive have yet to be fully exploited for humans. It is the drive to achieve those economies that keeps cities growing against the counter pressure of the social dangers of crowding and pollution.

With all its ability to conquer space and disease and to increase the natural fecundity of the earth, the high energy society has created for itself some monumental ecological hazards, of which three seem particularly important: the hazard of social disruption or sabotage by small dissident or irrational groups; the hazard of life-support failure through an uncontrollable onset of disease that attacks humans either directly or indirectly through their food crops; and the hazard of an abrupt collapse of living standards because of the economic exhaustion of nonrenewable energy resources. The high energy society is vulnerable to sabotage by small groups. A strike of garbage collectors, subway motormen, or drawbridge operators can paralyze the largest city in the world. Interdiction of petroleum supply by a very small number of the world's people can bring to their knees a much larger number of the world's people. A hydrogen bomb in the hands of a small nation or of a small group of international blackmailers could cause a mighty nation to quail. Plant and animal physiologists know that "conquering disease" is a perpetual race to build dikes against a relentless flood of organisms. Just as the potential losses from a disastrous flood mount as human occupancy and use of a "protected" floodplain grow, so do the potential human losses mount from a break in the elaborate dike system that high energy society has built to protect food crops, domestic animals, and human beings from the pandemic onslaught of disease organisms. The short life span of some of our tiny foes gives them an enormous advantage in this perpetual war of attrition; they can circumvent our chemical defenses, on which we rely heavily, by resistant mutagenesis. The third hazard of the high energy society, that of "running out" of nonrenewable energy resources before adequate replacements have been found, will be discussed in chapters 13 and 15.

It has been claimed that high energy societies will be ruled by democratic governments, not because freedom of choice is possible where there is surplus energy, but because of the need to regulate human relations in an industrial milieu by contract rather than by rules of status, and perhaps also because of the need for a high level of literacy and schooling. This thesis is far from proven. Germany under Hitler was a high energy, superbly industrialized society, as it was under the Kaiser, for that matter. And it was in Germany that what we call the high energy society got its early impetus. In Anglo-Saxon jurisprudence it is assumed that contracting parties are equal under the law and that a party cannot be bound to a contract he is coerced into. These rules are required for the maintenance of representative democracy, but not for the maintenance of a high energy society. As the energy systems supporting a high energy society get larger and more intermeshed, they render the society more vulnerable, not just to sabotage, but to control by a minority. Whoever controls the energy systems can dominate the society.

HEALTH, ENERGY, AND LONGEVITY

In no facet of human life are the benefits of energy more apparent than in health. Theodore Roszak (quoted by Daniel Bell, 1973, p. 404) has remarked that

> contemporary projections of pagan idylls neglect completely the diseases that wasted most "natural" men, the high infant mortality, the painful, frequent childbirth that debilitated the women, the recurrent shortages of food, and the inadequacies of shelter that made life nasty, brutish and short.

We need not revert to "pagan idylls" to see the same picture at closer range. It has been only within the last hundred years that men and women have ceased to live in almost constant fear of epidemics. In three years of the fourteenth century the plague killed at least a quarter of the population of Europe; it struck Venice twenty-three times between 1348 and 1576 and, in 1665, killed nearly a tenth of the population of London. Leprosy was a medieval scourge. In colonial America, epidemics of smallpox, yellow fever, measles, mumps, and diphtheria struck sporadically, and throughout the eighteenth century malaria was the most common debilitating disease. In 1793 a yellow fever epidemic in Philadelphia caused national, state, and local governments to disintegrate as almost the entire populace fled the city; 10 percent died. In the late eighteenth century about half of all deaths occurred before the age of ten years. In the early years of the nineteenth century, typhoid fever, typhus, tuberculosis, pneumonia, and scarlet fever became frequent causes of death. In 1807 the death rate in Philadelphia from tuberculosis alone was more than 300 per 100,000. Cholera became greatly feared; in 1832 more than 50,000 people fled New York City when cholera turned up there. Mortality rates during this period were higher among the poor than the prosperous and much higher among blacks than whites. In 1789 most white males in the United States were less than sixteen years old and the average life expectancy at birth was about thirty-five years. The reasons for such a grim picture of health (if anything, it was worse in Europe) are several, but all are related either to lack of energy or lack of knowledge, or both.

Except for the rich, until about 1825, with the introduction of cotton goods, underclothing was uncommon; the outer clothing was of heavy wool and leather, infrequently changed and laundered, often not taken off at night. Winter bathing was not feasible in rural areas and was uncommon at any time even in cities. In the winter water froze in bedroom pitchers and open fireplaces gave small comfort. Scabies and body lice were common. Shryock quotes (1960, p. 91) the diary of a Philadelphia Quaker lady who wrote in 1799 that, her family having installed one of the first shower baths, she had taken a shower and had become "wet all over" for the first time in twenty-eight years. Contaminated water supplies were common. There was no sewage system in any city in the United States until after 1800; only a few had sewage systems by 1850. Municipal water systems were introduced about 1825, after steam pumping systems became practical. Inadequate winter diets were common among the poor and rural populations. The crowded conditions of the early American cities worsened health statistics. From 1810 to 1859, the crude death rate in New York City rose from 1 in 46.5 to 1 in 27. In 1809, Dr. Benjamin Rush complained that Philadelphia was full of stagnant pools, that the docks had not been cleaned in more than thirty years, and that "privies are exceedingly offensive." By 1850, stoves had largely replaced fireplaces but in the north families lived mainly in one overheated, stuffy room during the winter. Harriet Beecher Stowe, as quoted by Shryock (1960, p. 92), wrote that it was little wonder "colds evolve into consumption."

Medicine in the United States in the late eighteenth and early nineteenth centuries was dominated by "bleeders and purgers" and the

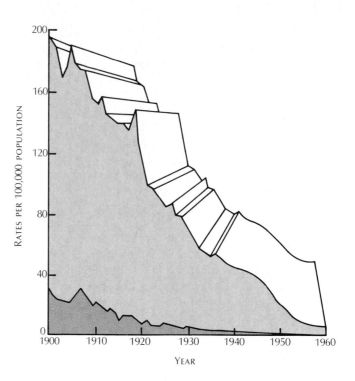

FIGURE 8.6
Declining death rates attributable to the infectious and communicable diseases of infancy and childhood, such as tuberculosis (upper curve) and typhoid fever (lower curve), are a major factor in the dramatic increase in life expectancy in the U.S. in this century. General factors associated with improved standards of living have contributed greatly to the decline in the incidence and the mortality of such diseases, although specific public-health measures have also been important. No immunization against tuberculosis has been adopted in the U.S., and the effectiveness of typhoid vaccine is questionable.

profession was infested with bumblers and charlatans. Hospitals were few and crowded. It was not uncommon to place two or three persons, related only by disease, in a single hospital bed and Benjamin Rush saw four in a bed in the Hotel de Dieu in Paris.

In the last half of the nineteenth century vaccination began to stem smallpox and quinine reduced the symptoms of malaria. A vaccine for yellow fever was not developed until 1939, and the southern third of the United States is still classified as a danger zone. Influenza is an epidemic disease of worldwide effect; the influenza epidemic of 1918 killed 20 million people, about 500,000 of them in the United States. Influenza vaccines still are only temporarily and partially effective. At the turn of the twentieth century, tuberculosis was the leading cause of death in the United States, but by the 1940s the discovery of streptomycin and other drugs reduced the incidence markedly. Poliomyelitis (infantile paralysis) was brought under control by the development and nationwide administration of a vaccine in 1954. In a span of about 100 years, infectious diseases in the high energy societies have been reduced from a constant threat to the individual and a scourge to society to a relatively minor position when compared with diseases of the heart, cancer, cerebrovascular diseases, and accidents as causes of death. In 1970 the life expectancy at birth of a citizen in the United States was seventy-one years. Medical authorities maintain that the general improvement of standards of living has contrib-

Life-Styles, Government, and Energy

uted greatly to the decline in the incidence and the mortality of infectious and communicable diseases of infancy and childhood. Although no immunization against tuberculosis has been adopted in the United States and the effectiveness of typhoid vaccine is questionable, death rates from both diseases fell dramatically between 1900 and 1960 (see figure 8.6). In high energy society nearly two-thirds of the deaths are now associated with the infirmities of old age.

THE NORTH AND THE SOUTH

In late colonial times in America it was as if two different English societies had been transported across the Atlantic. The Northern colonies reflected the working man's England—a growth society in which population had outrun production, in which there was much unemployment and misery, a society in which harshness and cruelty attended the production and protection of material wealth, and from which many of the less-favored sought refuge in austere beliefs such as those of the Puritans. The South reflected a more comfortable and confident English society, which brought the Anglican Church along as part of the baggage, not as the reason for emigration. Initial differences in social background, in reasons for emigration, and in the position of religion in daily life were to be accentuated by differences in the ways available to make a living and in the energy resources each region chose to develop and exploit.

The Puritans came to America seeking not freedom but independence. Within the Massachusetts Bay Colony they ruthlessly suppressed political and religious freedom. The first government in the North was a theocracy. New England colonial society was rigidly stratified. The lower classes were forbidden to wear finery, including any clothing made of silk. There was limited upward mobility. The life-style was almost morbidly religious and circumscribed; a child might smile, but before his adolescence he would have learned that in New England a smile was apt to be taken as a sign of the Devil at work. Puritan society, which for a century or more dominated New England, was a society at war with three enemies: nature, the Indians, and the Devil. The wilderness was evil, full of "fiery winged serpents" said Cotton Mather, and people who lived alone in it would become evil, for the moral man was a watched man according to the Puritans.

There was little slavery in colonial New England, not because of any great opposition on moral grounds, but because unskilled slave labor was of little use in an economy that was not based on large-scale hand agriculture and in which a variety of skills were needed. Fields were never very fertile in the glaciated North and for a long time the market for the abundant timber was small. Moreover, the winters were harsh, requiring sturdy dwellings and stocks of provisions. The New Englander was forced to look to the sea for much of his livelihood, and New England sailing ships became known in all the ports of the world.

In Virginia, Maryland, and the other southern colonies early American society was less rigid and more tolerant. One could wear whatever raiment he or she could afford. Upward social mobility was considerable. Most of the hand labor for the tobacco plantations in colonial times consisted of indentured white servants who paid for their passage from England by several years of work for the landowner who had sponsored them. Upon obtaining freedom, they frequently became landowners themselves, for good land was available. Thereafter, diligence, intelligence, and chance would determine one's material and social progress. The indentured servants, in New England, of which there were many, found it more difficult to become economically independent and many continued for wages what they had been doing under indenture—or they shipped to sea.

Today it seems somewhat ironic that the most farsighted, fundamentally democratic provisions of the United States Constitution were drafted by the Virginia delegation to the Constitutional Convention and that some of these provisions, especially those calling for a strong central government, were fought almost to the point of secession by the New England states, New York, and New Jersey, whose leaders were the States Righters of their day. The open face of good agricultural land in the South and the constant flow of English men and women from the indentured class to free society and economic independence probably account for the remarkable accent on individual liberty and the sense of nation that pervaded the South in late colonial times and left their marks in the Constitution.

Slavery did not become a major social and economic institution in the South until the successful Revolution cut off the supply of indentured servants from Britain; only then did the less skilled, less tractable, and less desired slave labor become an economic necessity on southern plantations. The introduction of slaves in large numbers and the gradual occupation of all good land reversed the social engine of the South; slavery, wealth, and political power came to be concentrated in the fertile lowlands, while poverty, political impotence and ethnic homogeneity were features of the hardscrabble highlands.

In both North and South the horse was the principal draft animal and transport engine: in both, fuel wood was almost the sole source of comfort and process heat; for both, night lighting was by oil lamps or pitch faggot. In the early years of the nineteenth century, the United States was ready to begin the transition from a low energy to a high energy society. The first step was industrialization. This took place mainly in the North, because it was there that water and wind power had been put to greatest use and that trade and commerce required the construction of canals and railroads. The South as well as the North had coal, iron, forests, and water power but had had no great need to exploit these resources.

The role of the free face of good agricultural land, mainly in the South in colonial times and then in the Ohio and Mississippi valleys in the early years of the United States, in the industrial development of the nation can hardly be over stressed. Throughout the colonial period and into the nineteenth century, according to Habbakkuk (1962, p. 11), American workmen were paid between 30 and 100 percent more than were English workmen, the wage floor being set by "the rewards and advantages of independent agriculture." The high cost of labor stimulated the adoption of labor-saving devices and systems. American machines typically were more poorly made, run at higher speeds, and replaced much more frequently than British machines of the same type. Whether or not such practices made American industry more competitive (in many cases it appears that they did), they nurtured the spirit of innovation and production efficiency, and led straight to Henry Ford's assembly line, the world's highest wages for industrial labor, and the intensive exploitation of inanimate energy sources.

LIFE IN THE UNITED STATES IN 1875

By 1875 the United States was a high energy society by contemporary standards, on a level with England and Germany. The West was being won, the continent was spanned by rails, Lake Superior iron and Pennsylvania coal were flowing to the steel mills of Pittsburgh, and the Midwest was becoming an agricultural cornucopia. The per capita energy consumption in the United States was already above the world average of 100 years later. But the American life-style of 1875 was very different from that of today. To get some appreciation of how far we have come since then, it may be useful to sketch the American way of life of 1875.

Most Americans lived in small, uncrowded towns, but cities were growing fast. Most workers still were farmers or made their living supplying farmers. Heating was a problem, even for the rich, and there was no air conditioning other than fans. Fuel wood was still the main source of heat in the home, and perhaps half of all the locomotives in the country still were woodburners. Central heating was common only in cities, where many buildings had steam radiators and many homes had hot-air furnaces; coal was the usual fuel for steam and hot-air generation. Homes were lit by kerosine lamps, which had replaced whale-oil lamps only about twenty years before. Streets and buildings in the cities were lit by gas manufactured from coal. Most housewives cooked on wood stoves, although a few had kerosine ("coal-oil") stoves.

FIGURE 8.7
The kitchen of the past contained few of the conveniences that are commonplace today. Photograph taken about 1890. (From Historical Pictures Service–Chicago.)

Fuel was a major item in the family budget. Whether wood, coal, town gas, or coal oil, it was not wasted. A typical kitchen range heated the kitchen (in many homes that was the only heated room during many days of the year), provided heat for cooking and, if it did not have looped pipes in the back of the firebox for providing hot water to a boiler, water for baths and washing was heated on top of the stove. The range was the working center of the home. The kitchen commonly served as a living room, dining room, and even bathroom (see figure 8.7). In those days a bathroom was apt to be just that; the other facility was outside.

Food was still subject to seasonal supply, although the expansion of the canning industry during the Civil War and the introduction of the steam-pressure canning process in 1874 had greatly improved the ability to store and transport food. Canned goods were expensive, however, and those who lived on the farm or in small towns, having eaten well during the latter part of the growing season, relied during

Life-Styles, Government, and Energy

the winter on home-canned fruits and vegetables (the Mason glass jar and lid had been invented in 1858), on smoked, salted, or canned meat and fish, and on those vegetables that would keep in "root cellars." In the North citrus fruit in the winter was for the urban wealthy, and vinegar was an occasional surrogate for lemon juice in pies. Food spoilage was a constant problem; in the summer, ice was generally available only in the cities, where the icebox was the sole means of home refrigeration until the advent of mechanical refrigerators shortly before World War I.

The telegraph, invented in 1837, was widely used for business and news purposes, but the principal means of distant communication between persons was still the letter. Alexander Graham Bell was perfecting the telephone, the first practical examples of which were to be made in 1877. The locomotive, the horse, and the steamship were the prime movers of transportation. Streetcars, wagons, and carriages of varied design were pulled by horses, except where steam trams ran in the streets. Walking was a respectable way of getting from one place to another. Travel was an exceptional, not an everyday, experience. But railroads were being built at a furious rate in 1875 America. When one needed to go from one city to another, there was no choice but the steam cars. That did not seem bad to people whose parents had to do all their traveling at the speed of horses and sailing ships.

It was a quiet age compared with ours, but it was neither healthy nor secure, and not very democratic, either. Infant mortality was high and adults ran many risks from endemic diseases such as tuberculosis, influenza, smallpox, diphtheria, typhoid fever, yellow fever, scarlet fever, cholera, measles, mumps, infantile paralysis, and syphilis. Medical competence, particularly in surgery, had grown since colonial times, but there were no X-rays, chemotherapy, or blood transfusions, and knowledge of disease vectors, hygiene, and nutrition left a great deal to be desired. Accidents in industry and transportation were common; working men frequently had fingers, arms, or legs missing. In addition to the constant dangers of epidemic disease, almost every house was a fire hazard. Most American residences and many commercial structures were built of wood. Overheated stoves, cracked chimneys, knocked-over kerosine lamps, a careless match—any of these could ignite a destructive fire. The great urban disasters of the nineteenth century were fires.

Fear of disease and fear of fire were often joined by fear of poverty. In 1875 not only comfort and social status, but the ability to avoid starvation or jail depended on one's wealth or income, and income often depended on vagaries of the market, on the interests of moneylenders, or on the favor of the boss. Working hours were long and working conditions could be hazardous and debilitating. Labor unions were growing and there was much unrest after the disastrous depression of 1873. Strikes against wage cuts and the displacement of workers by machinery were vigorously opposed by employers. Sabotage and violence made for much bitterness.

The woman's place was in the home, as manager of the household; she could not vote and found it very difficult to earn a living outside the home. Blacks were slaves no more, but many were worse off materially than they had been as slaves. Foreigners, invited to America as a source of cheap labor, came in great numbers, and were treated as inferiors by those whose ancestors had come on earlier boats.

THE NEW TECHNOLOGICAL REVOLUTION

In a brief period of twenty-seven years, from 1877 through 1903, a new technological revolution began. In America it succeeded the Industrial Revolution without a break. The Industrial Revolution was led by Great Britain; the technological revolution was led by the United States.

In 1877 both the telephone and the phonograph became practical realities; the first was invented in Boston, the second in New Jersey. In 1881, one of the world's first central electric power plants went into operation in New York City. In 1885 the world's first electric street railway opened in Baltimore; a year later the electrolytic process for producing aluminum was discovered by Charles Hall of Ohio. In 1892 the Duryea brothers of Illinois and Elwood Haynes of Indiana produced successful automobiles, and in 1896 Henry Ford built his first car. Finally, in 1903, man's first successful flights in an engine-powered airplane were made at Kitty Hawk, North Carolina, by the Wright brothers. Medical technology was also accelerating. Antitoxins for diphtheria and typhoid fever were discovered, Walter Reed demonstrated that mosquitoes were the vector of yellow fever, and Landsteiner discovered and classified human blood types, thereby making blood transfusion possible.

In half an adult lifetime the way had been prepared for high-speed communication and transport, for power technology in the home, for the proliferation of mobile power plants of a great range of size and power, for the effective concentration of business and commerce in the cores of cities, for the development of nuclear power, and for major advances in medical diagnosis and chemotherapy, in other words, for a very great change in the life-style of the average person in the society that made this transition. The life-style in a high energy society today is vastly different from the life-style in a high energy society 100 years ago, at the height of the Industrial Revolution, and it requires about fourteen times the useful heat and work per person. The gross energy consumption, however, because of the great increase in aggregate efficiency of the national energy system, has multiplied only three times.

SOCIETIES IN TRANSITION

No society ever began as a high energy society. All began from essentially the same base of energy utilization. Development into a high energy society requires a rate of increase in the availability of energy that substantially exceeds the rate of increase in population. When one looks at the energy history of Great Britain, Germany, North America, Japan, and the Soviet Union, one is struck by the fact that all have based their energy evolution primarily on mined energy (including the mined energy of soil nutrients). The low energy countries that appear today to have the best chance of making the transition are those with abundant fossil-fuel reserves. And to have much chance of remaining a high energy society when the fossil fuels eventually give out, there must be sufficient fossil-fuel capital now to finance the very expensive research and development required for adequate substitute systems based on nuclear transformations, conversion of solar energy, or utilization of geothermal heat.

Low energy societies in which large populations press closely on subsistence agriculture have little hope of breaking their energy bonds. Cottrell pointed out that "to reduce the number of farmers to the point where high-energy converters could be used on the farms would take an immense amount of employment in nonfarm production . . . to put Indian agricultural labor on . . . 25-acre farms would require 30 million new jobs in 10 years if in the meantime there were no change in the survival rate (1955, p. 129)." Even then the farms would be too small for efficient mechanization. The capital and energy costs of providing those new jobs outside agriculture would be very great. It was difficult for Cottrell to see how India or China could find enough energy to provide for farm mechanization and for employment of the displaced field workers. China now is attempting to do just that by expanding production of domestic coal, oil shale, and crude oil, by using this production to fuel growing industry, to manufacture fertilizer, and to trade for needed goods, and by striving to reduce the rate of increase of population. Cottrell persuasively argued (1955, p. 133) against the assumption that mechanized agriculture is so productive that eventually it will be practiced by all the peoples of the world, concluding that "unless a very marked decrease in fertility does occur, population growth will outstrip any increase in converters that is likely to occur (ibid., p. 168)" in large low energy societies, a point the Chinese appear to have accepted.

Since Cottrell's book first appeared in 1955, the Green Revolution has stimulated new hope that the low energy countries might pull themselves up by agriculture. Large increases in productivity are obtained, however, only at the cost of high investment in fertilizer, pesticides, and machinery, and at the cost of displacing agricultural workers who have nowhere to go and no way to earn their share of the increased production.

There are few occupations outside agriculture in the Third World and its population is increasing rapidly. Despite the Green Revolution, the increases in food production since World War II have been due mainly to expansion of cultivated area, which has now reached or surpassed economic limits. The labor-carrying capacity of agricultural land needs to be increased; new agricultural technology needs to increase yields, not save labor.

One of the common characteristics of the less developed countries is that the benefits of industrialization appear to lodge mainly in an elite minority, while the majority benefits hardly at all. In Mexico, for example, the top 10 percent of the population receives 42 percent of the total income, the bottom half of the population only 15 percent of the total income. Mechanization of agriculture drives the rural poor toward the few cities, whose industry cannot absorb them; but they remain, living in crowded, unsanitary shacktowns within the metropolitan areas, limning the disjunction between the two economies of the country.

Although many economists speak of "stages of economic development" as if there were an ordained sequence through which all nations must progress, at greater or lesser rates of advance (depending in part on their ability to understand and make use of economic principles), there is no persuasive evidence that such a law of progress exists. Indeed, there seems to be no good reason to conclude that economic progress is irreversible. The historical flow and ebb of civilizations is as logically interpreted in terms of changes in the ratio of population to the availability of energy as it is in terms of military campaigns, national morality, ethnic idiosyncrasies, or economic laws. Whenever and wherever the availability of energy is not increasing as fast as the population, or is decreasing while the population remains steady, then and there economic development is reversing, and the life-style of the affected population will begin to show some similarities with that of earlier times, with more human labor committed to agriculture, less meat in the diet, and less energy used in transportation.

JAPAN

The transition from low energy to high energy society took place slowly in western Europe and North America, although the pace quickened as the change advanced, especially in America. In Japan the transition and its concomitant revolution in life-style have taken place incredibly fast. The deliberate prose of a Japanese publication, *The Japan of Today*, published by the Japanese Ministry of Foreign Affairs in 1972, gives us a partial description of the changes that have taken place.

> Japan has been called the world's most rapidly changing society . . . from an agricultural society content with minimum standards of living 100 years ago into an industrial society beginning to enjoy the standards associated with the mass consumptive age . . . .
>
> At the end of World War II, Japan lost all of its overseas territories while its population soared beyond the 80 million mark . . . the food supply was at a minimum level . . . . Production facilities had suffered heavy damage . . . .
>
> Following the war . . . Japan's economy expanded 2 to 3 times as fast as other principal industrial nations . . . . A series of measures to democratize the nation's economic and social systems were carried out . . . the Civil Code was revised in 1947 to place women on an equal legal status with men in all phases of life and to abolish the old patriarchal character of the family . . . . The changes in the cultural and living environment . . . served to strengthen the tendency toward smaller family units . . . . The widespread practice of family planning has reduced the average number of children per couple to 2.2 in 1967, one of the lowest in the world . . . .
>
> The mode of living has also changed under the influence and widespread use of modern household appliances, together with the mass production of instant food preparations, a small but steadily expanding supply of frozen foods, readymade clothes and other wearing apparel . . . under the influence of the lighter household load and increased leisure time, more and more housewives are entering the labor force . . . .
>
> . . . people now travel the entire length of Japan for vacation and leisure activities, while the number of Japanese going abroad for pleasure has been increasing at a rate of 62 percent a year since 1965 . . . .
>
> Social welfare services have been greatly expanded . . . . [Although health standards have improved,] the Japanese people are now menaced by diseases such as apoplexy, high blood pressure, heart ailments, mental disorders . . . . These "diseases of civilization" are now the principal killers in place of the communicable diseases of the past.
>
> . . . a major change has been brought about in Japanese dietary habits, approaching the pattern of the Western Nations . . . . The excessive consumption of cereals has diminished with an increasing consumption of meat, milk and dairy products.

One wonders if the "excessive" consumption of cereals (rice) is being replaced by "excessive" consumption of animal fats and sugar. Other changes, not mentioned in the official publication, include the adoption of western habits of dress, a proliferation of juke boxes and transistor radios, and an increase in cigarette smoking. The crime rate, however, remains remarkably low for a high energy society.

INHUMANITY, ENERGY, AND DRUGS

Human society has come a long way in a short time. Man's inhumanity to man remains the greatest barrier to cultural evolution; the means and scope for expressing that inhumanity have changed with the evolution of energy. Slavery eighteen hundred years ago is described trenchantly in a passage from *The Golden Ass* of Lucius Apuleius, quoted by Ubbelohde:

> The Asse did greatly delight to behold the baker's art. O good Lord, what sort of poore slaves were there; some had their skinne black and blew, some had their backs striped with lashes, some were covered with rugged sacks, some had their members only hidden; some wore such ragged clouts that you might perceive all their naked bodies; some were marked and burned in the heads with hot irons, some had their haire half-clipped, some had lockes on their legges, some very ugly and evil favored, that they could scarce see, their eyes and faces were so black and dimme with smoke, and some had their faces all mealy. [1955, p. 87]

Christianity is supposed to have introduced a new scale of values with its emphasis on the worth of even the humblest human life, but Christian Spain and Portugal started importing slaves to meet their labor needs in the fifteenth century and, by 1700, England, France, Holland, Denmark, and the American colonies were also in the slave trade as competitors. The Industrial Revolution made slaves less needed and, beginning with Denmark in 1792, all the nations of Europe eventually prohibited the traffic and emancipated their slaves. During this same period, however, there was being created in the industrialized nations a class of economic slaves, including women and children, whose conditions of servitude sometimes approximated those of the former slaves. Elizabeth Barrett Browning's poem *Cry of the Children* (1843) called attention to child labor in Britain.

> For, all day, we drag our burden tiring
>   Through the coal-dark, underground
> Or, all day, we drive the wheels of iron
>   In the factories, round and round.

In some cases, children six years old were compelled to work twelve to sixteen hours a day. It was not until 1878 that the minimum legal age of employees in Great Britain was raised to ten years and the workday for children was limited to twelve hours. In the early years of the nineteenth century, children between the ages of seven and twelve constituted a third of the work force in factories in the United States. By 1900, almost 20 percent of all the American children between the ages of ten and sixteen were working for wages. It was not until 1938 that it became illegal to employ minors under sixteen years of age in

interstate or foreign commerce and under eighteen years in a hazardous occupation.

The increasing availability of energy, in other words, brought bad with good. The factory system of the Industrial Revolution produced goods in abundance at prices millions could afford; but it created a kind of stratified, undemocratic society based on economic slavery that repelled Karl Marx and many other observers. There were two ways out. One was to nationalize the means of production and have the new owners, the people, determine the conditions of labor and the schedule of production; the other was to build a new balance of power into the industrial society, so that no single interest (capital, labor, the state) could enslave the others. The nations with centrally planned economies took the first route; the nations with free-market or unplanned economies went the second road. Neither way was easy, and for neither group has the outcome proved entirely satisfactory. For both, the large problems of the future are similar.

The evolution of technology and related political structures has made it possible to reduce dramatically the ravages of disease and to triumph over the hostility of nature, but at the same time to kill one's fellow man in great numbers, efficiently in the gas chamber or inefficiently on the battlefield. Technology based on abundant energy has given man more options than he knows how to choose from, more power than it seems he can use with wisdom.

Historically, man is a creature of adversity. He likes challenge and novelty. Although mentally and morally lazy, he grows bored with security and sameness. Because high energy society has an inherent tendency toward individual security and environmental sameness, high energy man seeks diversion in sports, business, drugs, sex, rioting, and crime; he seems to sigh with relief when a real national crisis can be identified, whether the threat be to the economy as in the Great Depression, to national security as in World War II, or to life-styles as in times of energy shortage. One of the less lovely towns of the Industrial Revolution must have been Manchester; it used to be said that gin was the best way through it. At that time many hopeful people looked forward to the day when, because the evils of the world would have been abated and the benefits more equitably distributed, drugs such as alcohol would no longer be needed as a refuge from unpleasantness. Well, the evils were abated, at least the more personal ones such as illness and slavery, and the benefits not only were more equitably distributed but also were enormously increased. Did the use of drugs diminish? It did not. Drugs such as opium, morphine, and heroin, which could be purchased openly and legally in the United States throughout the nineteenth century, have become social problems only in the present century, and especially after its midpoint, when never in history had so many people "had it so good." Whereas in low energy societies drugs may be taken to forget, in high energy societies they appear to be taken to forget that there is nothing much to forget.

SCARCITY, AFFLUENCE, AND GOVERNMENT

It has already been noted that democracy is an improbable form of government in a low energy society. If one measures democracy by the degree of individual choice, this statement probably is also true for any society in which the availability of energy is increasing at a lesser rate than the population. It might be argued that the availability of resources in general is at issue, that shortages of iron and copper or aluminum would entail restrictions of freedom as well. So they would, but energy is more important, for without energy other resources do not exist, and with sufficient energy substitutions become possible. When the rate of availability of some energy sources such as motor fuel falls below that which is desired, some individual freedoms will be restricted as supplies are allocated, but when the rate of increase in the production of work and useful heat in the society falls consistently below the rate at which the population is increasing, or does not increase at all, serious abridgments of choice are called for and a truly democratic government will have a hard time surviving. Twenty years ago, Harrison Brown wrote:

> It seems clear that the first major penalty man will have to pay for his rapid consumption of the earth's non-renewable resources will be that of having to live in a world where his thoughts and actions are ever more strongly limited, where social organization has become all-pervasive, complex and inflexible, and where the state completely dominates the actions of the individual. [1954, p. 218–219]

SUPPLEMENTAL READING

Bell, Daniel. 1973. "Technology, Nature and Society." *American Scholar,* v. 42, p. 385–404.
    Essay on the social impact of the pursuit of economies of scale through technology.

Brecher, E. M., and others. 1972. *Licit and Illicit Drugs.* Mt. Vernon, N.Y.: Consumers Union.
    Thought-provoking information is included on the relation of drug usage to social conditions; see in particular chapters 1, 3, 21, 53, and 54.

Childe, V. Gordon. 1951. *Man Makes Himself.* London: Rationalist Press.
    The life-styles of early food gatherers, the events that may have led to the agricultural revolution, and the changes in life-styles brought about by agriculture and early urbanization are discussed lucidly.

Cottrell, Fred. 1955. *Energy and Society.* New York: McGraw-Hill.
    In chapter 10, Cottrell analyzes the distribution of consumer goods in relation to the energy status of society.

Durant, Will. 1950. *The Age of Faith.* New York: Simon and Schuster.
    Volume IV of the ten-volume *Story of Civilization* has some useful information on interest taking; see especially p. 630–633.

Engels, Frederick. 1892. *The Condition of the Working-Class in England in 1844.* London: George Allen and Unwin.
    Vivid descriptions of the squalor and misery of the urban proletariat in the boom days of the Industrial Revolution.

Langer, W. L. 1964. "The Black Death." *Scientific American,* v. 210, n. 2, p. 114–121. (Available as *Scientific American* Offprint 619.)
    Demographic and social effects of the plague in fourteenth century Europe.

Leaf, Alexander. 1973. "Getting Old." *Scientific American,* v. 229, p. 45–53.
    Contains an interesting description of the life-styles of three rural communities that are characterized by a high proportion of active elderly people; also discusses relation of certain physiological changes to the energy level of society.

Morison, R. S. 1973. "Dying." *Scientific American,* v. 229, p. 55–62.
    Relates the various causes of death in a society to its degree of industrialization.

Northrup, F. S. C. 1956. "Man's Relation to the Earth in its Bearing on His Aesthetic, Ethical, and Legal Values." In *Man's Role in Changing the Face of the Earth,* edited by W. L. Thomas, Jr., p. 1052–1067. Chicago: University of Chicago Press.
    Provocative essay contrasting Western democratic, technological society based on the law of contract with Eastern hierarchical nontechnological society based on the law of status.

Potter, D. M. 1954. *People of Plenty.* Chicago: University of Chicago Press.
    A fine analysis of the relation of economic abundance to life style and to the "distinctive American character"; this book is highly relevant to contemporary problems of resources.

Rush, Benjamin. 1948. *Autobiography,* edited by G. W. Corner. Princeton, N.J.: Princeton University Press.
    The author was a noted Philadelphia physician at the turn of the nineteenth century. His autobiography contains nuggets of social description.

Shryock, R. H. 1960. *Medicine and Society in America, 1660–1860.* New York: New York University Press.
    Contains much information on life-styles in America as well as on the common illnesses and feared diseases and how these changed with time.

Wertenbaker, T. J. 1927. *The First Americans.* New York: Macmillan.
    A valuable source of information on life-styles in the American colonies.

Chapter 9

THE GEOGRAPHY OF ENERGY

Had I been present at the creation,
I could have given some useful hints
for the better ordering of the universe.

—Alfonso X of Leon and Castile,
ca. 1275

SOURCES OF ENERGY IN THE WORLD

About 80 percent of the world's energy comes from fossil fuels, about 20 percent from dung and vegetable refuse (corn cobs and stalks, bagasse, chaff, and the like), about 1 percent from water power (mainly through hydroelectricity), and minor amounts from nuclear, geo-thermal, and wind power. In the dry lands that stretch from Morocco to India, it is estimated (Linton, 1965, p. 219) that the amount of dried dung burned as fuel each year is the equivalent of 500 million tons of coal; that corn cobs, corn stalks, and sugarcane trash in Latin America provide perhaps 250 million tons; and that wood fuel in for-ested regions contributes between 200 and 300 million tons. Table 9.1 and figure 9.1 show the distribution of the world's consumption of energy in 1971, based on national statistics as well as on unofficial estimates for types of energy consumption that were not recorded officially.

The geography of energy is the study of the global distribution of energy resources, the production, distribution, and utilization of those resources, and of the organizational structures for exploiting the resources. We will now consider the distribution of energy resources—arable land and food supply first, hydroelectric power next, and then the nonrenewable sources.

ARABLE LAND AND FOOD-ENERGY PRODUCTION

Less than 11 percent of the land area of the world is arable, that is, suitable for cultivation, and most of it is in the northern temperate zone. Iceland has none, Denmark is 63 percent arable, and there are

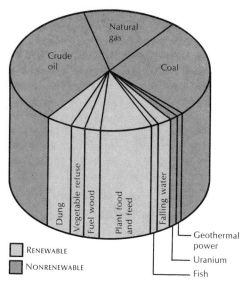

RENEWABLE
NONRENEWABLE

FIGURE 9.1
The world's consumption of energy resources in 1971.

Table 9.1 World Consumption of Energy (1971)

Source	Energy content (10^{15} kcal)	Percentage of total
Crude oil	22.7	34.8
Coal and lignite	17.1	26.2
Natural gas	10.9	16.7
Plant foot and feed*	6.2	9.5
Dried dung	3.3	5.1
Wood fuel	1.6	2.5
Vegetable refuse	1.6	2.5
Falling water	1.1	1.7
Uranium†	0.55	0.84
Fish	0.06	0.09
Geothermal power‡	0.04	0.06
Total	65.2	100.0%

NOTES: *Does not include feed for draft animals.
†Calculated on basis of 17 percent system efficiency.
‡Calculated on basis of 14 percent conversion efficiency.

thirty-seven nations, including Canada, Brazil, New Zealand, and Norway, in which less than 5 percent of the land is arable (see table 9.2).

Twenty-three nations have 74 percent of the world's arable land (table 9.3) and account for most of the world's grain and meat production. Total food production statistics, however, tell us little about the amount of food available to each person, one of the most basic measures of a nation's living standard or level. For that measure, we must do some calculations that, because they involve certain assumptions, will be outlined here.

Estimates of the daily per capita food supply in 132 nations are published by the Food and Agriculture Organization of the United Nations. These estimates are broken down for nine components: cereals, potatoes and other staples, sugars, meat, eggs, fish, milk, fats and oils, fruits and nuts. The food energy ingested daily in 1970 by the statistical person in each of these nations ranged from 1,750 kcal in Indonesia to 3,450 kcal in Ireland. The meat content of national diets, however, differed much more than the total caloric content, ranging from 6 kcal a day in India to 768 kcal a day in Uruguay. In chapter 5 we saw that it takes a lot of plant (feed) calories to produce a calorie of edible meat, and that the amount required differs according to the animal (hog, cow, sheep) and product (meat, milk, eggs) involved. Consequently, the nation that eats much meat puts more pressure on the energy production of its arable and pasture lands than does the

Table 9.2 Regional Distribution of Arable Land and its Relation to Population

Region	Total land area (10⁶ hectares)*	Arable land (10⁶ hectares)	Percentage arable	Mid-1970 population (10⁶ persons)	Percentage of world total arable land	Percentage of world total population	Arable land per person (hectares)
Europe†	493	148	30.0	462	10.4	12.7	0.32
U.S.S.R.	2,240	229	10.2	243	16.1	6.7	0.94
China	956	110	11.6	774	7.7	21.3	0.14
Asia‡	1,797	339	18.9	1,282	23.8	35.3	0.26
Africa	3,030	204	6.7	344	14.3	9.5	0.59
Oceania§	851	47	5.5	19	3.3	0.5	2.48
North America‖	1,968	220	11.2	228	17.8	6.3	0.96
Latin America	2,057	123	6.0	283	6.3	7.8	0.43
World	13,392	1,424	10.6	3,632	99.7	100.1	0.39

NOTES: *A hectare is a unit of area equivalent to that of a square 100 meters on a side; it equals 2.47 acres.
†Without U.S.S.R.
‡Without U.S.S.R. and China.
§Includes Australia, Polynesia (with New Zealand), Micronesia, and Melanesia.
‖Canada and U.S.A. only.
SOURCE: United Nations Food and Agriculture Organization, *Production Yearbook 1970*, v. 24 (Rome: 1971), p. 3, 51.

Table 9.3 Twenty-Three Nations with Most of the World's Arable Land

Nation	% of work force in agriculture	Arable land (10³ ha)	% of world total	Mid-1970 population (10⁶ persons)	% of world total	Arable land per capita (ha) (1970)	1970 grain* production (10⁶ ton)	1970 meat production (10⁶ pounds)	Fertilizer use (kg/ha of arable land)	Yield† Wheat	Yield† Rice
U.S.S.R.	32	228,500	16.1	242.76	6.7	0.94	137.70	19,615	35	14	36
U.S.A.	06	174,487	12.3	204.88	5.6	0.85	154.60	36,257	83	21	51
India	70	159,690	11.2	539.86	14.9	0.30	91.81	n.a.	11	12	17
China	63	110,300	7.7	773.66	21.3	0.14	130+	n.a.	n.a.	10	n.a.
Australia	10	44,354	3.1	12.51	0.3	3.55	10.89	4,284	25	12	62
Canada	09	43,404	3.0	21.32	0.6	2.04	20.62	3,273	17	18	n.a.
Brazil	50	29,800	2.1	93.39	2.6	0.32	25.43	5,149	20	10	16
Pakistan	74	28,400	2.0	114.18	3.1	0.25	30.13	n.a.	14	12	18
Turkey	74	25,231	1.8	35.23	1.0	0.71	14.87	1,175	18	12	38
Argentina	20	23,900	1.7	23.21	0.6	1.03	14.40	6,687	3	13	40
Mexico	52	22,500	1.6	49.09	1.4	0.46	12.88	2,061	24	28	29
Nigeria	79	21,800	1.5	55.07	1.5	0.40	1.62	n.a.	n.a.	n.a.	16
France	16	17,579	1.2	50.77	1.4	0.35	28.51	6,598	239	34	46
Burma	62	16,100	1.1	27.58	0.8	0.58	8.37	n.a.	4	8	17
Spain	38	15,700	1.1	33.78	0.9	0.46	9.37	1,985	79	11	62
Poland	36	15,100	1.1	32.53	0.9	0.46	6.55	3,165	162	23	n.a.
Ethiopia	89	12,900	0.9	24.63	0.7	0.52	1.68	n.a.	n.a.	7	n.a.
Niger	n.a.	12,200	0.86	4.02	0.1	2.28	0.03	n.a.	n.a.	12	20
Italy	24	12,221	0.86	53.57	1.5	0.23	15.62	2,898	101	23	53
South Africa	29	11,600	0.81	21.82	0.6	0.53	7.75	1,628	43	10	n.a.
Iran	49	11,300	0.79	28.66	0.8	0.40	6.37	608	8	9	33
Tanzania	95	10,700	0.75	13.27	0.4	0.81	1.06	n.a.	n.a.	9	13
Indonesia	67	10,570	0.74	115.40	3.2	0.09	19.36	n.a.	12	n.a.	20
Totals		1,057,400	74.2	2,571.1	70.7	0.39 (avg.)					

NOTES: *Without oats and rye.
†Quintals per hectare; a quintal is 100 kilograms.
SOURCE: United Nations (1971).

country that eats little meat. Differences in the *gross food-energy supply* among nations can be calculated accurately only where there exist adequate statistics on the consumption of feed, good estimates of the proportion of the feed supply obtained from pasture and woodland forage, and consistent measurements of the energy content of animal feed and animal products. Because such information for most of the countries of the world is not readily available, one must resort to a cruder method for comparison of gross food-energy supplies.

Fortunately, the United States Department of Agriculture has compiled and published figures on meat production in fifty-one countries and tabulated them in three main categories: beef and veal, pork, and mutton, lamb, and goat. Using multipliers based on statistics used in the United States (23 for beef and veal, 7 for pork, 37 for sheep and goat meat), one can calculate for each of the fifty-one countries a "national meat multiplier," which represents the approximate number of calories of plant food energy required for each calorie of edible meat produced in that country. For the United States, the calculation, based on 1970 production totals, is as follows.

$$(23) \, (22{,}272) \, + \, (7) \, (13{,}434) \, + \, (37) \, (551) \, = \, (m) \, (36{,}257)$$

$$m \, = \, 17.3$$

The second number in each of the pairs of the equation is the amount, in millions of pounds, of meat (beef and veal; pork; sheep and goat meat; total) produced in the United States in 1970. The symbol m stands for the number of plant calories it took to produce each calorie of meat, in other words, the "national meat multiplier" for the United States.

In comparisons of gross food-energy supply among countries, it does not matter how many of those 17.3 calories came from cultivated feed and how many came from pasture and woodland forage. The assumption that multipliers, based on practice in the United States, for individual animals can be applied to all countries certainly is a poor one, but is forced by lack of information. The beef multiplier has been scaled up arbitrarily from 21.5 (for recent practice in the United States) to 23 (United States practice in 1958) to fit better the slower maturing, leaner cattle of other countries. Meat multipliers have been assigned to those countries whose meat-production breakdown has not been published by the United States Department of Agriculture. Because there are regional patterns to the multipliers, conditioned by environment, religious tabus, and the degree to which agricultural production is subsidized by inputs of inanimate energy, such assignments are not as risky as they may seem. In the non-Moslem Far East, for example, where population presses on the land to such an extent that it would be very expensive, in social terms, to raise beef or mutton, we find cal-

> **BOX 9.1 The 7-to-1 Fallacy**
>
> Seven-to-one is a phrase that sticks in the mind. Applied to the conversion of plant food to animal products it can be very misleading.
>
> It does take about seven pounds of grain to add one pound of salable meat (at retail) to a beef animal in a feedlot, where it gets very little exercise and is protected by the administration of drugs and by human supervision. Even under these "ideal" conditions, the energy-conversion ratio is very different from the 7-to-1 weight ratio. Because a pound of grain contains about twice the energy in a pound of lean beef and about 50 percent more than that in a pound of fat beef (which is what comes out of feedlots), the energy-conversion ratio is more likely to be about 11-to-1. Remember, however, that considered here are individual animals for only a portion of their lives. When we look at the whole bull-cow-calf-steer energy system, incorporating the energy from grass and other wild forage and accounting for losses due to death, we get a view of the *system efficiency* of beef production and we find that the ratio of plant food eaten by the animals in the system to edible beef on the table (retail cuts) is closer to 23-to-1. If we estimate that 10 percent of the energy in beef as sold in the supermarket is trimmed off as unwanted fat by the consumer, the ratio goes over 25-to-1.
>
> Feedlots improve the system conversion efficiency and they do it quickly. Not only will a grass-fed steer have less fat, and therefore fewer calories per pound of meat, but it will take three years to achieve the weight that a feedlot animal can reach in fifteen to eighteen months.
>
> As demonstrated in the test, other animals have different conversion efficiencies. Beware the too facile application of that catchy 7-to-1 phrase!

culated meat multipliers that are low (for example, in the Philippines it is 10.1). Pork is the principal meat raised and eaten in this part of the world. In the rest of the world, the only country with such a low multiplier is Denmark, 10.4. In Moslem nations where pork is forbidden and where, except for southeast Asia, it is difficult to raise cattle or buffalo, we find the highest meat multipliers (for example, in Iran, it is 34.4). In the rest of the world, the only country approaching the Moslem multipliers is New Zealand (30.3), where both environment and tradition favor sheep raising.

From these calculated and estimated meat multipliers, the plant energy represented by the meat component of the daily diet can be calculated for each of seventy countries (see table 9.4); the plant energy

Table 9.4 Nations Ranked by per Capita Daily Gross Food-Energy Supply in 1970 (N = 70)

Nation	Net supply (kcal)	Meat production (lb./yr)	Meat eaten (kcal)	Meat multiplier	Gross plant energy represented by meat (kcal)	Gross food-energy supply (kcal)
New Zealand	3,320	838	665	30.3	20,150	25,868
Uruguay	3,020	317	768	25.0	19,200	23,326
Australia	3,220	377	665	27.0	17,955	22,843
Argentina	3,170	296	710	22.1	15,960	19,291
U.S.A.	3,290	187	643	17.3	11,124	16,240
Canada	3,150	167	672	16.5	11,088	15,905
Ireland	3,450	302	545	18.5	10,083	15,696
France	3,270	135	593	16.4	9,725	14,542
United Kingdom	3,180	86	524	17.4	9,118	14,173
Finland	2,960	100	285	18*	5,130	10,829
Belgium	3,150	168	483	13.0	6,279	10,783
Poland	3,140	101	332	13*	6,214	10,708
Switzerland	2,990	118	419	13.6	5,698	10,530
Israel	2,930	18	250	25.6	6,400	10,352
East Germany	3,040	125	497	13*	6,461	10,290
West Germany	2,940	140	444	13.1	5,816	10,290
Sweden	2,750	113	363	13.5	4,901	9,789
South Africa	2,730	91	254	25.2	6,400	9,650
Czechoslovakia	3,030	98	380	15*	5,700	9,590
Denmark	3,140	422	437	10.4	4,545	9,505
Netherlands	3,030	182	360	12.4	4,464	9,386
Norway	2,900	84	235	17.0	3,995	9,355
Austria	2,950	132	350	12.7	4,445	9,131
U.S.S.R.	3,180	85	240	19.5	4,680	9,040
Hungary	3,180	94	292	15*	4,380	8,761
Greece	2,900	52	173	23.9	4,135	8,661
Italy	2,950	56	218	17.4	3,793	7,882
Yugoslavia	3,130	85	246	15.6	3,838	7,700
Chile	2,520	67	195	22*	4,290	7,495
Spain	2,750	63	220	16.4	3,608	7,487
Colombia	2,190	51	182	23.2	4,222	7,164
Lebanon	2,360	n.a.	139	30*	4,170	7,128
Brazil	2,540	64	209	18.6	3,887	7,013
Cuba	2,500	n.a.	204	18*	3,672	6,805
Bulgaria	3,070	85	215	15*	3,225	6,762
Venezuela	2,490	51	132	22*	2,904	6,021
Mexico	2,620	48	140	18.9	2,646	5,726
Ecuador	1,850	n.a.	143	22*	3,146	5,453
Turkey	2,760	38	74	32	2,368	5,427
Jamaica	2,280	n.a.	128	20*	2,560	5,177
Kenya	2,240	n.a.	107	25*	2,675	5,139
Iraq	2,050	n.a.	90	30*	2,700	5,110
Taiwan	2,620	68	247	10*	2,470	5,089
U.A.R. (Egypt)	2,960	n.a.	54	30*	1,620	5,050
Portugal	2,730	50	106	16*	1,696	4,959
Ethiopia	2,150	n.a.	102	25*	2,550	4,882
Syria	2,450	n.a.	62	30*	1,860	4,813
Peru	2,200	24	97	22.2	2,153	4,708
Iran	2,030	25	68	34.4	2,339	4,695
Nicaragua	2,250	n.a.	71	23	1,633	4,693
Saudi Arabia	2,080	n.a.	73	35*	2,205	4,552
Dominican Republic	2,080	n.a.	95	20*	1,900	4,490
Afghanistan	2,060	n.a.	63	35*	2,205	4,236
Japan	2,450	20.5	73	11.5	840	4,172
Jordan	2,400	n.a.	53	30*	1,590	4,041
Zambia	2,250	n.a.	59	25*	1,475	3,842
Honduras	1,930	29	55	22*	1,210	3,675
Pakistan	2,350	n.a.	20	30*	600	3,628
Malaysia	2,190	n.a.	89	12*	1,068	3,536
Guatemala	1,950	28	44	23	1,012	3,491

Table 9.4 (continued)

Nation	Net supply (kcal)	Meat production (lb./yr)	Meat eaten (kcal)	Meat multiplier	Gross plant energy represented by meat (kcal)	Gross food-energy supply
Congo	2,160	n.a.	46	25*	1,150	3,349
Thailand	2,220	n.a.	83	11*	913	3,341
China	2,050	n.a.	134	9*	1,206	3,282
Philippines	1,990	31	100	10.1	1,010	3,183
Algeria	1,890	n.a.	43	25*	1,075	3,169
Haiti	1,930	n.a.	49	20*	980	3,019
Nigeria	2,160	n.a.	30	25*	750	2,981
Ghana	2,070	n.a.	26	25*	650	2,781
Burma	2,010	n.a.	38	12*	456	2,620
Ceylon	2,210	n.a.	8	15*	120	2,572
India	1,940	n.a.	6	15*	090	2,463
Indonesia	1,750	n.a.	20	25*	500	2,269

NOTES: *Estimated. n.a. = not available.

SOURCES: Data on net supply and meat eaten from United Nations (1971). Data on meat production from *Agricultural Statistics 1972* (Washington, D.C.: United States Government Printing Office).

represented by milk and eggs can likewise be calculated by use of the multipliers (5.4 for milk and 12.5 for eggs) previously calculated from United States practice. We do not know how much of the daily meat intake comes from poultry, but only in the industrial countries where the production of chicken meat has reached unusually high conversion efficiencies should the lack of this knowledge make much difference to the total. The objective is to calculate the total amount of plant (and fish) energy represented by the daily per capita food supply of each country (called Gross Food-Energy Supply in table 9.4). The calculated figures are not accurate, but they provide a much better means of international and inter-regional comparison than do simple food-energy supply figures.

The figure from the calculations shown in table 9.5 of 16,312 kilocalories per day in the United States is almost certainly conservative. If losses in transport, storage, and processing were added in, as well as the animal fats and oils, the final figure probably would be between 3 and 5 percent higher. Calculated in this way, gross food-energy supply ranges from a low of 2,269 kilocalories per person per day in Indonesia to a high of 25,838 in New Zealand.

The world pattern of gross food-energy supply (as calculated for table 9.4) shows vividly the unevenness of distribution (figure 9.2). In the lowest mapped category (3,600 kcal per capita per day and below) are China, India, and all of southeast Asia, about a third of Africa, and Haiti and Guatemala. Exactly one-half (1,816 million) of the world's people in 1970 lived in the countries represented by this lowest category, wherein the median daily supply (2,935 kcal per person) was less than one-seventh of the median (20,782 kcal per person) of the eight nations comprising the highest category, which represented only 7.5 percent (272 million) of the world population.

Table 9.5 Food-Energy Supply for the United States of America (1969)

Diet component	Food supply (kcal/day)	Multiplier	Gross plant energy represented (kcal/day)
Flour and cereals	656	1.0*	656
Potatoes and vegetables	180	1.0*	180
Beans, etc.	95	1.0*	95
Fruit	102	1.0*	102
Sugar	538	1.0*	538
Meat	633	17.3	10,950
Eggs	72	12.5†	900
Fish	26	1.0‡	26
Milk and butter	429	5.4	2,317
Fats and oils	525	1.0§	525
Coffee	23	1.0	23
Totals	3,279		16,312

NOTES: *Neglects losses in processing, transport, and storage and from refining, spoilage, and accident.
†Although this figure is now under 11 for the United States, a common multiplier of 12.5 was used for all countries, equivalent to United States practice in 1941.
‡Neglects aquaculture.
§Ignores fats and oils of animal origin.

SOURCE: Data for Food Supply from United States Department of Agriculture, *U.S. Food Consumption Price Expenditures* (ERS, Washington, D.C.: United States Government Printing Office, 1971).

FIGURE 9.2

Map of per capita gross food-energy supply in 1970. (Base map used with permission of The University of Chicago Department of Geography.)

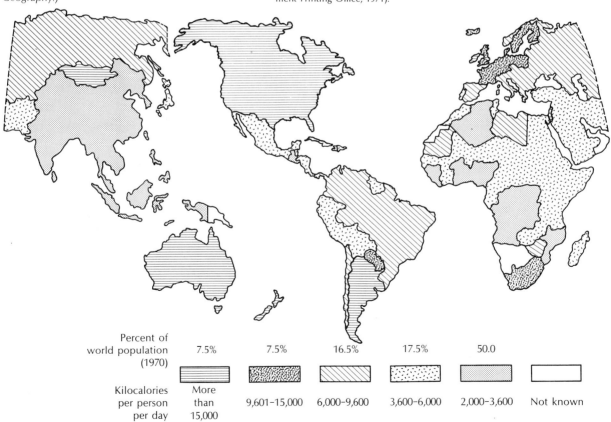

| Percent of world population (1970) | 7.5% | 7.5% | 16.5% | 17.5% | 50.0 | |
| Kilocalories per person per day | More than 15,000 | 9,601–15,000 | 6,000–9,600 | 3,600–6,000 | 2,000–3,600 | Not known |

The next-to-bottom category, still substantially below the world mean, includes about half of Africa, most of southwest Asia, Japan and Taiwan, Portugal, and almost a third of Latin America. The middle category (but not yet up to the world median) includes the Soviet Union, southeast Europe, about half of South America, three countries in Africa, and five or six in western Europe. The two upper categories, except for Mongolia, approximate the industrialized world of the late Victorian era and the grasslands that fed it.

HYDROELECTRICITY

Good sites for large power dams are distributed no more evenly throughout the world than are arable land or deposits of the fossil fuels. Deserts have no permanent streams, the arctic regions have few; most large rivers are in the temperate zones. Mountainous regions with steep, swift rivers contain numerous sites for relatively small dams, but few for large dams. Countries such as Switzerland and Norway, without much heavy industry and without much fossil energy, moved early to develop the power potential of their streams, consistent with their need to preserve floodplains for agriculture. Countries such as Germany, France, and the United States, with abundant coal deposits, were somewhat slower to build power dams but by now have put to use most of their prime sites. The unrealized water power potential in the world is still considerable (table 9.6) although most of it is in underdeveloped regions where a lack of productive use for the power to be generated has discouraged large capital investments in power dams. As we shall see, Latin America and Africa, which together have almost half the world's hydroelectric power potential, most of it undeveloped, are relatively deficient (with a few spectacular exceptions such as Libya, Nigeria, and Venezuela) in fossil-fuel resources.

Table 9.6 World Hydroelectric Power Development

Region	Potential* (10^6 kW)	Percentage of total	Development (10^6 kW)	Percentage developed
North America	313	11	59	19
Latin America	577	20	5	01
Western Europe	158	6	47	30
U.S.S.R., China, and other Communist countries	466	16	16	03
Southeast Asia and the Far East	497	17.5	21	04
Oceania	45	1.5	2	04
Middle East	21	1	n.a.	n.a.
Africa	780	27	2	0.3
Totals	2857	100	152[†]	5.3[†]

NOTES: *After Hubbert, 1962, p. 99.
    [†]Totals had increased to 210×10^6 kW and 7.4 percent by 1964.

Consequently, it would appear that any significant rise in living standards for most of the people of those two enormous regions will be based on the industrial use of hydroelectric power.

North America and western Europe have developed less than half their total hydroelectric power potential, but utilization of remaining sites will involve conflicts in land and water use, losses in power transmission from sites far away from use centers, and large capital investments. Because of these factors, it is almost certain that the total potential of these (and other industrialized regions) never will be developed fully. The spectre of reservoir siltation also overhangs the future of hydroelectric power. Unless an economic way is found to remove the silt that accumulates behind every dam, the useful life of a large power dam will be between 100 and 400 years. More than 90 percent of the world's hydroelectric power is produced by eighteen nations, more than 46 percent by only three (see table 9.7).

Tidal power is a form of hydroelectric power. There is a small tidal power plant in France, but elsewhere, the lack of tides of sufficient amplitude and the heavy investment required per unit of power output have deterred development.

When we look at the world distribution of electricity production (see table 9.8), including thermal, nuclear, and geothermal as well as

Table 9.7 Nations Making Greatest Use of Hydroelectric Power (1971) (N = 18)

Nation	Production (10^9 kWh)	Percentage of world total	Percentage of national total electricity produced	Percentage of national total energy produced
U.S.A.	269.580	22.0	15.7	1
Canada	160.984	13.2	74.4	10
U.S.S.R.	126.099	10.3	84.8	1
Japan	83.202	6.8	21.9	3
Norway	62.647	5.1	99.6	42
Sweden	52.027	4.2	78.2	12
France	48.726	4.0	32.7	3
Brazil	43.274	3.5	84.9	10
Italy	40.019	3.3	32.1	12
China	38.070	3.1	34.6	1
Spain	32.747	2.7	52.4	8
Switzerland	29.488	2.4	89.9	18
India	25.875	2.1	42.8	3
Mexico	16.802	1.4	53.6	3
Austria	16.778	1.4	58.4	9
Yugoslavia	15.644	1.3	53.0	6
West Germany	14.054	1.1	5.4	1
New Zealand	12.971	1.1	84.8	20
Totals	1088.987	89.0		
World	1224.167	100		

SOURCE: United Nations Department of Economic and Social Affairs, *World Energy Supplies 1968-1971*, statistical papers series J no. 16 (New York: 1973).

Table 9.8 World Electricity Production (1971) (10^9 kWh)

Region	Totals	Percentage of world total	Hydroelectric power	Percentage of region total	Fossil-fueled power	Percentage of region total
North America	1,934.331	37.1	430.564	22.3	1,461.332	75.5
Caribbean America	80.693	1.5	28.128	34.9	52.564	65.1
Other America	94.902	1.8	56.304	59.3	38.598	40.7
Western Europe	1,216.639	23.3	339.336	27.9	824.052	67.7
Western Asia	36,864	0.7	6.079	16.5	30.785	83.5
Far East	492.894	9.4	121.674	24.7	361.178	73.3
Oceania	77.631	1.5	25.995	33.5	50.462	65.0
Africa	94.322	1.8	26.588	28.2	67.734	71.8
Communist World	1,191.842	22.8	189.499	15.9	997.639	83.7
Totals	5,220.118	99.9	1,224.167	23.5	3,884.344	74.4

Region	Nuclear power	Percentage of region total	Geothermal power	Percentage of region total	Per capita consumption (kWh)
North America	41.887	2.2	0.548	0.03	8,455
Caribbean America	0	0	0.001	0	642
Other America	0	0	0	0	587
Western Europe	50.575	4.2	2.676	0.2	3,404
Western Asia	0	0	0	0	340
Far East	9.806	2.0	0.236	0.05	432
Oceania	0	0	1.174	1.5	3,979
Africa	0	0	0	0	263
Communist World	4.704	0.4	0	0	1,016
Totals	106.972	2.0	4.635	0.1	1,420

SOURCE: Production data from United Nations (1973).

hydroelectricity, we see that more than 60 percent is produced in two small areas of the world: the countries of Europe west of the Soviet bloc and in North America. These are the parts of the world that had, during the technological development of electric power generation and distribution, a great energy surplus to pay for the new systems, an industrial society capable of absorbing and putting to productive use a new energy form, a commitment to encourage the development of resources and the application of energy to work, and abundant fossil fuels as well as hydropower potential. If we add the Soviet bloc, we see that the rest of the world, including all of Latin America, Africa, Oceania, and Asia except the Soviet part, produced only 23 percent of the total electricity in 1971. The lesson seems apparent: electrification is a result, not a cause, of national affluence. Like meat in the national diet, electricity in the national energy consumption is a measure of living standards supported, in the main, by energy surpluses derived from the fossil fuels. Norway (see table 9.9), Sweden, Switzerland, and the other countries with little or no domestic resources of coal or petroleum are exceptions to this statement only insofar as they have not made productive use of *imported* fuel to drive ships, to run factories, and to create an energy surplus through trade;

Table 9.9 Per Capita Electricity Consumption in Selected Nations (1971)

1. Norway	15,344 kWh	22. Ireland	2,132 kWh
2. Canada	9,851	23. Spain	1,758
3. Sweden	8,412		
4. U.S.A.	8,312	WORLD AVERAGE	1,420
5. Iceland	7,777	24. Venezuela	1,261
6. New Zealand	5,363	25. Greece	1,199
7. Finland	5,009	26. Zambia	1,030
8. Switzerland	4,740	27. Argentina	1,003
9. Australia	4,689	28. Mexico	620
10. United Kingdom	4,601	29. Brazil	531
11. West Germany	4,344	30. Colombia	436
12. East Germany	4,101	31. Egypt	235
13. Kuwait	3,791	32. Saudi Arabia	144
14. Japan	3,622	33. China	140 (est.)
15. Belgium	3,499	34. India	109
16. U.S.S.R.	3,237	35. South Vietnam	74
17. France	2,877	36. Haiti	24
18. Israel	2,535	37. Burma	21
19. Italy	2,340	38. Indonesia	21
20. South Africa	2,221	39. Nepal	6
21. Poland	2,132	40. Yemen	3

SOURCE: United Nations (1973).

of course, these countries have, in addition, made profitable use of hydroelectric power. The recent addition of Kuwait to the ten greatest per capita users of electricity strikingly illustrates the thesis that electrification, like prolific agriculture, is dependent upon subsidy from another energy source, generally the fossil fuels.

COAL AND OTHER SOLID HYDROCARBON FUELS

We now turn to the chief source of energy for modern, industrial man, the fossil fuels. We shall consider first the solid fuels and then the fluid ones.

In the world distribution of beds of coal and lignite one can scarcely fail to note the general similarity to the distribution pattern of arable land. If anything, the coal measures show an even more striking preference for the temperate zones and for the northern temperate zone in particular (Figure 9.3). It is no coincidence that the two most powerful countries in the world produce 42 percent of the world's coal and lignite; together they have 76 percent of the world's reserves (see tables 9.10 and 9.11). In both countries, as well as in China, production has been increasing in recent years while production in western Europe and Japan has been decreasing. Four countries of the world, the U.S.S.R., the U.S.A., Canada, and China, have about 90 percent of all the coal and lignite in the world.

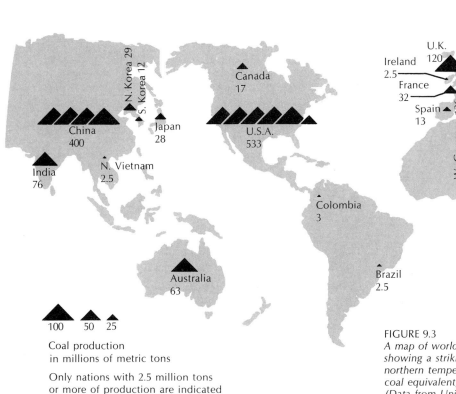

Coal production
in millions of metric tons

Only nations with 2.5 million tons
or more of production are indicated

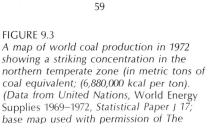

FIGURE 9.3
*A map of world coal production in 1972
showing a striking concentration in the
northern temperate zone (in metric tons of
coal equivalent; (6,880,000 kcal per ton).
(Data from United Nations,* World Energy
Supplies 1969–1972, *Statistical Paper J 17;
base map used with permission of The
University of Chicago Department of
Geography.)*

Table 9.10 World Coal and Lignite Production (1972)
(10^6 Metric Tons)

Nation	Production	Percentage of World Total
1. U.S.A.	537.3 ↑*	22.1
2. U.S.S.R.	481.0 ↑	19.8
3. China ⎫ North Korea ⎬ Mongolia ⎭	429.5 ↑	17.7
4. Poland	162.2 ↑	6.7
5. West Germany	135.8 ↓	5.6
6. United Kingdom	119.5 ↓	4.9
7. Czechoslovakia	79.3 ↑	3.3
8. India	75.8 ↑	3.1
9. East Germany	75.4 ↓	3.1
10. Australia	62.8 ↑	2.6
11. South Africa	58.6 ↑	2.4
12. France	31.5 ↓	1.3
13. Japan	28.1 ↓	1.1
Total	2276.8	93.7
World	2429.7	100.0

NOTE: *Production trend.
SOURCE: United Nations Department of Economic and Social Affairs,
World Energy Supplies 1969–1972, statistical papers series J
no. 17 (New York: 1974).

Table 9.11 World Reserves of Minable
Coal and Lignite (10^9 Metric
Tons)

Region	Reserves	Percentage of world total
1. U.S.S.R.	4,310	56.4
2. U.S.A.	1,486	19.5
3. Asia	681	8.9
4. Canada	601	7.9
5. Western Europe	377	4.9
6. Africa	109	1.4
7. Oceania	59	0.8
8. Latin America	14	0.2
Total	7,637	100.0

SOURCE: Paul Averitt, *Coal Resources of the United
States—January 1, 1967,* United States Geo-
logical Survey bulletin 1275 (Washington,
D.C.: United States Government Printing
Office, 1969).

PETROLEUM AND NATURAL GAS

Petroleum and natural gas, like coal and lignite, are formed in sedimentary basins. Economic deposits of the fluid hydrocarbons, however, are more widely disseminated and show a different distribution (see figure 9.4). The largest known reserves are in the Middle East, the region that also leads in production (figure 9.5). The only region in which consumption and production of petroleum are in equilibrium is that of the Communist countries; elsewhere supply exceeds consumption or consumption exceeds supply by large margins. From this pattern one could deduce, if one did not know already, that there is a great deal of petroleum moving about in the non-Communist world and that there may be differing or even opposed interests between the producing regions, such as the Middle East and Africa, on the one hand, and the consuming regions, such as western Europe, the Far East, and North America, on the other hand. As can be seen in table 9.12, refineries tend to be found in consuming regions rather than in producing regions. There are two reasons for this. First, it costs more to ship clean, volatile, and hazardous fuels such as gasoline and

FIGURE 9.4
A map of world oil and gas production (in energy units) in 1972 showing a distribution different from that of coal but still favoring the northern land masses (in metric tons of coal equivalent; 6,880,000 kcal per ton). (Data from United Nations, World Energy Supplies 1969–1972, Statistical Paper J 17; base map used with permission of The University of Chicago Department of Geography.)

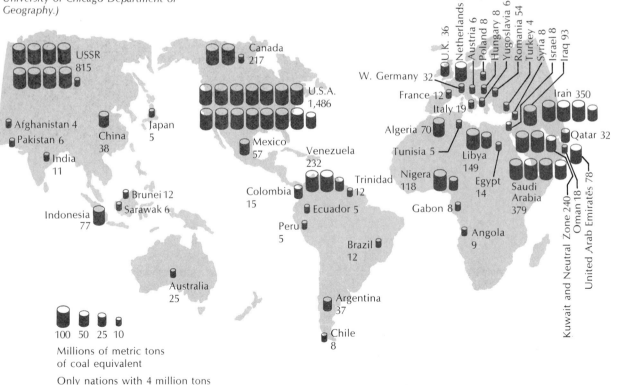

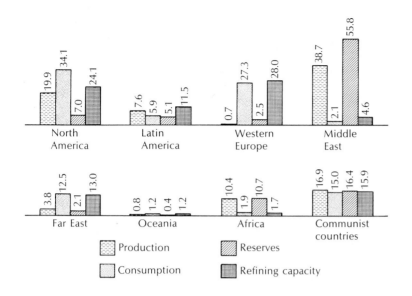

FIGURE 9.5
Oil production, consumption, reserves, and refining capacity in percentages of world totals, by regions.
(Based on 1973 data from Oil and Gas Journal *of December 31, 1973.)*

Table 9.12 Petroleum Production and Refining Capacity, by Leading Nations

Nation	Production		Crude oil refining capacity (January 1, 1975)	
	Barrels per day average (1974)	Percentage change from 1973	Nation	Barrels per day
1. U.S.S.R.	9,000,000 (est.)	+ 6.5	1. U.S.A.	14,216,287
2. U.S.A.	8,945,000	− 2.9	2. U.S.S.R.	8,800,000 (est.)
3. Saudi Arabia*	8,400,000	+11.7	3. Japan	5,133,840
4. Iran*	6,128,000	+ 0.6	4. Italy	3,952,660
5. Venezuela*	3,025,000	−10.0	5. France	3,341,600
6. Kuwait*	2,600,000	− 7.4	6. West Germany	2,986,723
7. Nigeria*	2,300,000	+15.0	7. United Kingdom	2,782,980
8. Iraq*	1,829,300	+ 2.0	8. Canada	1,877,550
9. Abu Dhabi*	1,750,000	+34.5	9. Netherlands	1,840,700
10. Libya*	1,700,000	−21.6	10. Venezuela*	1,531,715
11. Canada	1,682,000	− 6.5	11. Spain	1,165,000
12. Indonesia*	1,457,000	+10.0	12. Brazil	916,800
13. China	1,280,000 (est.)	+25.0	13. Netherlands Antilles	900,000
14. Algeria*	888,800	−19.2	14. Belgium	866,700
15. Qatar*	546,000	+ 5.4	15. China	850,000 (est.)
16. Mexico	513,500	+17.1	16. Singapore	845,650
17. Neutral Zone†	485,460	− 4.3	17. Iran*	789,000
18. Argentina	422,000	− 4.5	18. Mexico	760,000
19. Australia	370,600	+ 0.7	19. Australia	722,170
20. Brunei (Malaysia)	327,000	+ 0.6	20. Argentina	720,718
World	56,722,000		World	66,000,000 (est.)

NOTES: *Members of OPEC.
　　　†An area of disputed sovereignty between Saudi Arabia and Kuwait.
SOURCE: Gardner, 1974, p. 109, except for the figure for the U.S.S.R. and China, which are the author's estimates made from various news reports.

kerosine than it does to ship dirty, viscous, and relatively inert fuels such as crude oil and residual fuel oil. Second, refineries are designed for specific national or regional markets. Those in North America and Caribbean America, for example, are designed to produce the maximum amount of gasoline possible from a barrel of crude oil because the North American market has a high demand for gasoline. Refineries in western Europe, although their crude oil may come from the same source as that of an American refinery, are designed to produce much less gasoline and much more fuel oil.

In chapter 4 we noted that natural gas, although commonly associated with crude oil, is often found alone. The ranking of nations by petroleum and by natural-gas reserves (see table 9.13) illustrates this point again; some countries are richly endowed with both, whereas other countries have much more of one than the other. Ten of the top eleven nations on the list of proved reserves of petroleum are in Asia and Africa; six are Arab states, two are Communist states, and eight are members of the Organization of Petroleum Exporting Countries (OPEC).

OIL SHALE AND TAR SANDS

Oil shale, a fine-grained sedimentary rock containing solid insoluble organic matter genetically unrelated to crude oil that yields oil on

Table 9.13 Greatest Reserves of Petroleum and Natural Gas (January 1, 1975)

Nation	Petroleum (10^9 barrels)	Nation	Natural gas (10^{12} cubic feet)
1. Saudi Arabia[†*]	164.5	1. U.S.S.R.	812.0
2. U.S.S.R.	83.4	2. Iran[†]	330.0
3. Kuwait[†*]	72.8	3. United States	250.0
4. Iran[†]	66.0	4. Algeria[†*]	229.0
5. United States	35.3	5. Abu Dhabi[†*]	200.0
6. Iraq[†*]	35.0	6. Netherlands	94.8
7. Abu Dhabi[†*]	30.0	7. Saudi Arabia[†*]	55.0
8. Libya[†*]	26.6	8. Canada	52.5
9. China	25.0	9. United Kingdom	50.0
10. Nigeria[†]	20.9	10. Nigeria[†]	45.0
11. Neutral Zone[‡]	17.3	11. Venezuela[†]	43.0
12. United Kingdom	15.7	12. Australia	38.0
13. Indonesia[†]	15.0	13. Kuwait[†*]	32.0
13. Venezuela[†]	15.0	14. Iraq[*†]	27.5
15. Mexico	13.6	15. Libya[*†]	26.5
16. Canada	9.4	16. China	25.0
17. Algeria[†]	7.7	17. Norway	24.7
18. Norway	7.3	18. Brunei (Malaysia)	22.0
World	715.7	World	2,555.1

NOTES: *Member of the Organization of Arab Petroleum Exporting Countries.
†Member of the Organization of Petroleum Exporting Countries (OPEC).
‡An area of disputed sovereignty between Saudi Arabia and Kuwait.
SOURCE: Gardner, 1974, p. 108–109.

being heated to a sufficiently high temperature, occurs in large volumes in several areas of the world and constitutes an enormous potential resource. The principal deposits are in Brazil, central Africa, the U.S.S.R., and the United States. Oil shale rich enough to be on the verge of becoming an economic resource now that the price of crude oil has risen so sharply contains about 136 billion barrels of oil, of which 80 billion barrels are in the United States (see table 9.14). It is estimated (Duncan and Swanson, 1965, p. 30) that about 400 million barrels of oil have been produced from oil shale in foreign countries, mainly in Scotland, Estonia (U.S.S.R.), and Manchuria (China). Deposits have been exploited on a small scale in France, Sweden, Germany, Spain, South Africa, Australia, and Brazil. Present production is small and none of it is in North America.

Tar sands, unlike oil shale, contain soluble organic material that is a natural residual product of crude oil. They occur in Albania, Romania, U.S.S.R., Colombia, Venezuela, and the United States. The major world reserves appear to be in Canada (710 billion barrels in place), although both Colombia and Siberia may contain deposits equal to or greater than Canada's Athabasca tar sands. At present about 90 percent of the oil in the Athabasca sands that are currently being mined is recovered. Most of the Athabasca deposit lies under too much overburden to be recovered by stripping, the present method of mining, and the additional recovery problems involved in going underground are likely to reduce the recovery rate to between 35 and 50 percent. An optimistic estimate for the ultimate yield of the Athabasca sands is 300 billion barrels. But, because that total probably is about equal to all the crude oil that will ever be extracted from both Canadian and United States territory, it is well worth going after. Production from the Canadian tar sands, which in 1973 was about

Table 9.14 Shale-oil "Reserves" (10^9 Barrels)

Region	"Identified resources" or proved and indicated "reserves"		"Hypothetical resources" or inferred "reserves"	
	25–100 gpt*	10–25 gpt	25–100 gpt	10–25 gpt
Africa	100	small	‡	‡
Asia (inc. U.S.S.R. and China)	90	14	2	3,700
Europe	70	6	100	200
North America	418†	1,600	350	1,700
South America	small	800	‡	3,200
Total	678	2,420	>450	>8,800

NOTES: *Gallons recoverable per ton of shale.
‡Not estimated.
†Only 80×10^9 barrels of this amount are estimated to be recoverable economically with present techniques; consequently, all the other "reserve" figures in this table probably should be divided by five in order to obtain realistic estimates in terms of existing techniques of recovery.
SOURCE: Culbertson and Pitman, 1973, p. 500–501.

50,000 barrels a day, may be increased substantially in the next few years. Some oil is obtained also from tar sands in Albania, Rumania, and the Soviet Union.

NUCLEAR FUELS

The world's uranium reserves are concentrated in a few countries, principally the United States, Canada, South Africa, U.S.S.R., France, and Australia (see table 9.15). The present production reflects the distribution of known reserves. Information is not available on uranium deposits in the Soviet Union; there are believed to be large deposits in the Fergana area of central Asia, as well as in northern Siberia and central Kamchatka. Thorium is only now becoming an energy resource. The first thorium-cycle nuclear power reactor de-

Table 9.15 Non-Communist World Uranium Production and Reserves

Nation	1971 production (tons U_3O_8)	Reserves					
		at $5–$10 per lb		at $10–$15 per lb		at $15–$30 per lb	
		indicated	probable	indicated	probable	indicated	probable
U.S.A.	12,800	300	350	150	200	200	440
Canada	4,900	200	290	130	170	100	300
South Africa	3,800	205	15	65	35	55	70
France	2,700	45	20	5	10	n.a.	n.a.
Niger		12	13	13	n.a.	n.a.	n.a.
Spain		11	n.a.	4	30	15	250
Portugal		10	7	n.a.	12	n.a.	10
Australia	400	11	3	3	1	1	n.a.
Argentina		9	21	11	32	15	73
Sweden		n.a.	n.a.	350	50	150	200
Morocco		6	n.a.	11	n.a.	8	n.a.
Other non-Communist		17	24	28	16	n.a.	61
Total free world	25,020	826	743	770	556	588	1,404

SOURCES: Production data from Finch and others, 1973, p. 458.
Reserves data from DeCarlo and Shortt, 1976, p. 227.

Table 9.16 World's Thorium Resources (10^3 Tons ThO_2 Content)

Nation	Measured ore	Inferred ore	Total
India	300	250	550
U.S.A.	100	500	600
Canada	80	150	230
U.S.S.R.	100	100	200
Central and South Africa and Malagasy	50	50	100
Brazil	10	20	30
Malaysia	10	10	20
Denmark (Greenland)	15	n.a.	15
Australia	10	n.a.	10
Total	675	1,080	1,755

SOURCE: Shortt, 1970, p. 207.

signed for commercial use went into operation near Platteville, Colorado, in 1973. Many countries outside the United States, with meager or no supplies of fossil fuels or uranium, do have substantial deposits of inexpensive thorium (see table 9.16).

GEOTHERMAL ENERGY

By 1973 geothermal energy was being used in eight countries, mainly to generate electricity. Geothermal heat is also used directly for space heating, for air conditioning, for heating hothouses and soil for agricultural purposes, and in the processing of paper (New Zealand), diatomite (Ireland), and salt (Japan). Hot springs, long used for recreational and therapeutic purposes in Europe, Japan, Mexico, and the United States, represent another use of geothermal energy. Total installed generating capacity based on geothermal energy was 1,100 megawatts at the start of 1974, most of it in Italy, the United States, and New Zealand.

FIGURE 9.6
The geothermal regions of the world are in general the zones of active volcanism.
(From Oil and Gas Journal, *May 13, 1974, p. 103.)*

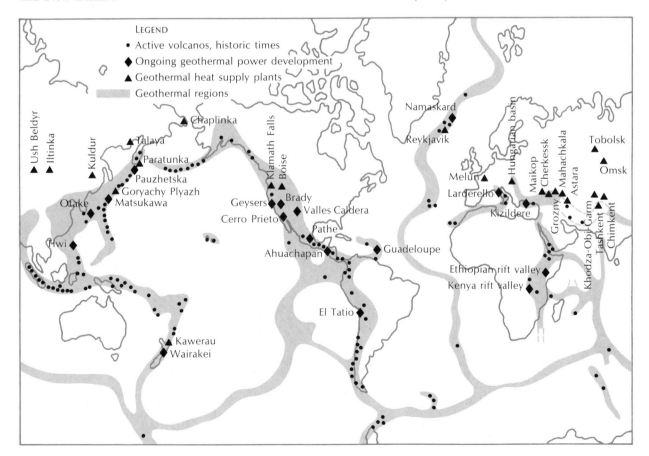

A wide range of guesses about the potential for geothermal energy have been published. Probable reserves, estimated on the basis of present technology and known reservoirs, although considerably larger than present production, do not offer a significant contribution to world or national energy needs. Should an economic technique be developed for extracting heat from dry rocks, or from the molten rock that exists at shallow depths in the world's volcanic system (see figure 9.6), geothermal power could meet the power demand of much of the world. The volcanic belts of the world do not favor the northern temperate zone as do the world's coal and petroleum fields. Given the economic feasibility (in energy-profit terms) of making electricity from either dry rock or magma reservoirs of heat, abundant power could be developed in southeast Asia, Latin America, and Africa.

CONSUMPTION AND USES OF ENERGY

The consumption of energy in the world is highly concentrated in the industrialized or developed regions: Europe, North America, Japan, and Australia. Energy in these regions comes overwhelmingly from the fossil fuels; a minor amount is in the form of hydroelectricity, and still less is nuclear; virtually none comes from refuse or other waste materials. Although heat energy derived from burning vegetable refuse and dung represents as much as a fifth of the world's total energy supply, it is spread over about 70 percent of the world's people, so that the per capita use of such fuels is very low. Most of the tropical world uses less energy in the form of fossil fuels and electricity (less than 4,000 kcal per day) than it does in gross food-energy consumption (see table 9.4).

The major uses of energy are for comfort heating and cooling, for process heating (including cooking and water heating), for work, and as food. In a low energy society, food, human, and animal power are major elements in the energy flow system; in a high energy society, they are minor. In a low energy society, the use of energy in industry and transportation are minor; in a high energy society, these are major components of the energy-flow system. The automobile is a striking illustration of this contrast (see table 9.17). In 1970, the industrialized nations of the Organization for Economic Cooperation and Development (OECD) owned 90 percent of the world's cars, an average of 241 cars for each thousand inhabitants, whereas the rest of the world had an average of only 6 cars for each thousand persons.

In the United States 50 percent of the crude-oil consumption is represented by automobile fuel; in Europe only 17 percent, and in the rest of the world, less than 5 percent. Grain fed to animals follows

Table 9.17 Automobile Population Densities

Nation	Cars/1,000 population
U.S.A.	434
Canada	313
France	244
West Germany	230
United Kingdom	211
Japan	85
World average	51

SOURCE: Leach, 1972, p. 5.

Table 9.18 Per Capita Energy Consumption of Selected Countries (1972)

U.S.A.	219,000 kcal	New Zealand	54,400 kcal
Canada	202,800	Italy	52,700
Kuwait	196,800	South Africa	52,200
Czechoslovakia	129,000	Israel	51,100
Belgium	121,900	WORLD AVERAGE.	37,400
Developed countries average.	117,000	Spain	33,300
		Argentina	32,600
East Germany	113,000	Greece	30,300
Sweden	108,200	Mexico	24,800
Netherlands	107,600	Iran	18,000
Australia	107,500	Portugal	17,100
Denmark	104,900	Peru	11,700
United Kingdom	101,700	China	10,700
West Germany	101,700	Brazil	10,000
Bahrein	96,500		
U.S.S.R.	89,900	Undeveloped countries average	6,800
Norway	87,400		
Poland	85,900	Egypt	6,100
France	78,300	India	3,500
Switzerland	68,200	Pakistan	3,000
Japan	61,300	Indonesia	2,500
		Nepal	300

NOTE: Fossil, nuclear, geothermal, and hydroelectric energy only; food and feed, fuel wood, fuel from animal and vegetable wastes are not included.
SOURCE: United Nations, 1974.

a similar distribution pattern. In 1962, the developed countries with unplanned economies consumed 180.9 million tons of feed grain, countries with planned economies, 61.8 million tons, and all the rest of the world only 24 million tons. The resulting meat production was 39.4 million, 17.2 million, and 18.5 million tons respectively. The per capita annual consumption of energy in the form of fossil fuels and electricity ranges from very small in some countries (see table 9.18) to very large, more than one hundred times greater, in others.

INTERNATIONAL ENERGY FLOWS

High energy society for the most part appears to have overrun its domestic energy supplies. Only four of the world's industrialized countries, Canada, Australia, the Soviet Union, and Poland were net exporters of energy in 1971. In figure 9.7, selected countries are ranked in order of the dependence of the individual inhabitant on *imported* energy. For each person in Belgium, Sweden, and Denmark, the equivalent of more than five tons of coal is *imported* each year; the average energy use in the world is less than half that figure. Seventeen of the top twenty importers are western European countries; fourteen of the top twenty import more than half of all the energy they con-

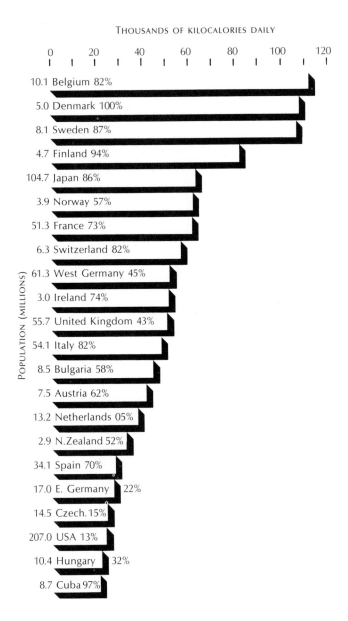

THOUSANDS OF KILOCALORIES DAILY

| | 0 | 20 | 40 | 60 | 80 | 100 | 120 |

10.1 Belgium 82%
5.0 Denmark 100%
8.1 Sweden 87%
4.7 Finland 94%
104.7 Japan 86%
3.9 Norway 57%
51.3 France 73%
6.3 Switzerland 82%
61.3 West Germany 45%
3.0 Ireland 74%
55.7 United Kingdom 43%
54.1 Italy 82%
8.5 Bulgaria 58%
7.5 Austria 62%
13.2 Netherlands 05%
2.9 N.Zealand 52%
34.1 Spain 70%
17.0 E. Germany 22%
14.5 Czech.15%
207.0 USA 13%
10.4 Hungary 32%
8.7 Cuba 97%

POPULATION (MILLIONS)

FIGURE 9.7
Dependency on imported energy of selected countries in 1971. Bars represent the amounts of daily per capita energy consumption represented by net energy imports. Percentages of energy dependency represent amounts by which total national energy production fell short in 1971 of meeting national energy demand.

sume. Thirty-four countries (each with a population of more than 4.5 million) depend on imports for 70 percent or more of their industrial energy (see table 9.19). These countries fall into two groups: a group of low energy agricultural nations and a group of industrialized nations in which the average per capita daily rate of energy consumption is between twenty and twenty-five times higher than the average of the other group. The first group needs imported energy if it is to

The Geography of Energy

Table 9.19 Nations Most Dependent upon Imports for Inanimate Energy

Nation	Population (millions)	Dependency (% of consumption)	1972 per capita daily energy consumption (kcal)
South Vietnam	18.8	100%	5,410
Upper Volta	5.7	100	225
Mali	5.1	100	430
Denmark	5.0	100	101,700
Haiti	4.9	100	530
Sudan	16.1	99	2,240
Kenya	11.7	98	3,110
Guatemala	5.3	98	4,900
Philippines	37.9	97	5,860
Thailand	35.3	97	5,750
Ethiopia	24.9	97	660
West Malaysia	9.5	97	9,350
Cuba	8.7	97	22,020
Madagascar	6.9	97	1,300
Tanganyika	13.2	95	1,360
Sri Lanka (Ceylon)	12.8	94	2,750
Finland	4.7	94	92,900
Malawi	4.6	92	1,000
Sweden	8.1	87	108,200
Japan	104.7	86	61,300
Uganda	10.1	86	1,240
Portugal	8.9	86	17,100
Italy	54.1	82	52,700
Belgium	10.1	82	121,900
Switzerland	6.3	82	68,200
Ghana	8.9	78	2,865
Morocco	15.2	76	4,200
Cameroon	5.9	75	1,890
France	51.3	73	78,300
Mozambique	7.6	73	2,710
Zaire	22.4	71	1,620
Spain	34.1	70	33,300
Greece	8.8	70	30,300

NOTE: Countries shown have populations of more than 4.5 million.
SOURCES: Population estimates from United States Bureau of the Census (1974).
　　　　Other figures from United Nations, 1974.

rise above a primitive level of subsistence; the second group needs imported energy to maintain the high standard of living it has attained.

In figure 9.8, selected countries are ranked in order of their per capita net *exports* of energy. In 1971, for each man, woman, and child in the United Arab Emirates, 1,333 million kcal of energy were exported—enough to have supplied all the energy needs (household, commercial, industrial, and transport) of 126 Italians, 87 Frenchmen, 63 Danes, or 30 Americans.

International energy flows consist primarily of food and petroleum. In energy content the petroleum shipments greatly exceed the food, but they are of almost equal importance to man. For both commodities there are distinctive flow patterns. The food that moves in ships is largely grain, sugar, soybeans, meat, and fish. Maize moves from the

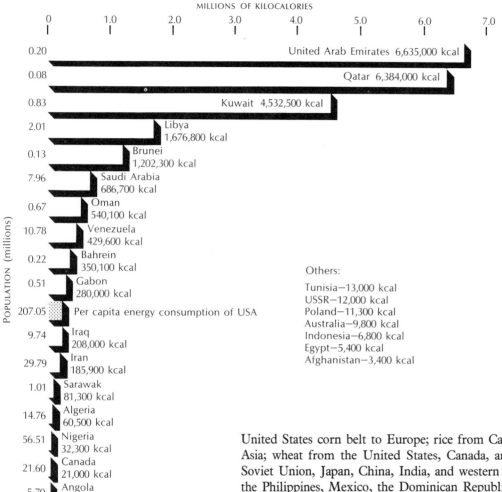

MILLIONS OF KILOCALORIES

Population (millions)	Country	kcal
0.20	United Arab Emirates	6,635,000 kcal
0.08	Qatar	6,384,000 kcal
0.83	Kuwait	4,532,500 kcal
2.01	Libya	1,676,800 kcal
0.13	Brunei	1,202,300 kcal
7.96	Saudi Arabia	686,700 kcal
0.67	Oman	540,100 kcal
10.78	Venezuela	429,600 kcal
0.22	Bahrein	350,100 kcal
0.51	Gabon	280,000 kcal
207.05	Per capita energy consumption of USA	
9.74	Iraq	208,000 kcal
29.79	Iran	185,900 kcal
1.01	Sarawak	81,300 kcal
14.76	Algeria	60,500 kcal
56.51	Nigeria	32,300 kcal
21.60	Canada	21,000 kcal
5.70	Angola	20,100 kcal

Others:

Tunisia—13,000 kcal
USSR—12,000 kcal
Poland—11,300 kcal
Australia—9,800 kcal
Indonesia—6,800 kcal
Egypt—5,400 kcal
Afghanistan—3,400 kcal

FIGURE 9.8
Main energy-exporting peoples. Bars represent daily per capita net energy exports in 1971.

United States corn belt to Europe; rice from California to southeast Asia; wheat from the United States, Canada, and Argentina to the Soviet Union, Japan, China, India, and western Europe; sugar from the Philippines, Mexico, the Dominican Republic, and Brazil to the United States and from Australia and the islands of the Indian Ocean to Europe. The United States produces about 70 percent of the world's soybeans and exports them to Europe, Japan, and Canada. The large flows of beef and butter are from Australia, New Zealand, and Argentina to Europe and the United States. Mutton and lamb are shipped mainly by Australia and New Zealand to Europe and Japan. Most of the fish catch of the world goes into the food supplies of the fishing countries, but appreciable amounts flow from nations such as Peru to the high energy countries (figure 9.9).

Rice and wheat move in substantial quantities from high energy to low energy societies, from those where grain raising is heavily subsidized by cheap fossil fuels to those where it is not. All the other foods of international commerce move mainly toward high energy societies; this is true especially of high protein foods, such as meat, fish, and soybeans. The world pattern of the flow of energy in food,

established during and by the Industrial Revolution, has not changed greatly during the past fifty years.

International oil flows, which at one time originated mainly in North and Caribbean America, now have a different pattern, reflecting the depletion of American reserves, great new discoveries in the Middle East and North Africa, and the rapid growth of oil consumption in western Europe and Japan. The rapid growth in production and consumption between 1958 and 1974 produced the global pattern of shipments depicted in figure 9.10.

Coal and natural gas move in limited amounts between continents. Coal is shipped from the United States and Canada to Japan and Europe, natural gas from Algeria to the United States and Europe, and from Indonesia and Alaska to Japan. Natural gas moves by pipeline from the Soviet Union to central and western Europe, and increasing amounts of electricity are being transmitted from Soviet Asia into Soviet Europe. Energy is the critical commodity of international exchange and the routes of its transport are of the highest strategic importance in a world still divided.

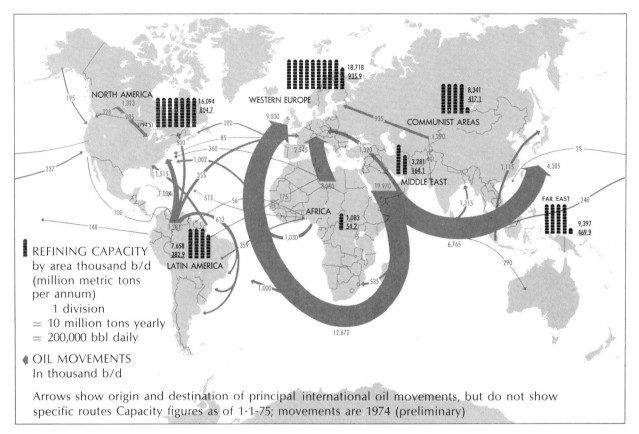

FIGURE 9.10
International oil flows in 1974. (Reproduced by permission of Petroleum Publishing Company.)

PETROLEUM IN WORLD POLITICS

The countries that for a thousand years or more were the energy-surplus masters of the earth have become energy-deficit nations. Western Europe, which used wind, and water, and coal to create the Industrial Revolution, now is heavily dependent on imported petroleum, although recent discoveries under the North Sea promise to reduce that dependency for some countries. The United States, once thought to have almost inexhaustible reserves of petroleum and natural gas, has become an energy-deficit nation; with industrialization, so have Japan and New Zealand.

As long as the foreign oil resources and the means of exploiting them were firmly in the hands of the major international oil companies, owned and controlled mainly by citizens of North America and western Europe, increasing dependence on foreign oil did not seem a serious threat to the stability of the high energy Western society. All that changed quickly in the third quarter of the twentieth century. The large international oil companies and the countries whose citizens control them (the United States, Great Britain, and the Netherlands) for some sixty years were the oil bankers and merchants of the world, assuring those countries not only of the energy they needed for economic growth and military strength but also of large repatriated profits from investments in overseas operations. For almost thirty years after World War II, these repatriated profits were substantially greater than the amounts the home countries were paying out for their increasing imports of petroleum.

Control of the market was the key to the growth of the big companies, and market control was built on control of refining, transport, and distribution facilities. With control of the principal markets, the big companies could dictate the price paid to the independent producer of crude oil, the amount the producing country received from each barrel taken, and the price to the consumer, except for taxes levied by the consumer's own government.

The big vertically integrated oil companies were formed in two different ways, which are exemplified by the histories of British Petroleum and Standard Oil of New Jersey. In 1912, Great Britain, the greatest sea power of the world and one of the major industrial nations, having decided to switch from coal to fuel oil for its navy and having no oil fields closer to home than Burma, bought a majority interest in the company that was to become British Petroleum and made it a quasi-official exploration and procurement arm of the government. British Petroleum developed into a worldwide, integrated oil company in which the majority interest still is held by the British Government. It was a different story in the United States, which had prolific oil fields at home and a multitude of producers vying for competitive advantage. In 1901, when Spindletop blew in, and later in 1932, when the giant East Texas field was discovered, it became painfully obvious that the finder-producer of crude oil in a free market during a time of surplus producing capacity faced economic extinction, that he could be as oil-poor as many of the farmers and ranchers of America have been land-poor during periods of surplus crop production. Attempts were made therefore to control both the oil market and oil production. The first objective (the *goal* was profit) was achieved by vertical integration—by obtaining control of the refineries of the country and by establishing a national network of wholesale and retail outlets; refineries and gas stations by the hundreds are beyond the means of the small oil producer and he is forced to do business with those who own them. The second objective was achieved by the authorization from the Congress and implementation by the oil-producing states of an interstate oil compact, which, in the name of conservation and through the police power of both the states and the federal government, limited oil production to quotas established by conservation commissions of the producing states. Successful attainment of both objectives led to an "orderly market" in which profits could be made even when the quantity of crude oil available far outstripped the demand for it. Although the hundreds of small "independent" oil producers whose access to refineries has been controlled by the major integrated companies often have chafed under enforced "prorationing" of their production, the governments of the oil-rich states have seen advantages in cooperating with industry to keep prices high: increased tax revenue, more stable employment, and prolongation of the period during which oil production could be an important part of the state's economy. At the same time, however, the consumer states, especially those without oil of their own, have begun to feel defrauded by the effective price control exercised by the big oil companies and the governments of the producing states and to suspect that such control, which they have perceived as being inimical to their own interests, has been maintained by unethical means.

Whereas vertical integration of the oil industry in Great Britain was the expression of the national interest of an oil-deficient country, in the United States it was an outgrowth of the producers' competition in an oil-abundant country. In western Europe, British Petroleum became the major oil company; Royal Dutch–Shell became a strong second. In the United States, the Standard Oil Company of New Jersey (now Exxon) emerged as the leader of a small pack of integrated companies that included Gulf, Texaco, Mobil, and Standard of California. The five American and the two European companies are known as the "Seven Sisters" of the international oil trade.

Outside the United States, Canada, and the Soviet Union (and now the North Sea countries), large oil reserves have been discovered mainly in countries that have been poor and politically weak. It has been relatively easy for the seven big international companies to gain and maintain control of those reserves. Back in 1929, the United States not only produced 70 percent of the world's petroleum but also was the largest consumer. To a large extent the oil business was an American business. As crude oil began to be discovered in larger and larger quantities outside the United States, in places where the costs of finding and producing were much lower, it became obvious that the domestic petroleum industry would be extinguished if the costs of producing oil in Saudi Arabia, for instance, were allowed to determine the world market price. Also, only the companies participating in the extraction of oil from the cheapest reservoirs would stay in business, and even they would suffer the loss of their assets in more expensive areas. Consequently, it was to the benefit of the big oil companies to maintain the world price at the highest level possible and, if this level were still below that required for profitable production in the United States, to get a protected market at a higher price level established at home, where expensive domestic oil could be produced at a profit, and some of the cheap foreign oil could be allowed entry (if produced by a company based in the United States) to produce a very profitable return. Until the 1950s, the distribution monopoly of the big companies worked to maintain a high world price. Then, as the world price weakened, the protected market in the United States appeared, guarded by a system of import quotas that maintained the domestic price of a barrel of crude about a dollar above the price at which it was available outside the United States. Both of these measures were based on the assumptions that supply would meet any foreseeable demand and that prices in a free market would decline. When the United States no longer had a surplus because of declining domestic production and rapidly rising consumption, the protected market ceased to exist and the import-quota system became meaningless.

Between 1929 and 1943, the world price of oil was pegged to the cost of production in the Gulf Coast region of the United States, no matter where the oil was actually produced, and the cost of transport from the Gulf of Mexico to the place of consumption was added. In this way both the producing nations and the consuming nations outside the United States were forced to pay tribute to the big oil companies. This control by the distributors and consumers of oil, which endured for about sixty years, no longer exists. How it came to pass into other hands is a classic story of the dynamics of energy resources.

The story begins in Mexico some forty years ago. American oil companies, successful in Texas, had extended their search and exploitation into geologically similar areas in eastern Mexico. They made some substantial discoveries, but in 1938 the Mexican government suddenly nationalized its petroleum resources. Acting perhaps more from outraged national pride, precipitated by the arrogant behavior of the foreign companies who refused to accept a government settlement of a labor dispute, than in the context of any long-range plan, Mexico was then faced with a task for which it was not prepared. The international oil industry was unanimous in its doubt about the ability of Mexico to operate such a complex and technologically developed industry. There were many self-serving predictions of failure. These predictions turned out to be very poor indeed. Mexico learned by doing. Petroléos Mexicanos (Pemex) operated existing fields, found new ones, and became the cornerstone of Mexican industrialization and economic growth, returning social benefits to the country far in excess of those that could have been expected under continued foreign exploitation. Mexico today is self-sufficient in petroleum.

Although the Pemex story was of minor consequence to the earnings of the international oil companies, its significance was not lost on other "backward" countries whose oil was being "developed" by those companies in later years. The formation of the Organization of Petroleum-Exporting Countries (OPEC) in 1960, however, appears in retrospect almost wholly a defensive move to try to halt erosion in world prices of crude oil rather than the first step in a plan to nationalize oil in the member countries, although OPEC did ask for new profit-sharing arrangements with the oil companies. Prices fell during the 1950s because of three developments that weakened the control of the Seven Sisters over the "free-world" market. First was the reluctant admission of some new players to the game. Most of these were American oil companies of the second rank who wanted to obtain foreign reserves because new reserves in the United States were becoming hard to find and costly to produce and who, therefore, found it increasingly difficult to compete with the big international companies, who kept finding easily oil that was incredibly cheap to produce. The list included Phillips, Sun, Cities Service, Standard of Indiana, Marathon, and later Occidental and Atlantic Richfield. These companies began to compete with the major companies for prospecting and production concessions and, when they found oil that was cheap to produce, they began competing in the world market.

The third factor to disturb the tranquility of the major companies world market during the late 1950s. The Soviets started to undermine the world price control structure of oil by selling and bartering Soviet oil with countries of western Europe as well as with Japan, Greece, Egypt, Cuba, and Brazil at prices below those maintained by the international companies. The Soviet exports led directly to reciprocal price cutting by the major companies, to smaller returns to the major exporting countries, and to the formation of OPEC.

The third factor to disturb the tranquility of the major companies during this period was Enrico Mattei, the genius of ENI (Ente Nazionali Idrocarburi), the Italian national petroleum company. Italy was a consumer country without oil of its own, either at home or abroad. ENI's task, as Mattei saw it, was to obtain oil for Italy as cheaply as possible, and that meant bypassing the international companies to deal directly with the producing countries. He bought and bartered for Soviet oil, and he made profit-sharing offers for concessions in oil-rich nations that were more favorable than those offered by the major companies. By 1956, profits were split evenly between the company and producing country. In 1958 Mattei said, "The people of Islam are weary of being exploited by foreigners. I intend to give them a more generous share of the profits and make them partners in the business of finding and exploiting oil resources." Through ENI, he offered 75 percent of the profits to the host country and was successful in obtaining agreements that enabled ENI to become a substantial producer of oil. The Italian company also offered technical and financial assistance to oil-*importing* underdeveloped countries who wished to explore for oil or to build refineries and thereby participate in the benefits of world oil. In the 1960s the Compagnie Française des Pétroles (CFP), the partly government-owned French company, and the Enterprise de Recherches et d'Activités Pétroliéres, ERAP, the wholly government-owned French company, further undercut the power of the international companies by making agreements with Iraq to take over revoked or confiscated concessions covering most of that country.

In 1960, the Organization of Petroleum Exporting Countries (OPEC) was formed. The original members were Venezuela, Libya, Saudi Arabia, Kuwait, Iraq, Iran, and Indonesia. Later Abu Dhabi, Algeria, Qatar, Nigeria, and Ecuador were accepted as members. For the first seven years of its existence, OPEC accomplished little more than to get organized, to recognize the range of interests and attitudes represented by its members, to plan strategy, and to attempt, with modest success, to halt the downward trend in oil prices and to obtain better profit sharing arrangements with the international oil companies.

It was thus during a period of increasing competition for production and markets, and of increasing pressure from the oil-exporting countries for a larger share of the ultimate value of their oil, that the 1967 Israeli–Arab war broke out. This brief conflict resulted in the closing of the Suez Canal, an event that was to have consequences for the industrialized world much more painful than the small increase in the cost of Persian Gulf oil hauled around Africa instead of through the canal. The increased transport cost was temporary, because the closing of the canal stimulated the design and construction of supertankers, of two hundred thousand tons or more. The supertanker of the 1970s cannot enter the Suez (or Panama) Canal, but has achieved such economies of scale that oil transport from the Persian Gulf around Africa to Europe is now cheaper than it would be in tankers small enough to traverse the canal.

But the closing of the Suez Canal revealed how dependent western Europe was on oil from the Persian Gulf and what power lay in the hands of any group that could control the flow of oil from there. Beginning with that revealing crisis of 1967, the power of OPEC rapidly increased, more favorable profit-sharing agreements were negotiated with the producing companies, the threat of confiscation became real, participation of the host country in the management of the industry was demanded, and the host countries wrested from the international companies the power to determine posted prices, the main measure of the producing country's revenue from oil. By 1974 new oil bankers were in power, and their key asset was the oil in the ground, instead of the control of transport, refining and market facilities. Oil reserves and production capability now were drawing money into the exporting countries and away from the consuming countries that for so long had garnered almost all of the profits from the cheap and convenient energy of oil, no matter where it had been produced.

In 1973 the Arab members of OPEC embargoed oil to the United States and the Netherlands and limited exports to western Europe and Japan, in order to bring pressure on the governments of those countries to lessen or renounce economic and political support of Israel. Reactions of the affected governments illuminated the power of those who control the flow of oil to the industrialized countries that import energy. The European Economic Community, refusing to honor the terms of a mutual aid pact, left the Netherlands to its fate—until the stubborn Dutch threatened to limit the flow of oil through its refineries to Germany. Japanese officials made propitiatory statements and renewed their efforts to make bilateral deals for oil at almost any price. It was every man for himself and a rewarding time for the sheikhs.

The income of the OPEC nations from their oil rose from less than $12 billion in 1971 to $30 billion in 1973 and about $90 billion in 1974. Trade deficits were either created or substantially enlarged for every industrialized country except the Soviet Union, Canada, and Australia. The industrialized consuming nations have been attempting to negotiate bilateral deals with producing nations, whereby services and goods will be exchanged for oil. In 1974 France agreed to build nuclear plants, factories, and pipelines in Iran in exchange for gas and oil exploration rights and started talks with Libyan and Iraqi officials on similar arrangements. Japan lent Iraq a billion dollars to build a refinery in exchange for rights to buy Iraqi oil, and OPEC openly encouraged other consuming countries to propose bilateral deals. A buildup of military strength in the OPEC nations is being financed by oil revenues. In 1974 and 1975 Iran and Saudi Arabia ordered fighter aircraft from the United States, France agreed to exchange Mirage jet fighters and other weapons for Saudi Arabian oil, and tiny Abu Dhabi bargained with the French to obtain thirty-five Mirage fighters in exchange for oil. As well as providing fighter planes, the United States is engaged in a long-range program to improve the Saudi navy and modernize the Saudi national guard, which protects the oil installations in that country.

A new era of barter between producers and consumers seems to be emerging. The National Iranian Oil Company (NIOC), for example, has been the instrument for agreements with Japan to finance a petrochemical plant and refineries; with Greece, India, and South Africa for shares in the refineries of those countries; with the Soviet Union for steel for a new mill, as well as with multinational companies (a NIOC/Standard Oil of Ohio sulphur plant, a NIOC/Goodrich caustic soda plant, a NIOC/Reynolds aluminum plant). Iranian oil is reported to be financing the building of twelve dams, a modern road network, nationwide electrification, a national telephone system, agricultural reform, a national education system, a machine-tool industry, and exploration for new mineral resources both in Iran and abroad. Iraq, through its national oil company, has worked out similar arrangements with Italy, Turkey, Hungary, Czechoslovakia, Spain, and the Soviet Union.

In Indonesia, foreign oil companies must contract with the national company (Pertamina), and the basic contract allows the foreign company to retain 35 percent of any oil discovered and requires that 65 percent of the remainder go to the national company. Pertamina in a few years has achieved the capability of contributing about half of the revenue of the Indonesian government. Libya, Nigeria, Saudi Arabia, Qatar, Peru, and Ecuador are other oil-producing countries that have national oil companies.

Although the OPEC cartel held together nicely while consumption was rising and demand was high, there are some who question its ability to remain unified and effective when the demand for oil levels off or even falls, since the internal conditions and problems of the member states differ so much. Four non-Arab members (Indonesia, Iran, Nigeria, and Venezuela) have 90 percent of the population represented by OPEC. In those countries oil income can be absorbed readily in welfare, education, and industrialization. For them, continued income is necessary. The Arab member states have small populations and ample financial reserves; they could afford to curtail oil production for years without harming their people or risking revolt.

NOT ONE WORLD, BUT FOUR

There are now four "worlds" into which the nations of the earth may be divided on the basis of access to energy resources.

1. *The high energy world of the unplanned economies,* most of which are now, for the first time in the history of industrialization, short of energy and dependent on continued imports of petroleum they no longer control to maintain their economic and social systems. They have high rates of energy consumption (above 50,000 kcal per person per day) and rather low rates of population growth (about 1 percent a year).
2. *The world of the planned economies,* in various stages of industrialization, self-sufficient in energy and able to use it for continued industrialization, to buy food, fertilizer, and high technology, and as an instrument of international political strategy. They have a wide range of energy consumption rates, from China (11,000 kcal per person per day) to the Soviet Union (90,000 kcal per person per day) and an overall rate of population growth slightly *under* that of the first world.
3. *The world of OPEC,* composed of unindustrialized countries that together have about 70 percent of the world's petroleum reserves. Calculated energy consumption rates in this world can be misleading, because they may represent mainly energy used in petroleum production, refining, and transport. Rates of population growth are high (around 3 percent a year).
4. *The rest of the world.* Energy consumption rates for most nations in this category are below 25,000 kcal per person per day, and for

many are below 5,000 kcal per person per day (if food, wood, and waste sources are not included). For this world, steep rises in prices for food and energy can create a crushing handicap to economic advancement. Rates of population growth are high (around 3 percent a year).

Half of the world's people consume only 8 percent of the world's inanimate energy supply; one-fifth of the world's people consume 72 percent. While this great disparity exists, the main fuel of the world, petroleum, is becoming expensive, long before most of the under-developed countries have acquired the capital, technology, and economy required to support the development of nuclear, geothermal, or solar energy.

A tabulation of world energy consumption by major regions (see table 9.20) shows what we already know, but in terms of regional balances. The Middle East, Caribbean America, and Africa produce a lot more energy than they use; western Europe, North America, and Japan (the Far East) use up the surplus of those other regions.

WORLD ENERGY RESERVES

Energy reserves depend upon the geophysical and geochemical constitution of the earth, the laws of physics, and the availability of economic recovery and utilization technology. These factors will be discussed more fully in chapter 13.

Reserves of nonrenewable resources are calculated in terms of quantities economically recoverable, not in terms of production or utilization rates. Reserves commonly are classified as *measured* or *proved*, *indicated* or *probable*, and *inferred* or *hypothetical*, each category representing a decrease in reliability of the figures given.

Table 9.20   World Energy Production and Consumption (1972) (10⁶ Metric Tons of Coal Equivalent)

Region	Production	Percentage of total	Coal and lignite	Crude petroleum	Natural gas	Consumption	Percentage of total	Per capita consumption (kg)	Approximate per capita surplus or deficit (kg)*
North America	2,323	30.7	554	744	960	2,661	35.9	11,531	−1,210 (def.)
Caribbean America	326	4.3	5	275	42	158	2.1	1,227	+1,560 (surp.)
Other America	82	1.1	5	54	16	126	1.7	759	− 290 (def.)
Western Europe	586	7.8	338	25	170	1,436	19.4	4,000	−2,250 (def.)
Middle East	1,225	16.2	7	1,177	39	95	1.3	857	+9,330 (surp.)
Far East	263	3.5	122	101	22	565	7.6	482	− 270 (def.)
Oceania	93	1.2	64	21	5	93	1.2	4,275	− 60 (def.)
Africa	441	5.8	64	359	15	134	1.8	363	+ 860 (surp.)
Communist countries	2,227	29.4	1,271	584	347	2,149	29.0	1,800	+ 70 (surp.)
World	7,566	100.0	2,430	3,340	1,616	7,410	100.0	1,984	+ 45 (surp.)

NOTE: *A kilogram of coal (equivalent) contains 6,880 kcal.
SOURCE: United Nations, 1974.

The Geography of Energy

The terms *potential* and *speculative* are also used. Reserves of renewable resources are calculated in terms of unused capacity or rate of potential production.

Apart from proved or measured reserves and actual production capacities, there generally is a range of figures for reserves to choose from. An informed choice may require a good deal of background study of the various methods used to arrive at figures for potential reserves. Chapter 13 goes into this problem in some depth. Here a tabulation of selected figures has been made (tables 9.21 and 9.22).

The figures for nonrenewable reserves have been calculated in terms of energy content and are put into two categories, *reasonably assured* and *probably discoverable*. In general, the first category will include reserves classified by others as *probable* or *indicated* as well as those classified as *proved* or *measured*, while the second category will include reserves or resources called *hypothetical, inferred,* or *possible* by others, but not those called *speculative*.

Averitt (1973, p. 140) gives 9,500 billion tons as the world total of *identified resources* of coal, a term that encompasses both *measured* and *indicated* coal reserves and assumes complete recovery. Because it seems probable that a cumulative recovery factor of only 60 percent will be achieved, I reduced the total of 9,500 billion tons to 5,700 tons of *reasonably assured reserves* and then split that number into two parts: 3,200 billion tons of bituminous and anthracite coal and 2,500 tons of lignite and brown coal, in accordance with the ratio of estimates published in the proceedings of the World Power Conference of 1968 in Moscow. The former tonnage was multiplied by 26 million Btu, the latter by 20 million Btu; the two results added yield 133.2×10^{18} Btu or 33.3×10^{18} kcal for the energy content of the reasonably assured world coal reserves. The calculation for reasonably assured reserves in the United States was made similarly from Averitt's estimate of 550 billion tons for *identified resources*. The totals for quantities that are probably discoverable (and economically recoverable for use) are based on Averitt's estimates of *hypothetical resources* but multiplied by 0.50 as a recovery ratio since much of this coal, if it exists, will be found at depths requiring underground mining.

Totals for *reasonably assured* petroleum supplies in the world and the United States are based on figures of proved reserves published by the *Oil and Gas Journal* (December 30, 1974, p. 109–110). The total amount *probably discoverable* in the United States is based on Hubbert's estimate (1972, p. 20) of 200 billion barrels ultimately recoverable and the amount *probably discoverable* in the world is based on the average ($1,725 \times 10^9$ barrels) of the two figures for the total recoverable oil postulated by Hubbert (1969, p. 196) as determining the range for ultimate-recovery estimates.

Reasonably assured reserves of natural gas are based on estimates made by the *Oil and Gas Journal* (December 30, 1974, p. 109–110).

Table 9.21 Nonrenewable Energy Reserves

Resources	Reasonably assured		Probably discoverable*		Total	
	Quantity	10^{18} kcal	Quantity	10^{18} kcal	Quantity	10^{18} kcal
World						
Coal	$5,770 \times 10^9$ tons	33.3	$3,665 \times 10^9$ tons	23.8	$9,365 \times 10^9$ tons	57.1
Crude oil	627.9×10^9 barrels	0.9	897×10^9 barrels	1.3	$1,525 \times 10^9$ barrels	2.1
Natural gas[†]	$2,033 \times 10^{12}$ cu. ft.	0.6	$5,967 \times 10^{12}$ cu. ft.	1.5	$8,000 \times 10^{12}$ cu. ft.	2.1
Shale oil	136×10^9 barrels	0.2	$2,672 \times 10^9$ barrels	3.8	$2,808 \times 10^9$ barrels	4.0
Tar sand	390×10^9 barrels	0.6	818×10^9 barrels	1.1	$1,208 \times 10^8$ barrels	1.7
Uranium (burned)[‡]	$1,596 \times 10^3$ tons U_3O_8	0.2	$1,299 \times 10^3$ tons U_3O_8	0.2	$2,895 \times 10^3$ tons U_3O_8	0.4
Thorium (converted)[§]	675×10^3 tons ThO_2	3.9	$1,080 \times 10^3$ tons ThO_2	6.3	$1,755 \times 10^3$ tons ThO_2	10.2
Uranium[‖]	$2,184 \times 10^3$ tons U_3O_8	19.0	$2,703 \times 10^3$ tons U_3O_8	23.5	$4,887 \times 10^3$ tons U_3O_8	42.5
Thorium[‖]	?	>5.9	?	>9.4	?	>15.3
Geothermal traps	—	0.1	—	?	—	>0.1
United States						
Coal	330×10^9 tons	1.98	515×10^9 tons	3.35	845×10^9 tons	5.33
Crude oil	34.7×10^9 barrels	0.05	65×10^9 barrels	0.09	100×10^9 barrels	0.14
Natural gas[†]	247.3×10^{12} cu. ft.	0.08	453×10^{12} cu. ft.	0.12	700×10^{12} cu. ft.	0.19
Shale oil	80×10^9 barrels	0.11	964×10^9 barrels	1.36	$1,044 \times 10^9$ barrels	1.47
Tar sand	None	None	4×10^9 barrels	0.01	4×10^9 barrels	0.01
Uranium (burned)[‡]	273×10^3 tons U_3O_8	0.04	450×10^3 tons U_3O_8	0.07	723×10^3 tons U_3O_8	0.11
Thorium (converted)[§]	100×10^3 tons ThO_2	0.58	500×10^3 tons ThO_2	2.90	600×10^3 tons ThO_2	3.48
Uranium[‖]	520×10^3 tons U_3O_8	4.53	$1,000 \times 10^3$ tons U_3O_8	8.70	$1,520 \times 10^3$ tons U_3O_8	13.23
Thorium[‖]	?	>0.87	?	>4.35	?	>5.22
Geothermal traps	—	0.01	—	?	—	>0.01

NOTES: 10^{18} = one quintillion.

*And economically recoverable for use.

[†]Includes natural-gas liquids.

[‡]At 1 percent burnup in a burner reactor (using only U-235).

[§]At 40 percent burnup in an almost-breeder reactor.

[‖]At 60 percent burnup in a breeder reactor.

Table 9.22 Renewable Energy Resource Reserve Capacity

Type of energy resource	Availability of energy	World	United States
Food and feed	Actual	6×10^{15} kcal	1.1×10^{15} kcal
	Potential	*	*
Water power			
Fluvial	Actual	243×10^9 watts (1967)	51×10^9 watts
Fluvial	Potential	2857×10^9 watts	161×10^9 watts
Tidal	Potential	64	
Total	Reserve capacity	2678×10^9 watts	110×10^9 watts
Wind power	Actual	small	small
	Potential	modest	modest
Vegetable energy (except food)	Actual	small	small
	Potential	modest	modest
Refuse energy	Actual	small	small
	Potential	modest	modest
Solar heating	Actual	small	small
	Potential	large†	large†
Solar power	Actual	0	0
	Potential	small to very large	small to very large
Geothermal power	Actual	0	0
	Potential	modest to very large	modest to very large
Fusion power	Actual	0	0
	Potential	zero to extremely large	zero to extremely large

NOTES: *Depends on energy subsidy to agriculture; world production could be increased considerably by more intensive use of fertilizer, but U.S. production could not.
†Harrison Brown (1954, p. 185) calculated that more than 20 percent of the total energy needs of man could be supplied by solar heat, by the use of heat pumps in temperate climes, and solar collectors in areas of high insolation; this is an attainable goal.

The figures for *probably discoverable* natural gas supplies are based on Hubbert (1969, p. 193). Both figures include natural-gas liquids.

Shale-oil reserves are based first on an estimate made by the United States Bureau of Mines (Schramm, 1970, p. 198) of 80 billion barrels recoverable by 50 percent extraction from shales that contain between 30 and 35 gallons per ton, are less than 1,000 feet below the surface, and can be mined by the conventional room-and-pillar method, and second on estimates by Culbertson and Pitman (1973, p. 500) of identified and hypothetical shale resources. Because shale oil has yet to be produced at a profit in the United States, only the highest grade deposits are here considered as reasonably assured reserves. The figure reached by the Bureau of Mines represents about 20 percent of the estimate made by Culbertson and Pitman of identified oil-shale resources in the United States from rock containing more than 25 gallons per ton. The 20 percent figure was therefore applied to the world's total given by Culbertson and Pitman (ibid., p. 501) for identified shale-oil resources to obtain a figure for reasonably assured reserves of 136 billion barrels. For the probably recoverable resources, 50 percent of the oil in all the shale deposits containing more than 25 gallons per ton and 20 percent of the oil in all shale containing

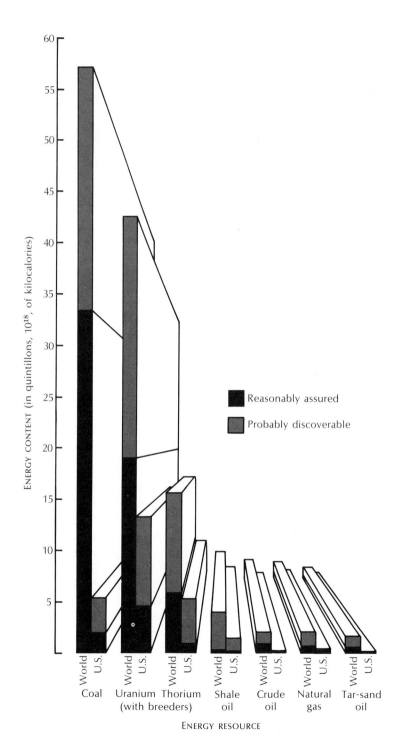

FIGURE 9.11
Although the heights of the columns in this diagram are debatable, the order of magnitude is a matter of general agreement. If present day nuclear power plants are not replaced by breeder power plants (or almost-breeders, such as the Canadian CANDU design), the uranium column will be shortened to about 3×10^{18} kcal and, if plutonium is not allowed to be recycled as fuel, to about 1×10^{18} kcal. Without breeders, the thorium column would be greatly reduced.

between 10 and 25 gallons per ton, as estimated by Culbertson and Pitman (ibid., p. 500-501), were added together; from the result, the totals for reasonably assured supplies were subtracted, giving total figures for probably recoverable amounts for the world and the United States.

Tar-sand reserves were estimated by a very crude method. The Canadian tar sands are calculated to contain 710 billion barrels of oil in place, of which about 300 billion barrels may be economically recoverable with existing and foreseeable technology. This ratio applied to the estimate by Phizackerley and Scott (1967, p. 551) of 915.2 billion barrels for the world's tar-sand resources gives a figure for the world's reasonably assured reserves. Probably discoverable reserves were obtained by adding estimates of 700 and 800 billion barrels respectively for the content of Siberian and Colombian deposits not included in the 915.2 billion total and assuming a cumulative 50 percent recovery ratio. The figure for probably discoverable reserves in the United States is that estimated by the United States Bureau of Mines (1970, p. 15).

Uranium reserve figures are from the United States Atomic Energy Commission as cited by DeCarlo and Shortt (1970, p. 277). The energy content of uranium "burned" in light-water reactors is taken as 580×10^9 Btu per ton of uranium oxide. The energy content of uranium used in breeder reactors is taken as sixty times that figure and in addition, the \$15-\$30 per pound resources are counted in the reserve base. Thorium reserve figures are those estimated by Shortt (1970, p. 207). The energy content per ton of thorium oxide (ThO_2) "converted" in a heavy-water reactor is taken as $23,200 \times 10^9$ Btu (40 percent burnup) and of ThO_2 used in breeder reactors as $34,800 \times 10^9$ Btu (60 percent burnup).

Geothermal-energy estimates are based on White's (1965, p. 14).

In print, such a tabulation appears more accurate than it is, but it does give a basis for comparing the energy potentials of the various natural energy funds. With present technology, coal is the major asset in the world energy bank, as it is also for the United States. With foreseeable technology (the breeder reactor and efficient *in situ* extraction methods for shale oil and tar-sand oil) nuclear-fuel reserves become equivalent to those of coal and the available energy in oil shale and tar sand becomes larger than that in the remaining reserves of crude oil and natural gas. The energy to be expected from geothermal traps is negligible in comparison.

When we turn to renewable energy resources, we must deal in terms of unrealized capacity or rate of availability, in other words, with power. Even calculations of diet, although we do not think of them that way, are power calculations, because they deal with rates of energy flow. Here we are no more concerned with efficiency of energy utilization than we were when calculating the reserves of nonrenewable energy, but only with potential inputs to human energy systems. For instance, we could calculate the potential for increase in the world's food-energy supply rate, basing it on certain assumptions about applications of fertilizer, effectiveness of pesticides, and genetic advances, but the result would not tell us how many people might be fed on that supply. But again, the effort is useful in obtaining perspective. Except for food, *power* sources are interchangeable. At present, water power and vegetable energy (fuel wood, dung, other wastes) are the most used in the world. Solar heating is used, but on a very small scale. With available technology, the greatest potential lies in solar heating, which could supply over 20 percent of man's energy needs. With existing and anticipated technology, rather small increments could be obtained from plants grown to provide energy, from refuse burned, gasified, or electrolyzed, and from wind. Only with provident technology can solar power, geothermal power (based on terrestrial heat flow rather than thermal traps), or fusion power provide energy flows for man's use, yet any one of the three, given a technological breakthrough, could become a major source of power.

FOOD RESERVE CAPACITY

From statistics on land use published by the Food and Agriculture Organization of the United Nations (1971), it would appear that the total amount of cropland in the world could be more than doubled, because only 44 percent of the potentially arable land is being used for crops. But that figure is very misleading. The best lands have long been used for crop production and much of the potentially arable land is in desert and tropical regions where there are serious limitations on productivity. The largest uncultivated areas are in tropical Africa and South America. The soils of tropical forests are notoriously deficient in nutrients and those of arid and semiarid regions require irrigation. Disease vectors such as the tsetse fly are a major barrier to food production in tropical Africa. Massive injections of capital would be required to achieve and sustain productivity in such lands. The oceans are often looked to as a potential reservoir of food, but the world fish harvest has ceased to expand, and there is disturbing evidence of overfishing in some of the major fisheries. At present, a potential doubling of the world's total food supply seems an optimistic estimate of the food-supply reserve potential. The problem is not that of a finite limit but the one familiar to the student of other natural resources, the exponentially-rising cost of exploiting marginal supplies. Sustained productivity of the world's marginal food lands and of the world oceans will demand not only an increasing subsidy from other energy sources but also effective worldwide controls to prevent overuse of land and overharvesting of fish and other seafood.

To increase crop yields on land now cultivated requires more fertilizer, insecticides, motorized equipment, and motor fuel. Nations that do not have sufficient indigenous supplies of inanimate energy already are finding it difficult to pay for imported fuels. Economists in the Indian government, for example, estimated in early 1974 that 80 percent of India's export earnings in that year would go to pay for imported oil; yet the exports themselves were imperilled by the rapidly rising price of the oil needed to produce and transport them.

SUPPLEMENTAL READING

Brobst, D. A. and W. P. Pratt. 1973. *United States Mineral Resources.* United States Geological Survey, professional paper 820. Washington, D.C.: United States Government Printing Office.

Information is included on coal, geothermal resources, nuclear fuels, oil and gas, and oil shale.

Campbell, Robert W. 1968. *The Economics of Soviet Oil and Gas.* Baltimore: Johns Hopkins.

Chapter 1 is an informative review of the development of the Soviet energy economy.

Engler, Robert. 1961. *The Politics of Oil.* Chicago: University of Chicago Press.

The author contends that the oil industry is a private government controlling most of the petroleum resources of the world and has "totalitarian potentials."

Jensen, W. G. 1970. *Energy and the Economy of Nations.* Henley-on-Thames, England: G. T. Foulis.

A succinct review of the economics of energy, as well as the demand and supply of it, that leads to an argument for the establishment of national and international energy policies.

Lutfi, Ashraf. 1968. *OPEC Oil.* Beirut: Middle East Research and Publication Center.

Rather brief treatment of the structure and operations of the international oil companies and of the early aims and activities of OPEC.

Mulchansingh, U. C. 1972. *Energy and the Third World.* Mona, Jamaica: University of the West Indies, Geography Department.

Based on a good deal of pertinent data, this mimeographed report sets in perspective the energy problems of the underdeveloped nations.

Rouhani, Fuad. 1971. *A History of O.P.E.C.* New York: Praeger.

The author was the first Secretary General of OPEC; this book, although somewhat outdated by recent events, is vital to an understanding of OPEC.

Slocum, Marianna. 1974. "Soviet Energy: An Internal Assessment." *Technology Review,* v. 77, n. 1, p. 17–33.

Raises questions about the magnitude of Soviet fossil-fuel reserves and shows that the costs of developing Siberian hydrocarbon resources will be high.

Tanzer, Michael. 1969. *The Political Economy of International Oil and the Underdeveloped Countries.* Boston: Beacon Press.

An excellent, documented analysis of the growth and development of the international oil companies, of the symbiosis between the major companies and their home governments, of the economics of international oil, and of the problems underdeveloped countries have with oil companies; separate chapters on India, Mexico, China, and Iraq.

United States Bureau of Mines. 1970. *Mineral Facts and Problems.* United States Bureau of Mines bulletin 650 (1970 ed.). Washington, D.C.: United States Government Printing Office.

Summaries of information on production, technology, reserves, uses and outlook for mineral commodities, including the mineral fuels.

William Tenney

Chapter 10

THE SOCIAL ECONOMY OF ENERGY USE

Work is of two kinds: first, altering the position of matter at or near the earth's surface relatively to other such matter; second, telling other people to do so. The first kind is unpleasant and illpaid; the second is pleasant and highly paid.

—Bertrand Russell, 1935

PHYSICAL, PECUNIARY, AND SOCIAL ECONOMY

The Greek *oikonomia* meant household management; the original meaning of the English *economy* was identical. Today, it is recognized that the management of man's household will decide the future of sentient life on this planet. It might therefore be useful to go back to the original meaning of economy, but to use it in a broader sense, as we now use ecology. In order to manage a household well, one needs to know its structure, its inhabitants, the resources available to it, and the cost of those resources—cost measured not in money but in work, in "life-minutes," in opportunities foregone, in wear and tear on the household, and in the diminishing availability of resources in the future. One needs to know the inhabitants—not only their physical needs but also their human wants, not only their capacity to cope with penury or to enjoy affluence but also the limits of their self-control under stress.

If the house is Earth, its resources are energy and materials. Even food is energy and materials. The only resource that comes from outside the house is solar radiation. Man's physical well-being and material comfort depend upon his access to, and use of, his resources. That part of household management that deals with the relations among the house, resources, and man is the *physical economy*. Its units of measurement are those of energy, time, mass, and distance.

The inhabitants of the house are man, other animals, and plants. Man is the manager of the household. Called a political animal by Aristotle and regarded as an economic animal by Adam Smith, man is above all a social animal. Loving, playing, striving for status, warring, engaging in ritual—all are important aspects of his life and welfare. The part of household management that deals with social values is the *social economy.* Its units of measurement include privileges and signs of status, such as uniforms, medals, office carpets, media exposure, titles, votes, prices above intrinsic values. Social costs include the costs of keeping peace within the society, of maintaining a tolerable environment, and of coping with the physical damage and health casualties of economic activities.

We must recognize a third part of household management, the *pecuniary economy.* Money (or gold) may be regarded as wealth or as a symbol of debt. Where money is regarded as wealth, its accumulation becomes a prime social goal and its possession an attribute of status and a lever in political activity. Napoleon once remarked that religion was what kept the poor from killing the rich. Cynics might say that the vision of widespread material wealth in a growth society has gradually replaced religion in serving this function. Money seeking and the use of money as the prime measure of value tend to create an economy of their own—often at odds with both the physical and the social economy.

In industrial society, the word *economy,* especially in its verb and adjective forms, *economize* and *economical,* carries the connotation of doing things the frugal way. In the context of household management, however, it would mean doing things the best way (for the members of the household), and the best way may not always be the frugal way.

The three kinds of economy are involved every day in the decisions we make about the use of energy and materials. The physical economy of energy use tells us how much energy we use, in what forms, at what efficiencies, for what purposes, and how much useful heat and work we obtain. The social economy of energy use tells us what social costs and benefits attend each use of energy and gives us a separate means of comparing alternative modes and uses. The pecuniary economy of energy tells us how much energy costs in its several forms, at different places and in varying quantities, as compared with other goods and services. The pecuniary cost may not reflect all the social costs; for example, the cost of coal commonly has not included the cost of restoring mined land. The pecuniary cost may not be consistent with physical economy; for example, large users of power may pay a lower unit cost even when they add disproportionately to the energy expense of meeting peak loads. The physical economy of energy use might suggest that we use high-sulfur coal in some power plants because it is the most efficient fuel available; but social economy will tell us that the social costs, in adverse effects on environment and health, are unacceptable.

As an illustration of the three kinds of economy, consider the choice of transportation modes for a commuter who could use his car, a municipal bus, or a bicycle. Analysis of the net propulsion efficiency (physical economy) would show the bicycle as the best economic choice. An analysis of the dollar costs and benefits (pecuniary economy) might show the car as the best choice, because it is faster than either the bicycle or the bus (time saved in transport can be put to gainful use) and in direct operating cost perhaps less than the bus fare. However, an analysis based on social costs and benefits (social economy) might suggest strongly that the consumption of nonrenewable resources and the environmental effects of air pollution, noise, and crowding would be minimized if the bus were used, and that the slight loss of the individual's time would be of no great cost to society.

Revealed in this simple illustration is the fact that the criteria and measurement of value get increasingly difficult to work with from physical through pecuniary to social economy. In physical economy, we encounter few problems that demand judgment. The main one is how to compute the energy cost of goods, such as nylons, and services, such as selling. In pecuniary economy, the major difficulty is putting monetary values on social costs, such as the loss of a family homestead beneath the waters of a reservoir, and benefits, such as the increases in literacy from public investment in education. In social economy, we face a lack of accepted units of measure as well as the enormous question of goals and standards. If social efficiency is to be measured in terms of the physical needs of mankind, the bright lights of Las Vegas reflect a social cost, but if social efficiency is to be measured in terms of freedom of individual choice, those bright lights become a social asset and vie for positive valuation with other unessential but desired goods and services.

Although it is the most difficult of the economic conceptions to work with, social economy may be the most important, for it is the measure of social adaptation to environmental constraints. In the long run, communities, societies, and nations are held together more by shared social values than by shared material interests or imposed political force.

In what follows, it will be assumed that individuals in any society share the goals of life, liberty, and the pursuit of happiness, although they may have accepted different social strategies to achieve these goals. As we shall see, such a postulate is required before one can discuss social efficiency, social benefits, and social costs.

SOCIAL BENEFITS OF ENERGY USE

Many of the social benefits of energy use were brought out in chapter 8. Longer life and better health come from more and better food energy, as well as from medical skills and measures for control of disease developed through the allocation of surplus energy to research, education, and public health services. Choice in life-style, opportunity for leisure, and release from physical or economic slavery come from energy surplus. These are primary benefits. Secondary benefits include the acquisition of knowledge, which depends upon the provision of leisure. Knowledge in turn may be used to extend the tangible benefits of energy use.

The definition and valuation of social benefits differs from one society to another. For example, consider the valuation of leisure. Sahlins (1972) argues that Stone Age hunters and gatherers should be called affluent because they were satisfied with the products of a work-day of less than four hours and were not frustrated by unfulfilled wants. They appear, as do some contemporary primitive societies, to have been indifferent to easily obtainable additional goods, perhaps because their use of resources, other than the basic food supply, served social ends exclusively. This was not a welfare-state society; the energy surplus was taken in leisure for the able-bodied; the weak, the halt, and the aged were not supported.

Another type of society in which freedom from drudgery was highly valued was that of the Benedictine and Cistercian monasteries (and of the larger populations they influenced) during the Middle Ages and the Renaissance. A number of technological advances made by the monks, especially in the use of water power, released labor and increased productivity. Lewis Mumford praises the monks for the way they used the labor saved.

> Rewarding work they kept for themselves: manuscript copying, illumination, carving. Unrewarding work they turned over to the machine: grinding, pounding, sawing. In that original discrimination they showed their intellectual superiority to many of our own contemporaries who seek to transfer both forms of work to the machine, even if the resultant life proves to be mindless and meaningless. [1967, p. 269]

All western Europe appropriated the labor-saving devices of the monks and imitated their regularity and efficiency, if not their principles of equality, chastity, sharing, and support of the infirm and elderly. The social benefit of leisure is described by Mumford.

> How great this release [from labor] was can be discovered by the number of holidays the medieval worker enjoyed. Even in backward mining communities, as late as the sixteenth century more than half the recorded days were holidays; while for Europe as a whole, the total number of holidays, including Sunday, came to 189. . . . Nothing more clearly indicates a surplus of food and human energy, if not material goods. [Ibid., p. 271]

Perhaps unfortunately, the lay populace that readily adopted or successfully imitated the monks' virtues of work, order, and regularity failed to heed the utility of continence in preserving their surplus of leisure, and gradually the holidays became workdays and work again became drudgery. Gradually, the ideal of surplus goods was substituted for the ideal of surplus time, and it became more desirable to amass wealth than to share leisure.

There are clear social benefits to a society that is able to maintain an energy surplus by the efficient use of available resources. These social benefits do not depend so much as has been imagined on the absolute rate of energy use, but more on the wants of the society and the extent to which those wants can be satisfied through the use of energy. Wherever the rate of energy availability is nearing marginal limits, the number of persons to be satisfied becomes critically important. Renaissance society rejected the Benedictine idea that the energy saved by the use of water power should be used to support prayer and spiritually rewarding (not productive) work. The great church edifices of the High Middle Ages bear eloquent testimony that medieval society had directed its energy into spiritual and aesthetic works. One of the striking social characteristics of the Renaissance was the concentration of wealth in the hands of certain families, who gradually took over from the Church the role of patron of the arts. Art then passed from being a high form of the rewarding work that was the social benefit of reducing drudgery to being a product of ostentation, paid for out of the surplus goods accumulated by a relatively few people. A new social emphasis on the production of more goods than needed, superimposed on a strong population growth rate interrupted only by two traumatic episodes of plague, reduced the labor surplus more quickly than population growth alone would have. Leisure, enjoyed by many Europeans in the High Middle Ages and the Renaissance, eventually became a privilege only the rich enjoyed—and they obtained their leisure at the expense of other people.

How our values affect our language may be illustrated by the necessity to emphasize here that the term *labor surplus* as used above does not refer to unemployed persons but to an excess of human musclepower over what was needed for a reasonably healthy existence. In modern industrialized society, unemployment is regarded as a social evil and even *underemployment* is pejorative. In an older time unemployment was a major social goal, and the maximization of underemployment a prime social strategy.

There are assumed or postulated social benefits of economic growth and exploitation of resources that may not in fact exist, at least beyond a certain stage in the industrial development of a society. Adam Smith's "invisible hand" has acquired a more deceptive form in the "doctrine of positive externalities," which holds that private production and selling will have beneficial social effects apart from the actual producing and selling. This doctrine has been the principal justification for public expenditures in support of production and trade. As Garvey emphasizes, "public intervention in the market seemed fully consistent with the prevailing economic doctrine, so long as it enriched the incentives for private selfishness (1972, p. 30)." Only recently has a feature of modern industrial society that has long disturbed biologists and resource geographers become evident to some economists: "external diseconomies"—in other words, social costs—may increase faster than economic and population growth, and positive externalities (social benefits) may become negative as growth continues.

SOCIAL COSTS OF ENERGY USE

The social costs of energy use depend to a great extent on the rate of use and the kinds of energy used. First, the basic social cost of energy use is *labor*. It takes physical labor to avail oneself even of tropical fruits and to keep on the shady side of a rock; it requires labor to hunt and fish, to plow and harvest, to herd and milk, to mine and lumber. Labor requires expenditures of time as well as of energy; time spent in labor is time that cannot be devoted to other pursuits. The principle that labor, when applied to energy resources, can be returned many times in new forms is the basis and hope of modern society.

Another social cost is *social degradation,* of which slavery is the extreme form. The use of energy produces divisions or strata in society characterized by differing degrees of control over energy. Those low in the demographic column or social pecking order may lead lives of misery. The inequitable distribution of the benefits of energy use may thus be a social cost.

Then there is *warfare,* which involves the use of energy and is widely regarded as a social cost, although one could argue that some wars may have led to social benefits or at least prevented greater social costs.

Pollution is a social cost, whether esthetic (in decreasing the enjoyment of life in a society), demographic (in decreasing the functional effectiveness of a society), pathogenic (in decreasing the physical health of a society), or ecological (in decreasing the life expectancy of a society). Ecological pollution may be the ultimate social cost of energy use, the cost that threatens the extinction of a society. Decreased life expectancy of a society may arise from the exhaustion of the energy resources on which the society depends, an overwhelming flood of disease organisms, very high-energy accidents or warfare, massive civil violence that entrains a pervasive collapse of social sanity, or irreversible deleterious changes in the life-support system.

The ecological costs of using *renewable* resources accrue partly from lack of stewardship, but mainly from overpopulation. Soils are depleted of their nutrients, land is subjected to accelerated erosion because of timber cutting and overgrazing, floods increase in virulence, reservoirs and canals silt up, fish and game disappear. As productivity becomes more a goal than a strategy, agriculture becomes more monocultural and more dependent on sources of external energy for continuous infusions of fertilizer, pesticides, soil conditioners, and power, and consequently is more vulnerable to pestilential attack and to failure of the external energy system. The inanimate portions of the life-support system, especially the transport, communication, and control networks, become more intricate and more susceptible to catastrophic disruption by sabotage, warfare, a general decrease in public sanity, or interdiction of energy supplies. Even while the system works well, future social costs will accrue from the increasing capacity to process resources beyond the ability of human populations to absorb the products and from the growth of those populations beyond the capacity of the earth and human ingenuity to provide resources to process or sinks for waste disposal.

The use of fuel wood for energy might be thought of as entailing negligible social cost, until one recalls the early environmental degradation in Cyprus, Italy, France, and England brought on by deforestation spurred by the hunt for firewood and the later need for charcoal. Today the Sahara Desert is advancing northward over land being denuded in a frantic search for fuel (it also appears to be spreading southward). The population of North Africa has grown from 5.5 million in 1900 to more than 35 million today. Animals, farmers, and fuel gatherers exert a fierce pressure on the semiarid environment; Campbell estimates (1973, p. 672) that about one kilogram of wood fuel per person is used daily and observes that woody shrubs commonly are pulled up roots and all.

In an industrial society, the direct social costs of energy *use* will outweigh the direct social costs of energy *production*, because more people are affected by the consequences of energy use than by the consequences of its production. Most of the fossil fuel produced is burned in or near cities; air pollution from the stacks of large stationary power plants and from the exhausts of automobiles and trucks is concentrated in urban areas. Waste heat discharged from such power plants may or may not constitute pollution (a social cost), but it too is concentrated in urban areas.

Esthetic degradation is a real social cost, however limited it may be to the perception of certain groups within total populations, and however its valuation may wax and wane with changes in the social environment. That offshore drilling platforms, strip mines, and high-tension lines are widely regarded in North America as esthetic disadvantages of energy use, whereas in the Soviet Union and Mexico they seem to be viewed as visible reminders of national economic progress and technological achievement, means that *here* they entail esthetic costs that *there* they do not.

Contemporary problems as well as the presentiment of future crises have raised questions about the social economy of energy use in high energy societies. Is the automobile, which uses up about one-sixth of the energy input to the American economy, a socially efficient machine as it is being used? Is agriculture in the United States, highly productive in terms of the individual agricultural worker, efficient in social terms? What is the social value of advertising? Do we have adequate mechanisms for comparing the social impact of alternative technological proposals involving energy use? Will the single-minded pursuit of pecuniary economic efficiency ultimately lead to irreversible and potentially catastrophic social costs? What can we do to control future energy use so that its social efficiency will improve?

The imminent destruction of the world's tropical rain forests, felled for timber, for slash-and-burn agriculture, and to clear spaces for other human activities, and the weakness of contemporary efforts to preserve a biological diversity in genetic banks and ecological preserves are reflections of the kinds of problems that societies based on the concept of economic efficiency seem unable to cope with—in particular, the problem of foregoing immediate, calculable, economic benefits in order to prevent the long-term accrual of incalculable, but potentially severe, social costs.

The social costs of energy use may require that we forego some indirect benefits. For example, the development of energy-conversion facilities in the interior American West, for converting the kerogen of oil shale into oil, or coal into gas or methanol, could easily and quickly use up all the unappropriated water in the region, leaving none to be used in developing other industry or supporting communities other than those necessary to the energy conversion.

"BENEFITS" OF ENERGY USE IN EMPLOYMENT

If the creation of employment opportunities is to be regarded as a social benefit, a tenet that *is* part of the conventional wisdom in the growth society, regardless of the planned or unplanned nature of the economy, the exploitation of inanimate energy resources must be regarded as a mixed blessing, for its effect is to displace labor. In all of its application, most strikingly in agriculture and transportation, inanimate energy decreases the necessity for human labor. Conventional wisdom, in contending that opportunities for productive labor are increased when inanimate power replaces manual labor, ignores the fact that beyond some optimum level, difficult to define and measure but nonetheless real, the production of goods and services becomes a social cost rather than a benefit. At that point, the maintenance of "full employment" requires the wasteful use of both human and inanimate resources and is no longer a rational social goal. The growth of the automobile industry, for example, has been sustained by planned obsolescence and the creation of wants that far exceed needs. In periods of economic retrenchment, it becomes clear that Americans can make do—perhaps for years—with the cars they already have and that the replacement market is largely artificial. Here is a trenchant illustration of the social waste engendered by a society fixated on the "need" for growth and jobs.

DISTRIBUTIONAL ANALYSIS OF SOCIAL BENEFITS AND COSTS

Social costs and benefits lend themselves to distributional analysis almost as readily as do pecuniary costs and benefits. The social benefits obtained from the intensive use of petroleum products are spread throughout the United States; but the social costs are paid by urban populations beset by air pollution and the relatively few communities affected by refinery effluents and occasional oil spills. The social benefits of burning low-sulfur coal are enjoyed mainly by the citizens of certain Midwest cities, while the social costs are felt in the northern Rocky Mountains. The social benefits of burning the high-ash coal of the Colorado plateau are enjoyed only in a few cities of New Mexico, Arizona, Utah, and southern California, while the social costs are felt in the areas where the mines and power plants are.

The distribution of the costs and benefits, both social and pecuniary, from the construction of public hydroelectric power projects long has been a subject of study and controversy. Particularly questioned has been the investment of federal funds in projects whose benefits accrue mainly to the people in the areas to which the generated power is distributed. The prevailing hypothesis has been that the social benefits from an injection of cheap power in one part of the nation, especially if that part's social health, measured by income, educational levels, and the availability of health services, has fallen below the norm, will spread throughout the regions in which the projects are situated. Such federal investment has been regarded, in other words, as a necessary form of insurance against social decay that, even though apparently localized, would weaken the national social organism. So few objective studies of the effects of large public power projects have been made that any generalization is impossible, but it appears likely that some large regional investments such as the Tennessee Valley Authority and Bonneville Power Administration have achieved the social goal, whereas many smaller projects have provided little more than a temporary infusion of construction money into a depressed area.

The distributional analysis of social costs and benefits is not skewed in the time dimension by discount rates that cause today's benefit to be valued more highly than tomorrow's as is the analysis of pecuniary costs and benefits. For this reason alone, a social economic analysis provides a much better basis for forecasting, planning, and decisions about the welfare of society than does pecuniary economic analysis. A method of making decisions in which the discount rate of expected returns or benefits is increased according to arbitrary estimates of the instability of the social structure in question may contribute to that instability, so that the estimates of instability may become self-fulfilling prophecies because the pecuniary economic analysis led to decisions that decreased social stability. For example, the strategy of retail merchants to charge higher prices in a ghetto because of an estimated higher risk, or the strategy of an oil company to demand a quicker return from operations in a country deemed unstable for investment purposes, may augment the very risk or instability that the higher discount rate implied in those decisions was meant to counter.

THE SOCIAL COSTS OF FOOD PRODUCTION

The social benefit of an adequate supply of food is health. But a system that delivers an adequate supply of food, especially a highly mechanized system, may incur significant social costs. Quantities of chemicals may be applied to tilled land in order to restore soil nutrients and guard against pests. Some of these additions may seep into groundwater or be washed into surface streams, causing water pollution. Poor tillage and overgrazing may lead to soil erosion. In some areas irrigated agriculture is depleting groundwater that cannot be restored in any period of time meaningful to man. Feeding grain and legumes to livestock to produce food for humans is extravagant where an alternative use of the feed or the land on which it is grown would yield a social benefit and causes an enormous production of animal wastes (one cow or steer produces as much sewage as approximately thirteen humans), which, when concentrated in feedlots, poses a vexing problem of disposal and pollution. The unnecessary use of energy, as in many frozen foods and in open refrigerators, is a social cost when energy is scarce, or whenever nonrenewable resources supply the energy used. Chemical additives designed to make food taste better, last longer, or look nicer may involve hidden health costs. A recent technological triumph, the non-food food, designed to sell but not to nourish, has been called an incentive to malnutrition. A system that makes a wide variety of food available to most of the members of a society, especially if it is a high energy society, allows, if it does not promote, poor habits of eating and the related social costs.

Finally agriculture, especially mechanized agriculture, simplifies ecosystems; it creates vast fields covered by one strain of one species of a food crop bred by man. Such monoculture is highly vulnerable to small variations in climate and to prolonged anomalies of weather. The Irish potato famine, in which 1.5 million persons starved to death, was a vivid illustration of the magnitude of the social cost of reliance on a simple agricultural ecosystem.

SOCIAL BENEFITS AND COSTS OF PETROLEUM

The social benefits and costs of petroleum will be discussed in some detail because it is now the crucial nonrenewable world resource; its control and use will be major factors in world economics and politics for the remainder of this century. The social benefits of petroleum are twofold. First, it represents a great natural subsidy, because it can be found and extracted from the ground with a very minor expenditure of energy compared to the energy it contains. A barrel of crude oil that contains almost 1.5 million kcal of recoverable energy may cost less than 1 percent of that to get out of the ground. Although 5 percent may be expended in transporting the oil to the country in which it is to be used and 10 to 11 percent will be used in refining, about 80 percent will remain to be used. The second benefit is petroleum's versatility. Not only are petroleum products easy to transport, but also they can be used almost anywhere, in a wide range of converters, for a great many purposes.

Whatever the profits of the oil companies, their products were a cheap source of energy to the industrial countries. The amazing rebound of western Europe after the destruction of World War II and the astonishing emergence of Japan as one of the world's industrial giants were due mainly to the cheap fuel that subsidized the rebuilding and expansion of national industrial systems. The postwar industrial boom was powered by the natural subsidy of petroleum, much of it coming from countries that participated only marginally in the economic advances of the industrial world.

In the planned economies and in Mexico the natural subsidy of oil has been converted to social benefits by government direction. There is a range of ways in which the natural subsidy can be allocated. All sectors of the economy may be provided with petroleum products at cost; one or more sectors whose rapid development is considered essential may be provided with petroleum products *below* cost while the other sectors are charged more; or all sectors can be charged more than the cost in order to create a monetary surplus to be allocated for specified social needs. Part of the subsidy may be distributed at the source by deferring the mechanization of petroleum production and distribution in order to provide jobs for more workers. In any case, the entire natural subsidy from indigenous oil is available to the producing country. Any surplus oil can be sold abroad for economic or political advantage or held in the ground for future needs. The possibility of curtailing production of a depleting natural resource in order to obtain its social benefits for a long time is perhaps the most important difference between public and private exploitation of nonrenewable resources. In private exploitation, the future value of a resource in the ground is strongly discounted; the result is a heavy pressure for immediate production and as much profit as soon as possible.

The versatility of petroleum is a great advantage in the industrialization of a country and in creating the possibility of improved lifestyle. It can fuel trucks, tractors, and small electrical plants and thus link remote agricultural villages to trade and market centers, increase the harvest of crops, and bring the benefits of electrification and education to rural regions. Compared with oil, coal is cumbersome and hydroelectricity requires too much investment.

The social *costs* of petroleum begin with the fish killed by explosions during seismic prospecting and end with the exhaustion of resources; they include ecological damage from well blowouts, tanker spills, and pipeline leaks; health damage from air pollution, largely by automobiles and trucks, but also from refineries and furnaces; the waste of associated natural gas by flaring and the loss of future benefits by inefficient extraction and wasteful use; the costs of paying tribute in times of scarcity, real or artificial, to foreign producing nations, to oil companies, or to both; the esthetic degradation of oil fields and cities; the ecological damage of waste heat; and the potential unquantifiable costs of the accumulation in the atmosphere of carbon dioxide, and of population growth rates and living standards based on the quicksand of a nonrenewable resource.

SOCIAL ANALYSIS OF THE AUTOMOBILE

Throughout the world, the automobile is a symbol of economic achievement and personal freedom. Its social benefits include those of mobility. Because of it larger cities are possible, with resulting economies of labor division and scale. Dispersion of retail centers and factories is possible, with advantages of location and transport efficiencies, as well as pollution abatement. A wider choice of where to go and what to do is available to individuals. The automobile also has psychological benefits, in pride of ownership, the feeling of freedom, and the sense of power. The car is also an outlet (not without social costs) for displacement behavior—the frustrations of office or family can be relieved behind the wheel.

Economies of scale in primary distribution networks for goods and services can be achieved where all consumers have access to automobiles and can go to Supermarket Plaza. Urban sprawl is made possible by the automobile. Yet, if it did not bring great social benefits, it would not have become such a pervasive feature of the high energy landscape. Although the *physical* economy of the total distribution network, when one considers the inefficient use of myriads of automobiles, leaves a great deal to be desired, its social efficiency is hard to deny—it allows people who work in a city to live in the quasi country (where the air is cleaner, the nights are quieter, and the threat of violence dimmer), and it gives them (generally) more choice and greater convenience than they would find in the city, as well as a headstart for the beach or the mountains on weekends. Again we have a contrast between conclusions drawn from different kinds of economics. Physical economics might say "Live in the city." Social economics says "Live in suburbia." The answer from pecuniary economics will depend upon the cost of energy.

The social costs of the automobile include the psychological and physical attrition of those who use it under stress; the amenity costs of highways, freeways, and flatworm communities strung out along sleazy main streets; the shortened and damaged lives of those killed and injured in automobile accidents; the diversion to the automobile of money that might otherwise be available for more socially valuable goods or services; the imponderable costs of the rapid depletion of nonrenewable resources, especially those of energy; the cost of dependence on energy and other resources in foreign hands; costs of air pollution from automobile exhausts and water pollution from refineries and tanker spills; and the social cost of the automobile's considerable contribution to the weakening or displacement of the family as the fundamental brick in the social edifice.

Calculation of the social efficiency of the automobile involves weighting all these benefits and costs. One would expect almost as many different answers as there were people making the calculation. Society, however, bypasses formal value weights and calculations, making them implicitly by its decisions in the political arena and the market place.

The interaction of social economy and pecuniary economy is well illustrated by the automobile. In the United States the automobile is a necessity. When its social costs are perceived as being too high, it is the machine itself that is tinkered with. The alternative solution of replacing the automobile as a means of transport involves unacceptable social as well as very large pecuniary costs. Consequently, the machine is adjusted to reduce its casualty costs (seat belts and padded dashboards are added), its pollution costs (emission control devices are installed), and its resource costs (it is made smaller). Its use is curtailed to some extent by allowing or imposing higher fuel prices and by subsidizing mass transit systems, but both adjustments entail social costs of their own that limit their acceptability. The financial burden of reducing a social cost is difficult to allocate equitably or successfully through the market mechanism. Attempts to reduce the social cost of the automobile by increasing the price of the car and its fuel may improve the urban environment and conserve scarce resources, but the burden falls much harder on the poor than on the rich; accordingly, there is less pressure to clean up the urban air and motor fuel is placed under price controls.

The prospect that the internal-combustion engine fueled by gasoline fairly soon will make the automobile an unacceptable urban citizen has led to several studies of the economic impact of conversion to an automobile that does not pollute. Not unexpectedly, much of the impact turns out to be a social cost expressed in dollars—the cost of retraining mechanics. Ayres estimates (1969, p. 28) that 1.5 million mechanics would need to be retrained at an approximate cost of $10,000 each—a total of $15 billion.

SOCIAL BENEFITS AND COSTS OF COAL

Coal is the drudge of the energy world. Drab and unloved, coal powered the Industrial Revolution and provided the affluence that led to the growth of technology and to the discovery of more versatile and cleaner energy systems based on other energy resources. Today, in those countries that have large coal reserves, the unglamorous fuel is being reappraised and is seen to have great potential for contributing to the needs of high energy society. Because of the work of mining, mechanical, chemical, electrical, and combustion engineers, coal is potentially much more versatile than it was when it was the mainstay of the industrializing world. Coal now can be converted into gaseous or liquid fuels that can be transported by pipeline and tanker or used near the conversion plants. It can be burned in power plants at the mine to produce electricity that can be transmitted hundreds of miles over ultra-high voltage lines. It can be hauled in unit trains to be burned in distant power plants. The social benefits still to be derived from coal in North America are enormous. It can provide for the maintenance of a high energy society as domestic reserves of oil and gas are gradually exhausted and as the effort is intensified to develop adequate alternatives to nonrenewable energy resources.

The social costs of using coal are many. Some, because of their visibility, tend to be exaggerated. Others, because they are not obvious, tend to be overlooked. All can be reduced by thoughtful planning and careful conservation measures.

The social costs of underground coal mining include subsidence, acid mine drainage, and fires. Subsidence is caused when rock collapses into abandoned mine openings. Acid mine drainage occurs where groundwater, flowing through the shattered coal that remains underground after mining ceases, takes sulfur into solution from pyrite in the coal and leaves the abandoned mine as dilute sulfuric acid, toxic to plants and fish. Uncontrolled fires, both underground in abandoned mines and in surface waste piles, damage health, property, and the environment. The Bureau of Mines reported 110 underground fires and 460 burning coal-refuse banks in 7 states in 1969; the great majority were in Pennsylvania and West Virginia. Accumulations of carbon monoxide in buildings are a health hazard from mine fires. As lower-grade coal is mined and higher-grade coal is required for burning, a growing percentage of the total material hauled out of the mine is being removed as waste and piled near the mine.

The social costs of strip-mining coal can range from severe and long-lasting to mild and transient. In hilly country, unreclaimed strip-mined areas can suffer heavy erosion, leading to long delays in the reestablishment of vegetative cover and to downstream siltation and acidification. In the flat lands of Illinois and Indiana, fully reclaimed mined lands support heavier crops than they did before mining, because of the enrichment of the restored topsoil by nutrients from below. Strip mines in Texas are being restored to full agricultural productivity within a year after mining ceases. The average annual rainfall and the steepness of slopes are the major factors affecting the reclaimability of strip-mined lands. For some of the semiarid Rocky Mountain areas underlain by coal and oil shale sure means of restoring the vegetative cover have yet to be found. The social costs of strip mining in Appalachia are not likely to be the same as those in the Rocky Mountains. In Appalachia rainfall is high, but steep slopes are common and much of the coal has a high sulfur content. Much of the coal stripping in Appalachia has been marginally economic and the imposition of reclamation costs would have put some operators out of business. In the Rocky Mountain coal areas that are suitable for stripping rainfall is low, but steep slopes are not common and the coal has a low sulfur content. In the Rockies, the present and projected stripping operations will be little affected if reclamation costs run to several thousand dollars an acre. A coal seam 10 feet thick contains 14,000 tons per acre; a reclamation cost of $3,500 per acre, which may be sufficient for complete reclamation, will add 25 cents per ton to the cost of coal that sells in Chicago for between $12 and $15 a ton and that has a value of $3 or $4 a ton at the mine.

The social costs of coal mining include the costs to the miners of accidents or disease, such as "black lung," which affects only underground coal miners; the costs of living in dingy, poor communities, and of being subject to the shifting markets for coal as well as to the economic limits of the seams. Between 1965 and 1972, 1,412 lives were lost in underground coal mining in the United States, a rate of 0.61 deaths per million tons of coal, more than five times greater than the death rate for surface or strip-mining of coal. In addition, much more coal is left behind in underground mining (an average of 43 percent), probably never to be recovered. If the environmental impact of strip mining can be kept temporary, there is clearly a social profit in mining coal that way.

SOCIAL BENEFITS AND COSTS OF NUCLEAR POWER

The social benefits and costs of atomic energy or nuclear power have received much more public attention in recent years than those of any other form of energy. A brief review here may be helpful as a prelude to the discussion of energy decision making in chapter 12. Excluded from this discussion will be the military uses of nuclear energy.

The social benefits of nuclear power include replacement or supplementation of depleted fossil-fuel resources, cessation of the air and water pollution caused in the fossil-fuel cycle of electricity generating plants, and provision of electricity for nations and regions that have little or no access to cheap fossil fuels. For some countries, the benefits can be summed up as "clean" power in large quantities at reasonable cost from domestic resources. Nuclear power plants produce no carbon dioxide, and thus do not contribute to the accumulation in the atmosphere of that gas, regarded by some scientists as a long-range threat to life on earth.

The social costs of nuclear power include the hazard of a catastrophic reactor accident (technically not a risk, unless it can be quantified), thermal pollution from power plants, radioactive pollution from power plants and fuel-reprocessing plants, perpetual surveillance of high-level radioactive waste materials, and "the plutonium peril." The hazards to human health and well-being are somatic (directly and immediately affecting the individuals exposed to damaging doses of radioactivity), carcinogenic (causing cancer in exposed individuals after a delay of between five and twenty years), and genetic (affecting adversely the descendants of exposed individuals).

Any exposure to radioactivity, whether natural background or manmade, is harmful to humans. Background radiation, low at sea level on or near the ocean and high in elevated regions particularly if the rocks are granitic, may have little or no somatic or direct effect on exposed individuals but almost certainly is responsible, by induced mutagenesis, for some "natural" deformities and cancers. In contemporary high energy society, by far the largest exposure to harmful radiation comes from the medical use of X rays; then comes natural or background radiation and, finally, a very minor source, nuclear power. Short of widespread nuclear warfare, this division of sources probably will be maintained into the next century. The possibilities of reducing medical exposures far outweigh the probable increases in exposure from the nuclear power industry, so that any increases in the latter could more than be compensated for by reduction in the former.

The proliferation of nuclear power plants and their satellite plants for fuel enrichment and fabrication and for reprocessing the spent fuel carries, however, the possibility of increasing local levels of radioactivity that could range from a small regional increase in the incidence of cancer and birth defects to a catastrophic level. In addition, the small quantities of dangerously radioactive wastes now produced can be expected to increase with the growth of nuclear power. Since these waste materials must be kept out of the biological environment for between six hundred and a thousand years, the problem of perpetual surveillance or secure disposal is serious.

The most efficient fuel for the breeder reactor, which carries the burden of hope for a vast extension of the energy resources available to man in the next half century, is plutonium, the fuel that is produced or "bred" in a breeder reactor that uses uranium as its primary fuel. Plutonium not only is one of the most toxic substances known but has a half-life of about twenty-four thousand years, which means that any particular packet or vault containing plutonium will be hazardous for half a million years, even if no more plutonium is added. With about five kilograms (eleven pounds) of plutonium and some easily available technological knowledge and apparatus, an atomic bomb can be made. If breeder power plants based on the uranium fuel cycle proliferate, there will be rapidly growing quantities of plutonium in transit between the fuel-reprocessing, fuel-element fabrication, and breeder power plants. The possibility that enough plutonium to make a bomb could be diverted clandestinely from this transport system by a group bent on blackmail of previously unimaginable scale constitutes what has been called "the plutonium peril." Proliferation of breeder and almost-breeder power plants based on a thorium fuel cycle would create the same hazard, uranium 233 being usable in weapons.

Safe and "permanent" disposal of radioactive waste involves a social cost that cannot yet be measured. No permanent repository for highly radioactive waste has been constructed in any country now accumulating such wastes. Proposals have been made for disposal in the deep ocean, under the Antarctic ice cap, in outer space, and in natural underground salt formations. The last-named has received the greatest attention in the United States, where a permanent repository in Kansas was identified but abandoned when the site (not the geologic environment) was shown to be defective. At present, the disposal problem appears less serious and the cost of safe disposal less onerous than the ecological and social problems involved in other portions of the nuclear-fuel cycle.

SOCIAL BENEFITS AND COSTS
OF HYDROELECTRIC POWER

The social benefits of hydroelectric power were evangelic visions of the first half of the twentieth century. They were assumed and proclaimed, both in the United States and the Soviet Union. That the reality was somewhat inferior to the vision should in no way obscure the usefulness of the vision. In the United States, hydroelectric power was a vital part of successful social experiments such as the Tennessee Valley Authority and the Bonneville Power Administration. Subsidized by all citizens, cheap and clean hydroelectric power revitalized economically stagnant areas and brought better living to the citizens of those areas. In the Soviet Union, hydroelectric power has aided the development of resources and occupation in Siberia and can be vital to the electrification of the European part of the Union should coal supplies start to fail. In neither nation, however, does hydroelectric power now supply as much as 5 percent of the total energy needs.

The social costs of hydroelectric power are, with two exceptions, minor:

1. Dams and reservoirs cover land that might be used for agriculture or other purposes.
2. Water is wasted by increased evaporation.
3. The weight of the water in a dam may trigger local earthquakes.
4. Rivers no longer flood and deposit fertile silt and the river channel downstream from a dam is more susceptible to erosion.
5. Wild life is displaced and salmon runs depleted.
6. Phreatophytic plants and disease vectors (of, for instance, schistosomiasis) are encouraged.
7. The dam will eventually be silted up and useless.
8. The electricity transmission lines occupy land and may disfigure the landscape.

Reservoirs in arid regions cause serious water losses by increased evaporation. The annual evaporation from Lake Powell and Lake Mead, for example, is more than 10 percent of the total flow of the Colorado River. In an arid region, such as the American Southwest, evaporation reduces the amount of water available for agriculture, mining, manufacturing, and people, but is not charged for in the price of the power produced at Hoover Dam. The second important cost, and in the course of time the major one, of hydroelectric power is the cost of the depletion and exhaustion of the power potential of dams as they fill with silt. Water mills use falling water as a renewable, inexhaustible resource; hydroelectric dams and power plants do not. A feasible means of desilting clogged reservoirs may be developed. But the cost of the desilting is likely to be large, no provision is made in present pricing practice to accumulate a fund to defray that cost, and thus that cost—or alternatively, the cost of developing an alternate power source—is a deferred social cost, deferred by the generations reaping the benefits of "cheap" hydroelectric power to the generations who will inherit the hydraulic ruins.

ADJUSTING SOCIAL EFFICIENCY TO RESOURCE-USE EFFICIENCY

Social efficiency sometimes accords with the physical efficiency of energy use and sometimes does not. In an economy of scarcity, the two tend to converge; for example, social efficiency in a low energy society may require that all cultivated plant food be reserved for human consumption and not fed to animals. Social efficiency, in the sense of supporting a given culture, also may lead to the inefficient use of animals—even in a low energy society—because of religious beliefs and tabus. In an economy of abundance, social efficiency may diverge greatly from resource efficiency. For example, large amounts of meat, thermal electricity, and gasoline may be produced and consumed in ways that represent low resource efficiencies but that are socially efficient in a high energy society.

Mechanized agriculture, in which crops and livestock production are subsidized by the injection of inanimate energy, reduces the number of people who can live directly off the land. While the supply of inanimate energy increases faster than population, and while agricultural production from the land available can yet be increased by greater use of fertilizer and pesticides, by the development of highly productive plants and animals, and by more effective use of farm power, there is a net social gain from mechanized agriculture. But where or when population increases faster than the energy supply or than total agricultural production, which is subject to the law of diminishing returns, agricultural production is apt to decrease and the social costs of mechanized agriculture begin to outweigh its social benefits. Where there is widespread unemployment, it may be more efficient socially to use idle hands productively than to use machines efficiently.

A worldwide problem in adjusting social efficiency to resource efficiency has long been exemplified by the relation of upstream producer to downstream consumer. Because of the physical economies inherent in streams and other bodies of water for transport and power, agricultural, forest and mineral raw materials may be produced in the upper reaches of river basins, but rarely (the upper Mississippi basin is an exception) are they processed or consumed there. The river subsidizes the delivery of cheap raw materials to processors downstream and the ocean subsidizes the delivery of raw materials and finished products to consumers overseas. Economic and political power is concentrated in the communities downstream, which then tend to exploit the producers upstream. River basins in many parts of the world have thus become arenas of economic and political conflict between these two groups; in some cases, cultural differences exacerbate the dissension created by conflicting economic interests.

As the industrial world developed, the producers upstream came to be represented by economic and political colonies, while the consumers downstream were the industrializing "home" countries. A contemporary analogy of the problem is seen in the international flow of crude oil. The underdeveloped oil-exporting countries represent the producers upstream; the highly developed oil-importing countries the consumers downstream. Social efficiency as a goal, whether we are considering the economy of a single river basin or of the entire world, would require resolution of this conflict, so that the ultimate social benefits of the resources upstream are conserved and shared among all those who contribute to their value and not preempted by the consumers downstream.

When resources are abundant the consumers downstream commonly find it easy to control the benefits of resource use, especially if the resource can be obtained from any of several producers upstream. When resources are scarce the shoe may be on the other foot. A transfer of power may occur even during the abundance of resources if the major producers upstream can form an effective cartel and can withstand a boycott by consumers better than the consumers downstream can withstand an embargo by producers. This kind of power transition occurred in the international oil trade between 1971 and 1974. Two kinds of nations may be hurt very badly in such a power exchange: the developed nation that depends mainly on imported oil to fuel its economy and maintain its social system and living standard (Japan is a striking example) and the underdeveloped nation that depends heavily on cheap imported energy for its release from the low energy treadmill (India is an example). Although oil deliveries to such countries may not be shut off, the rise in price that accompanies an effective cartel and is a mechanism employed by the producers for enforced sharing of the benefits of oil consumption can reverse abruptly the economic and social advance of both groups of nations. In the long run, such adverse effects may be seen as previews triggered by man of what may happen, more slowly, to all nations that persist in building their economy and their society on nonrenewable energy resources.

The social efficiency of using coal to generate electric power if nuclear fuels, natural gas, or imported oil are available is questionable. A comparison made by the Council on Environmental Quality (see table 10.1) shows that coal causes by far the most harm to the environment. Even so, coal may yet be able to compete with expensive oil and nuclear power if the effects on the environment of mining the coal can be minimized at some reasonable financial cost.

Table 10.1 Comparative Environmental Impact of 1,000-Megawatt Electric Energy Systems Operating at a 0.75 Load Factor With Low Levels of Environmental Controls or With Generally Prevailing Controls

System	Air emissions			Water discharges				Solid waste			Land use		Occupational health		Potential for large-scale disaster
	Tons ($\times 10^3$)	Curies ($\times 10^3$)	Severity	Tons ($\times 10^3$)	Curies ($\times 10^3$)	Btu's ($\times 10^{13}$)	Severity	Tons ($\times 10^3$)	Curies ($\times 10^4$)	Severity	Acres ($\times 10^3$)	Severity	Deaths	Workdays lost ($\times 10^3$)	
Coal Deep-mined	383	–	5	7.33	–	3.05	5	602	–	3	29.4	3	4.00	8.77	Sudden subsidence in urban areas, mine accidents
Surface-mined	383	–	5	40.5	–	3.05	5	3,267	–	5	34.3	5	2.64	3.09	Landslides
Oil Onshore	158.4	–	3	5.99	–	3.05	3	n.a.	–	1	20.7	2	.35	3.61	Massive spill on land from blowout or pipeline rupture
Offshore	158.4	–	3	6.07	–	3.05	4	n.a.	–	1	17.8	1	.35	3.61	Massive spill on water from blowout or pipeline rupture
Imports	70.6	–	2	2.52	–	3.05	4	n.a.	–	1	17.4	1	.06	.69	Massive oil spill from tanker accident
Natural gas	24.1	–	1	.81	–	3.05	2	–	–	0	20.8	2	.20	1.99	Pipeline explosion
Nuclear	–	489	1	21.3	2.68	5.29	3	2,620	1.4	4	19.1	2	.15	.27	Core meltdown, radiological health accidents

NOTES: Severity rating key: 5 = serious, 4 = significant, 3 = moderate, 2 = small, 1 = negligible, 0 = none.
n.a. = not available

SOURCE: Reprinted from The Council on Environmental Quality, *Energy and the Environment–Electric Power* (Washington, D.C.: United States Government Printing Office, 1973), table 3, p. 14.

THE EXPONENTIAL IMPERATIVE
IN SOCIAL ECONOMICS

The exponential imperative, which was discussed in chapter 5 as a major constraint on the achievement of absolute objectives such as perfectly clean air and water, cheap power for everyone, and the indefinite use of nonrenewable resources, plays a similar role in social economics. For instance, a social cost of energy use that may be minor and acceptable when little energy is used, may become major and unacceptable when much is used. The expedient methods of disposing of radioactive wastes from the early nuclear power and fuel-reprocessing plants illustrate this point. As nuclear facilities proliferate such disposal practices can no longer be condoned because their potential social cost rises exponentially with each increment of waste delivered to the biosphere.

Surface mining in the United States may offer another example. Rapidly increasing demand for coal and industrial and agricultural minerals, coupled with the more rapid advance in the efficient technology of surface mining than of underground mining, is causing surface mining to increase at such a pace that it may be four times as intensive in the year 2000 as it was in 1972, when for the first time more coal was produced from strip mines than from underground mines in the United States. If the present rate of increase continues, by 2000 some twelve hundred square miles of land surface will be torn up each year to get at the valuable material beneath and between 1972 and 2000 over twenty thousand square miles, an area very nearly the size of Massachusetts, Connecticut, and New Jersey combined, will have been disturbed by surface mining. Whether or not the related social costs will be outweighed by the social benefits derived from the mined materials is not here the question; that the costs are increasing exponentially is difficult to deny. Still another example is urban air pollution, which has grown exponentially because of the synergistic effect of a population implosion into urban areas, a rate of growth of the automobile population that is double that of the human population, and the increasing rate at which people in the cities are using their cars.

In a democratic society with a market economy, government subsidies and operational programs are major ways of trying to obtain social efficiency while allowing economic efficiency to dominate private decisions. This system emphasizes the deferral rather than the prevention of social costs and often results in the costs being assessed to a group rather different from the one that benefited from whatever engendered the costs. Subsidies have stimulated the development of transportation networks, rural mail delivery and electrification, agricultural production, exploration for minerals and their production, flood control, and hydroelectric power production. One of these subsidies is known as the depletion allowance.

THE SOCIAL ECONOMY OF
THE DEPLETION ALLOWANCE

The depletion allowance (no longer available to large oil companies) is a tax provision about which much has been said but little is known. It has been called a tax break for oil companies, obtained and maintained by the wealth and political muscle of this industry. That charge is misleading because a depletion allowance applies to all mineral production, even—under certain circumstances—to groundwater. It has been claimed that the depletion allowance has allowed individuals and corporations to get by without paying any federal taxes. That claim is untrue because the depletion deduction cannot exceed 50 percent of taxable income. It is stated that the depletion allowance is simply a subsidized incentive for minimal exploration and production. That statement is untrue because depletion has been defended, since its incorporation in the first federal income tax law in 1913, as a return of capital, similar to depreciation except that, in the case of mineral resources, the value of the wasting asset can never be determined accurately until the last ton of ore or barrel of oil is out of the ground. It is argued that the financial effect of the allowance is to enrich producers at the expense of consumers. That argument is debatable because much of the tax saving to mineral and oil producers in an economy where they compete for the available market is passed through to the consumer, so that the major pecuniary effect may be to lower the price of gasoline and other products of mined materials.

What is undebatable about the depletion allowance is that it has had the *effect* of stimulating exploration and production of mineral resources, including oil and gas. It is a bonus for success in finding and exploiting nonrenewable resources. It can be obtained only by producing those resources and selling them at a profit. The higher the profit, the greater the bonus. The depletion allowance encourages the exploration and discovery of mineral deposits (a social benefit), accelerates the production and depletion of those resources (a social benefit or cost depending on the stage of economic development of the nation and the degree of physical depletion of the resources), and rewards price fixing if that can be achieved (a financial cost to the consumer and a social cost to the nation because it works against any sharing of the social benefits of the natural subsidy).

If one puts aside the return-of-capital argument and looks at the depletion allowance simply in terms of social economy, one sees that its retention during a period of scarcity that drives up prices of energy resources can be justified only if it is believed that an incentive, in addition to high prices, is needed to find more reserves quickly and if it is reasonable to assume that large undiscovered, economically recoverable "reserves" actually exist.

BENEFIT-COST TRADEOFFS

There is no such thing as a free social benefit. The traditional "free goods" of the economist—air, water, soil fertility—are no longer free, if they ever were. A high energy economy based on fossil fuels is in fundamental conflict with an ideal of clean air, clean water, undisturbed scenery, and the maintenance of fragile or complex ecosystems. We can have such environmental amenities only by paying for them through increased energy costs in goods and services, through decreased energy consumption by the individual, through increased taxes to pay those environmental and social costs that are not included in the prices of goods and services, or through some combination of the three. All these consequences decrease disposable income and lower the material standard of living.

It may be technologically feasible to mine by underground methods all the coal needed to fuel power plants and to make coke and to remove almost all coal's impurities before burning it. It may be possible to design and reroute oil pipelines and tankers so that the danger of oil spills is reduced to insignificance and to replace all gasoline, kerosine, and diesel-powered vehicles by vehicles powered by steam, hydrogen, methanol, or electricity. It might even be theo-

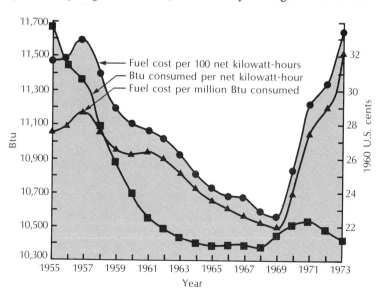

FIGURE 10.1
Fuel efficiency and fuel cost in the electric utility industry of the United States, 1960–1973.

The Social Economy of Energy Use

retically possible to restructure our cities so that all travel into and within them would be by a clean and technically efficient system of mass transport. But to achieve such possibilities would be so costly that our material standard of living would be lowered greatly. In fact, our average standard of living already is being lowered by substantial increases in food and other energy costs, due in part to measures to protect the environment.

Concern for the quality of the air has sharply increased the cost of electricity because of a scarcity of fuel acceptable for burning under new air quality standards, increases in transport costs for fuel (because in many instances it must now come from much farther away), the cost of removing most of the sulfur from residual fuel oil made from "sour" crude, and the cost of intensified stackgas cleaning. The demand for clean-burning natural gas has been rising while reserves have been falling to the extent that some utility plants have had to switch to fuel oil and thus to incur painful cost increases. Concern for aquatic ecology has helped raise the cost of electricity by forcing the use of cooling towers for water from power plant condensers. Concern for the environmental health and safety of coal miners has reduced the productivity of underground mining and contributed to the rise in the cost of coal used by power plants.

Without the continuing subsidy of natural-gas consumers by natural-gas producers, imposed through federal control of wellhead prices on interstate gas, electricity prices would be rising even faster.

Concern for air quality has been increasing the cost of automotive transport by decreasing the efficiency of automobiles, increasing the cost and weight of the vehicle, and increasing the refining and distribution costs of producing low-lead and lead-free gasoline. The number of car-miles per constant dollar obtainable by the American citizen driving a 1974 model was 30 percent below the 1967 average. To a society that had regarded the automobile as an almost living symbol of the standard of living of the family, the community, and the nation, that had interpreted "the land of the free" to mean the land where almost everyone had an automobile, this decrease in personal transport per dollar was indeed a decrease in living standard.

The demand for a safer, cleaner, and less disturbed environment has raised the cost of energy in the United States, thus adding a constraint on our ability to obtain energy. The exponential imperative will prevent us ever attaining absolute safety, absolute cleanliness, or absolute ecological tranquility. In trying to get the last bit of sulfur oxide or particle of fly ash out of stackgas, in trying to remove the last small doubt about the safety of power reactors, or in trying to drill wells or dig mines without making holes in the ground, we come to a point where the work expended for each increment of improvement, having grown exponentially, becomes very large when measured against the incremental value of the product; in physical or economic terms, it will never be practicable to go the last percentage point, to be absolutely clean, safe, and tranquil.

The same considerations apply to food production. We put almost ten units of man-directed energy into the food production system in the United States for each unit of food energy placed on the table. Like automotive transport, the food production system in this country is inefficient and has been growing more so. Not only the monetary costs but also the social costs are rising. Reducing the social costs by banning persistent pesticides, by measures to protect innocent wildlife from broadcast poisons, by decreasing the linearity of fields and increasing the biological diversity of crops inevitably raises the monetary cost of farm products. About 8 percent of the cost in the United States of putting food on the table represents inanimate energy expended in the food system. As energy costs rise, partly in response to reductions in social costs, they add to food prices. This percentage is increasing, as more agricultural chemicals are used per unit of farm production and as more electricity is required in food processing and storage. As prices rise, the consumer makes substitutions. Not all kinds of food are equally expensive. Protein costs more than fat. Meat protein costs more than vegetable protein. Frozen foods cost more than canned foods. Protein is necessary; frozen foods are not. With rising food-energy prices, a rejection of expensive protein and frozen foods reflects choices made by consumers between convenience and taste on the one hand and cost and health on the other. These are the kinds of decisions that have to be made by governments in trying to adjust the conditions of energy use so that reasonable choices between environmental sanity and the quality of human life, between the social costs and the social benefits of energy use, will be encouraged rather than prevented.

THREE VARIABLES AND THE THREE ECONOMIES

The social balance sheet on energy for any human community depends ultimately on the skill with which the interaction between people and available energy is managed. Energy use, environmental sanity, and the quality of life are interdependent variables. Each variable has at least one real limit. The limit on energy use is a maximum, measured as the rate of availability of useful energy that can be sustained during any chosen time span. The limit on environmental sanity can be thought of also as a maximum, either of pathogenic pollution or of unleashed disease organisms, but in either case determined by the ability of the human species to persist. The quality of life, too, has a maximum and that is a number of persons—determined either as the greatest number that can be supported by the available food supply or as the greatest number that the social mechanism can tolerate before it destroys itself.

No static analysis of the balance of energy, environment, and happiness can be more than a still photograph from an old movie. The elements are dynamic and constantly changing, more so in those societies that regard change as progress than in those that regard change as hazardous. Rates of change and their directions or senses in relation to the limits of the interdependence are the significant elements in any useful analysis of it. There is now a strong trend toward including social costs in the prices of goods and services and trying to deal with social benefits and costs in the mathematics of traditional economics. This trend can be useful—as long as it does not succeed too well. The only social costs of energy use that can handily be incorporated into the price of energy are those related directly to the maintenance of environmental quality: the cost of reclaiming strip-mined land, of reducing pathogenic automobile exhaust emissions and power plant effluents, of siting energy plants and handling energy wastes safely, of reducing the use of certain pesticides and other additives in food production, and of improving the environmental health and safety of workers in industries that supply energy. These are not by any means all the social costs of energy use. The social costs of automobile accidents, of increasing the environmental instability of producing food crops, of the slow accumulation in the environment of man-induced radioactivity and carbon dioxide, of the increasing rate of waste heat production, of social stresses related to socially inefficient uses of energy, of the gradual depletion of soil fertility, and of the dislocations occasioned by the exhaustion of nonrenewable energy resources are difficult or impossible to quantify. If they could be quantified and put into dollars and cents, traditional economic analysis might very well lead one to the conclusion that the benefits from some of man's present activities will be exceeded by social costs within a period of time meaningful to those living now. It must be remembered, however, that traditional economic analysis strongly discounts future benefits but not future costs. Consequently, doomsday conclusions from a procedure that somehow incorporates all the social costs of energy use might be unjustified or at least premature. Social and economic studies may be of little use if they ignore physical and financial parameters. But much conventional economic analysis fails because it ignores physical and social parameters. In both the free world and the planned world, standard economic analysis commonly is based on a limited range of assumptions about social efficiency and physical economy, notably, in the free world, the construct of "economic man" and the assumption that the market economy is socially efficient and, in both worlds, the assumption that the availability of resources is not a limiting factor in economic planning. It might be useful to substitute other assumptions from time to time and to test their validity as rigorously as possible.

In physical economics, there is no discount rate; benefits and costs, whether accruing now or in the future, are the same. In pecuniary economics, the value of future benefits always is discounted and the value of future costs may be ignored, especially if they are difficult to "put in dollar terms." In social economics, the question of discounting future benefits and costs hinges on the historical perception of the society itself. In many low energy societies, the discount rate is low, both the past and posterity being honored in action as well as word. In high energy society, the social discount rate is high, the past and posterity getting lip service but little else.

Projected social benefits (or low social costs) of exotic energy sources account for much of the attraction of solar, fusion, geothermal, and wind power. That these can prove to be expensive social benefits (or cost reductions) seems to be easily ignored. The high cost would be in the work and materials required to establish an energy system based on one or more of these sources. If we can exploit (at a net work profit) any of them on a large scale, however, we shall have escaped from the two most threatening social costs of present energy systems, environmental poisoning and social collapse due to the exhaustion of nonrenewable resources.

SUPPLEMENTAL READING

Ayres, Edward. 1969. *The Economic Impact of Conversion to a Nonpolluting Automobile.* Springfield, Va.: National Technical Information Service (PB 199-925).
Concludes that, except for retraining mechanics, the impact of conversion would not be great.

Campbell, Ian A. 1973. "Ephemeral Towns on the Desert Fringe." *Geographical Magazine,* v. 45, p. 669-675.
Relates population growth in North Africa and the resulting increase in fuel and food demands to the northward advance of the Sahara.

Cook, Earl. 1975. "Ionizing radiation." In *Environment— Population, Pollution, Resources,* edited by William Murdoch, 2d ed., p. 297-323. Sunderland, Mass: Sinauer Associates.
Reviews the potential social costs of man-induced radiation and examines some of the related decision problems.

Council on Environmental Quality. 1973. *Energy and the Environment—Electric Power.* Washington, D.C.: United States Government Printing Office.
A good summary of the environmental impact of various electricity generating systems; especially valuable are the quantitative evaluations of the effects.

Foreman, Harry, ed. 1970. *Nuclear Power and the Public.* Minneapolis: University of Minnesota Press.
Papers of a symposium; pertinent to social economy are the papers by Auerbach, Eisenbud, Stannard, Green, Ramey, and Commoner.

Garvey, Gerald. 1972. *Energy, Ecology, Economy.* New York: Norton.
Several of the social costs of energy use are put into an analytical frame that emphasizes "depletive waste," the solvency of the ecosystem, and accumulation of more waste than can be assimilated.

Hubbert, M. K. 1972. "Man's Conquest of Energy: Its Ecological and Human Consequences." In *The*

The Social Economy of Energy Use

Environmental and Ecological Forum 1970-71. Oak Ridge, Tenn.: United States Atomic Energy Commission (TID-25857), 186 p.

Succinct treatment of ultimate consequences in terms of options available.

Kapp, K. William. 1950. *The Social Costs of Private Enterprise.* Cambridge, Mass.: Harvard University Press.

A pioneering work in the field of social economics; it questions the narrow scope of traditional economic analysis, as well as the social efficiency of the market.

Kneese, A. V., and R. C. d'Arge. 1969. "Pervasive External Costs and the Response of Society." Reprint no. 80. Washington, D.C.: Resources for the Future. Presents economic models to deal with "pervasive externalities," in other words, social or environmental costs.

Kubo, A. S., and D. J. Rose. 1973. "Disposal of Nuclear Wastes." *Science,* v. 182, p. 1205-1211.

In which attention is paid to means of containing the potential social costs of inadequate radioactive waste management.

Mostert, Noel. 1974. *Supership.* New York: Knopf.

The actual and potential damage to the world oceans that may be done by supertankers is carefully and convincingly set forth.

Mumford, Lewis. 1967. *The Myth of the Machine—Technics and Human Development.* New York: Harcourt, Brace and World.

Contains a perceptive discussion of the Benedictines' role in the development and proliferation of water power technology in western Europe—by an author who regards the High Middle Ages as Paradise Lost.

National Academy of Sciences. 1974. *Rehabilitation Potential of Western Coal Lands.* Cambridge, Mass.: Ballinger.

Coherent, lucid exposition of the environmental impact of surface mining in a region of scant water and of the small amount of knowledge about the probabilities of success in rehabilitation.

Nephew, Edmund A. 1973. "The Challenge and Promise of Coal." *Technology Review,* v. 76, p. 21-29.

A review of potential social benefits and costs.

Ridker, Ronald G., ed. 1972. *Population, Resources, and the Environment.* Washington D.C. Commission on Population Growth and the American Future, United States Government Printing Office.

Contains papers by Commoner and by Ehrlich and Holdren on the social costs of energy use, the former ascribing a major causal role to economic growth, the latter contending that population growth is the more important.

Richards, Paul W. 1973. "The Tropical Rain Forest." *Scientific American,* v. 229, n. 6, p. 58-67. (Available as *Scientific American* Offprint 1286.)

Emphasizes an impending social cost still not much appreciated outside the community of biologists.

Russell, Bertrand. 1935. *In Praise of Idleness.* London: George Allen and Unwin.

In the trenchant and witty title essay of this sparkling collection, Lord Russell argues persuasively that work is the great social disadvantage of the Western world.

Sahlins, Marshall. 1972. *Stone Age Economics.* Chicago: Aldine-Atherton.

An anthropologist describes societies in which work has been regarded as a social cost and in which the spirit of the gift transcends and replaces economics.

Sporn, Philip. 1972. "Possible Impacts of Environmental Standards on Electric Power Availability and Costs." In *Energy, Economic Growth, and the Environment,* edited by Sam H. Schurr, p. 69-88. Baltimore: Johns Hopkins.

A respected utility engineer and manager translates reduced environmental costs into increased financial costs.

Study of Critical Environmental Problems. 1970. *Man's Impact on the Global Environment.* Cambridge, Mass.: MIT Press.

Contains a number of sections dealing with effects on the environment that are related to energy: carbon dioxide and particles in the atmosphere, increased heat flux, oil in the ocean, radioactive wastes, and a separate section, p. 287-306, on energy products.

Symposium on Environmental Aspects of Nuclear Power Stations. 1971. *Environmental Aspects of Nuclear Power Stations.* Vienna: International Atomic Energy Agency.

Packed with information on expected social benefits and potential social costs of nuclear power.

Willrich, Mason, and J. B. Taylor. 1974. *Nuclear Theft: Risks and Safeguards.* Cambridge, Mass.: Ballinger.

A system of safeguards can be developed that will keep the risks of theft of nuclear weapon materials at very low levels (whatever that means).

William Tenney

Chapter 11   ENERGY AND
THE UNITED STATES

. . . The mad rush of the last
hundred years has left us out of breath.

—Eric Hoffer, 1967

AMERICA'S ENERGY ENDOWMENT

The North American continent was richly endowed with energy resources: great forests for fuel wood as well as lumber; abundant flowing
streams to turn mills and turbines; large and widely distributed
accumulations of petroleum and natural gas; numerous coal seams;
significant deposits of uranium, thorium, and oil shale. Every natural
energy resource of which man has made use was available in abundance
in North America. Although not distributed evenly throughout the
continent, energy resources were widespread enough that in only
limited areas was human occupation held back by lack of fuel and
power. Probably 80 percent of the land area now included in the
United States and Canada was covered by forest in the seventeenth
century. Water power in the eastern part of the continent was abundant
in terms of eighteenth and early nineteenth century needs; in addition
to power, waterways provided a transport network that facilitated
colonization, early trade, and industry. In every region there was coal
(see figure 11.1) the fuel that subsidized America's quick transition,
in the latter half of the nineteenth century, to an industrialized high
energy society. In the early part of the twentieth century, it became
clear that America's heritage of crude oil and natural gas was likewise
generous and widespread (see figure 11.2). Oil shale (see figure 11.3),
tar sands, and uranium, and thorium deposits (see figure 11.4) are more
localized than the other energy resources but contain more potential
energy and may provide much of the continent's energy in the next
century. Geothermal heat pockets are found mainly in the western
United States.

Regions most favorably situated to make use of solar radiation for
heating, cooling, and power are those with high annual solar insolation,

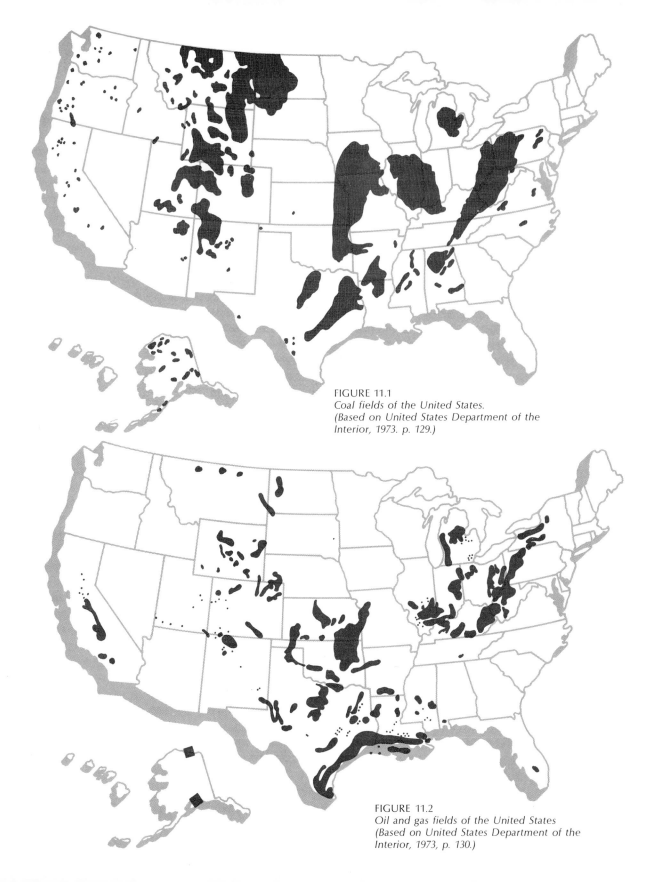

FIGURE 11.1
Coal fields of the United States.
(Based on United States Department of the
Interior, 1973. p. 129.)

FIGURE 11.2
Oil and gas fields of the United States
(Based on United States Department of the
Interior, 1973, p. 130.)

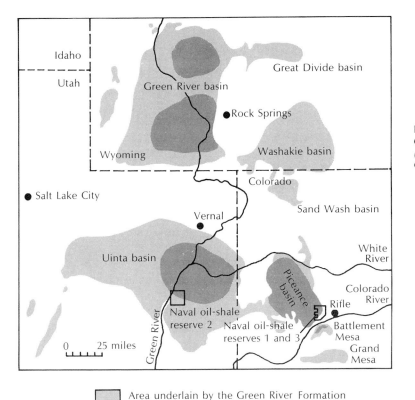

FIGURE 11.3
*Oil shale deposits of the United States.
(Based on United States Office of Oil and
Gas, 1968, p. 76.)*

Area underlain by the Green River Formation
in which the oil shale is unappraised or low grade

Area underlain by oil shale more than 10 feet thick,
which yields 25 gallons or more oil per ton of shale

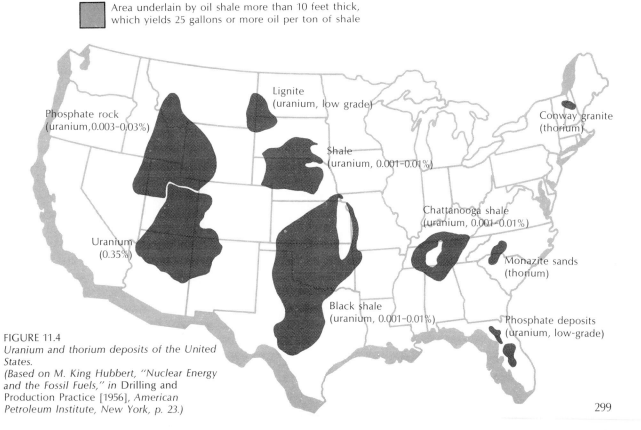

FIGURE 11.4
*Uranium and thorium deposits of the United
States.
(Based on M. King Hubbert, "Nuclear Energy
and the Fossil Fuels," in* Drilling and
Production Practice [1956], *American
Petroleum Institute, New York, p. 23.)*

in other words, those areas with few cloudy days and high annual temperatures—desert areas and certain warm coastal areas that prevailing winds keep relatively free of cloud cover. The United States has many favorable areas for solar collectors, particularly in the arid Southwest.

HISTORICAL PATTERN OF ENERGY USE IN AMERICA

For 250 years after the first settler landed, fuel wood was the principal source of inanimate energy, and the horse was the main source of animate energy. During most of this period, wind was the propelling force for most water transport. In the past hundred years, however, dramatic changes have taken place. In the United States, fuel wood gave way to coal, then the coal industry ceded the largest share of its market to the oil and gas industry; human muscle power became quantitatively unimportant, although more highly valued; hydroelectric power, a valuable form of energy because of its renewability and cleanliness, has never achieved more than a minor share of the national energy input; nuclear power, growing rapidly, as yet provides only about 1 percent of the national total.

Per capita and total energy consumption in the United States have increased greatly since 1890. More significant, however, is the increase in the output of useful heat and work (see figure 11.5); this output rose steadily through the nineteenth century, while per capita gross

FIGURE 11.5
Energy consumption in the United States in 1974 was 18 times higher than it was in 1870, but the useful heat and work produced in 1974 was more than 80 times that produced in 1870. The difference was caused by an increase in efficiency from 8 to 36 percent.

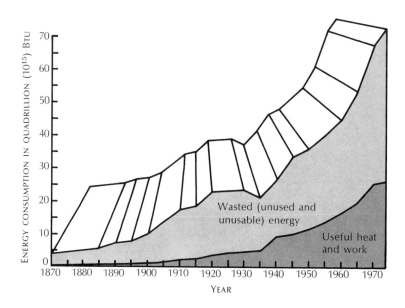

300

energy consumption actually *declined* between 1800 and 1880. The reason for this rather surprising decline lies in a progressive increase in the aggregate efficiency of energy use; in 1890, for example, about the same amount of energy was consumed for each individual in the country as had been consumed in 1800, yet the useful output from each unit of energy input was quite a bit greater, and so the material level of living in the United States, because the efficiency of energy use had increased faster than the human population, was higher in 1890 than it had been in 1800.

COAL FUELS THE EMERGENCE OF AN INDUSTRIAL NATION

Coal gave the United States an enormous energy subsidy for industrialization and strongly influenced the demographic evolution of the nation. Coal, converted to coke, smelted the ore to make the steel for railroads, tall buildings, massive factories, great bridges, and millions of machines. Coal, burned in locomotive fireboxes and under the boilers of steamships, replaced inefficient wood and made possible greater power and speed in transport. Coal, glowing in pot-bellied heating stoves, kitchen ranges, and furnaces, provided comfort and cooking heat in homes, stores, offices and workshops.

Nearness to coal fields—in Pennsylvania, Alabama, and Illinois—was a major factor in siting steel mills, which in turn were important in siting steel-using factories such as those manufacturing locomotives, trucks, automobiles, machine and other metal tools, and heavy equipment for construction, farming, and mining. The sites of steel mills and their satellite industries determined the shape of a considerable portion of the national transport network; coal and iron ore came to the steel mills by train and ship, steel moved from mill to factory by train and truck, and finished products were carried from factory to user by train, truck, and ship. Patterns of industry, population, transport, commerce, and even politics were influenced greatly by the geographical distribution of coal.

The importance of coal as an energy source in the second half of the nineteenth century is illustrated by the early energy history of California, a state deficient in coal that quickly used up its readily available fuel wood. Until the transcontinental railroad was completed in 1869, coal was brought to California by sailing ship from coastal mines in Oregon, Washington, and British Columbia, and as well from Pennsylvania around South America. The Sierra foothills were stripped of timber in the quest for fuel, mine timbers, and lumber. With completion of the transcontinental railroad coal from fields in Utah and Wyoming started to flow to California. Development of

California's prolific oil fields near the turn of the century was stimulated by a continuing need for fuel for heating and for steam generation.

Many of the European immigrants to the United States after 1850 found their economic foothold in coal mining or in the steel mills. Many present-day communities in places as far apart and as environmentally diverse as Wilkes-Barre, Pennsylvania, Louisville, Colorado, and Roslyn, Washington reflect this early movement of European immigrant groups into coal-mining regions.

PETROLEUM AND NATURAL GAS TAKE OVER

Coal was available so widely and at such an enormous energy profit, and the national system for distribution and utilization of coal so well developed, that only where coal was expensive, as it was in California, was there much demand for petroleum as fuel, after it began to be discovered in North America in the 1850s. In 1860, with only a few producing wells in the United States and Canada, crude oil was a resource whose supply clearly exceeded the demand for it. The consequent low price stimulated primitive refining research aimed at producing kerosine for lamps more cheaply than it could be produced from coal. "Coal oil" or kerosine made from coal had begun to displace the whale oil used for interior lighting in the industrialized parts of the world. Not only was whale oil rising in price as the sperm-whale population of the world was being reduced, but also better light could be obtained from kerosine with less smoke and (to some) a less objectionable smell. When it became clear that kerosine could be made from crude oil more cheaply than from coal, a market for petroleum was assured, but that market did not increase spectacularly until the invention of the Welsbach mantle in 1883, after which the light produced by kerosine was so much better than that produced by sperm-oil lamps, candles, or kerosine lamps with wicks, that the demand for kerosine for lighting tripled within a few years. The "age of kerosine" lasted about 50 years, during which period much gasoline, a necessary byproduct of kerosine production, was discarded as a waste product. Then came the automobile population explosion and the beginning of national electrification. The former caused demand for gasoline to soar, the latter reduced the need for kerosine, but increased the demand for fuel oil. In sum the two developments greatly enlarged the market for petroleum products.

Natural gas, much of it unavoidably produced with crude oil, was wasted in the oil fields on a profligate scale until the rapid growth, during and following World War II, of regional and national pipeline systems for its distribution. Its production and consumption then increased almost as fast as distribution networks could be constructed. Together, natural gas and petroleum by 1950 provided more than half the gross energy input to the economy.

ADVENT OF THE AUTOMOBILE AND ELECTRICITY

In 1900 there were eight thousand automobiles in the United States. In 1920 there were 8 million. In 1970 there were almost 90 million. Gasoline consumption by automobiles has increased faster than the automobile population, because the average automobile has put on a great deal of weight over the years and has become less efficient in the use of its fuel. Crude oil, the supply of which exceeded demand for some sixty years, began to be more profitable to find and produce in the 1920s as demand for gasoline and fuel oil grew. It was in this period that the fruits of vertical integration and market control by the major integrated oil companies, begun as a defense against perennial oversupply, began to be harvested, when a certain measure of artificial scarcity could be created and maintained. The automobile and oil were to change the face and mind of America as much as steel and coal had changed its skeleton and vital functions.

Electricity was a laboratory curiosity until the 1880s, when coal-fired central generating plants began to be constructed. Then the advantages of electric lighting, cooking, water heating, and power for machines became available for the first time on a mass scale and a rush was on to electrify the nation. The incremental step from a society whose nocturnal activities took place in the smoky, flickering glow of oil and glass-chimneyed kerosine lamps to one whose homes, work places, and streets were brightened by pressurized gas lamps was now followed by a quantum leap to the brilliant almost-day nights of electricity. Even today one may experience these two abrupt transitions, when traveling at night in an underdeveloped country, from a remote candlelit hamlet to a kerosine- or gasoline-lamped village and then to the visual shock of a coruscant electrified city.

Subsidy of the automobile and of electrification by government hastened the spread of both. From about 1910 it became an article of the conventional wisdom in the United States that public investment in highways would always be more than returned in increased regional and national productivity. It took more than twenty years longer for the same concept to be fully applied to electric generating facilities. The lag was because vested economic interests (private utilities) effectively delayed and limited the government's entry into electricity production, whereas those economic interests that opposed highway subsidies, the railroads, were overwhelmed by the interests that campaigned for them, automobile and truck manufacturers, oil companies, cement companies, and thousands of local businessmen. It took a depression to get the government into the large-scale construction of dams for electric power, with the dams being looked upon as subsidies for employment as well as regional economic stimulators.

THE BUDDING GIANT OF NUCLEAR POWER

Electricity generation by heat derived from the kinetic energy of nuclear fission is a modern version of turning swords into plowshares, since nuclear fission was first used in weapons. In 1957 the first commercial nuclear power plant went into service at Shippingport, Pennsylvania. Since then the growth rate of nuclear generating capacity has been high, although total installed capacity in 1974 was still about one-quarter that of hydroelectric generating capacity in the country.

The expansion of nuclear power based on reactors that use only the U-235 in the atomic fuel appears limited by the availability of cheap uranium ore. Although the fuel accounts at present for 10 percent or less of the cost of the power produced, a quadrupling in fuel price, which may well occur before the end of the present decade, would increase the cost of nuclear electricity by 30 percent. If the cost (in constant dollars) of delivering coal to conventional coal-fired power plants does not increase proportionately, nuclear power plants of the present design will find themselves at a considerable cost disadvantage. An efficient breeder reactor power plant will almost eliminate fuel as a cost factor in the price of the electricity produced and might produce electricity more cheaply than most coal-fired plants. The decision to build breeder power plants, however, will be decided on the basis of social rather than pecuniary economics. Opposition to the construction of nuclear power plants has been strong in some states, almost nonexistent in others. The 1975 distribution of commercial nuclear power plants operating, under construction, or planned (see figure 11.6) in the United States probably reflects both an expected lack of comparably priced alternatives and the degree of acceptance of nuclear power by local residents.

Other countries, notably France and the Soviet Union, are moving ahead vigorously with programs to develop breeder reactors. The program in the United States probably will not produce a commercial breeder power plant until about 1990; after that, however, the growth of nuclear power production could be spectacular, if questions of environmental and social risks have been resolved.

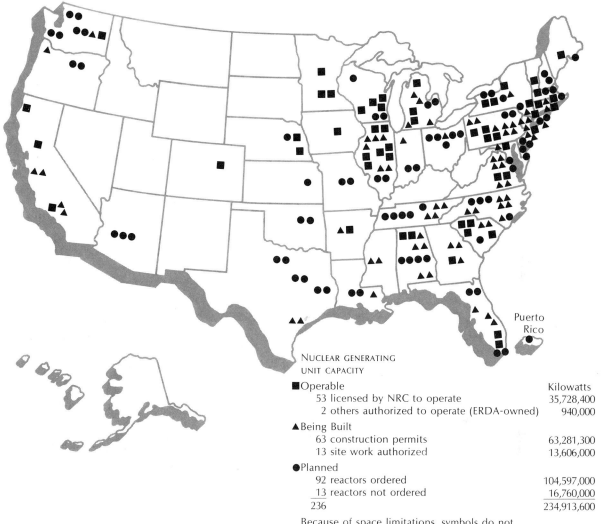

NUCLEAR GENERATING
UNIT CAPACITY

	Kilowatts
■Operable	
53 licensed by NRC to operate	35,728,400
2 others authorized to operate (ERDA-owned)	940,000
▲Being Built	
63 construction permits	63,281,300
13 site work authorized	13,606,000
●Planned	
92 reactors ordered	104,597,000
13 reactors not ordered	16,760,000
236	234,913,600

Because of space limitations, symbols do not
reflect precise locations.

DEMOGRAPHIC CHANGES AND ATTITUDES RELATED TO ENERGY USE

The adolescent United States dreamed of destiny and conquest; but the destiny was not political and the conquest was not military. The course of the American empire was westward, "across the wide Missouri." Its scouts were hunters, trappers, prospectors. Its combat troops were loggers, miners, and railroad builders. Its occupation forces were farmers, stockmen, tradesmen, railroad and sawmill workers. The Manifest Destiny of the United States was hegemony

FIGURE 11.6
Nuclear power plants in the United States at first tended to be situated long distances from sources of coal, natural gas, and fuel oil. With the increasing scarcity of acceptable fossil fuels, however, nuclear plants are invading even Texas.
(Based on map dated June 30, 1975, issued with a news release by the Grand Junction Office of the United States Research and Development Administration, July 25, 1975.)

over the rich North American continent and the building of an industrial and technological society surpassing any society the world had yet seen.

In the popular phrase of the frontier era, "winning the West," the verb is peculiarly appropriate to American attitudes of the period. The West and its resources were to be won as fruits of a campaign in which valiant effort would be assisted by the unseen hand of a Providence prejudiced in favor of those who helped themselves without waiting to be helped. A form of providential assistance in which nineteenth century America placed great faith was an abundance of natural resources. The United States Congress, for illustration, refused to accept evidence presented to them by John Wesley Powell, explorer of the Colorado canyons and director of the United States Geological Survey, that water in the interior West would inevitably limit agricultural development. Powell's plea that homestead laws for the arid regions be adapted to water scarcity fell on deaf ears. The wealth of timber, coal, iron, and other resources already exploited, the spread of productive agriculture, and a rapid increase of industrial production led to confidence in the boundless resources of the continent as well as in the almost unlimited ability of Americans to develop or exploit those resources for the greater material welfare of the nation and its citizens.

The strength of the unifying concept of a Manifest Destiny unfolding as the hostile environment of the frontier was conquered and as its resources were transformed into the beauties of productive farms and busy cities, and the corresponding weakness of conservation philosophy throughout the nineteenth century reflect not only the momentum of an industrializing society with an abundance of resources on which to feed but also the work ethos and material ambition brought to the United States by millions of European immigrants. These immigrants were farmers and laborers escaping harsh economic conditions and seeking opportunity to gain material security. They eagerly cut, dug, sawed, plowed, and blasted their way across the American West. They had little patience with Thoreau's transcendental philosophy or Rousseau's and Francis Parkman's noble savage. They transformed Thomas Jefferson's agrarian moralism into an interest-based Populist political movement and they ignored the warnings of George Perkins Marsh about the adverse consequences of thoughtless exploitation of natural resources. This mass immigration to America included little intellectual leaven; the educated European of gentle or noble birth seldom found anything attractive in American society except its money. The geographic frontier disappeared in the early years of the present century, but frontier attitudes toward resources persisted. The effects such attitudes have had on energy decision making will be discussed in chapter 12.

CONTEMPORARY ENERGY PRODUCTION AND CONSUMPTION IN THE UNITED STATES

A few of the United States produce most of the energy that all consume. A conterminous group of five states (see table 11.1)—Texas, Louisiana, Oklahoma, New Mexico, and Kansas—produces more than 90 percent of the nation's natural gas; the same group produces almost 75 percent of the domestic total of crude oil. Appalachia and the Ohio Valley dominate the coal-production statistics; a contiguous group of five adjoining states, extending from Mexico to Canada along the Continental Divide produces most of the country's uranium (see table 11.2), as well as substantial amounts of the fossil fuels. The only major

Table 11.1 Distribution by Major Producing States of Fossil Energy (1971)

Natural Gas			Crude Oil			Coal		
State	Cubic feet $\times 10^6$	Percentage of national production	State	barrels $\times 10^3$	Percentage of national production	State	Tons $\times 10^3$	Percentage of national production
Texas	8,550,705	38.0	Texas	1,222,926	35.4	Kentucky	119,389	21.3
Louisiana	8,081,907	35.9	Louisiana	935,243	27.1	West Virginia	118,258	21.1
Oklahoma	1,684,260	7.5	California	358,484	10.4	Pennsylvania	81,562	14.5
New Mexico	1,167,577	5.2	Oklahoma	213,323	6.2	Illinois	58,402	10.4
Kansas	885,144	3.9	Wyoming	148,114	4.3	Ohio	51,431	9.2
California	612,629	2.7	New Mexico	118,412	3.4	Virginia	30,628	5.5
Wyoming	380,105	1.7	Alaska	79,494	2.3	Indiana	21,396	3.8
West Virginia	234,027	1.0	Kansas	78,532	2.3	Alabama	17,945	3.2
Arkansas	172,154	0.76	Mississippi	64,066	1.9	Wyoming	8,052	1.4
Alaska	121,618	0.54	Montana	34,599	1.0	Montana	7,064	1.3
Mississippi	118,805	0.53						
Colorado	108,537	0.48						
U.S.A.	22,493,012		U.S.A.	3,453,914		U.S.A.	560,919	

SOURCE: Production data from United States Department of the Interior, *United States Energy Fact Sheets 1971* (Washington, D.C.: United States Government Printing Office, 1973).

Table 11.2 Distribution by Major Producing States of Energy (1971)

Hydroelectricity			Uranium			Total Energy		
State	(kWh $\times 10^6$)	Percentage of national production	State	Pounds of uranium oxide	Percentage of national production	State	(Kcal $\times 10^{12}$)	Percentage of national production
Washington	71,429	26.8	New Mexico	10,567,000	43.1	Texas	4,313	25.5
California	39,045	14.7	Wyoming	6,986,000	28.5	Louisiana	3,592	21.2
Oregon	34,305	12.9	Colorado	2,536,000	10.3	New Mexico	1,135	7.9
New York	25,178	9.5	Utah	1,445,000	5.9	Wyoming	878	5.2
Alabama	9,912	3.7	U.S.A.	24,520,000		Oklahoma	802	4.7
Montana	9,595	3.6				West Virginia	798	4.7
Tennessee	9,420	3.5				Kentucky	769	4.5
Idaho	7,469	2.9				California	722	4.3
Arizona	6,621	2.5				Pennsylvania	529	3.1
North Carolina	5,910	2.2				Illinois	413	2.4
U.S.A.	266,320					Kansas	381	2.3
						Ohio	347	2.1
						Colorado	289	1.7
						Virginia	190	1.1
						Utah	179	1.1
						Alaska	148	0.87
						Indiana	141	0.83
						U.S.A.	16,916	

SOURCE: United States Department of the Interior (1973).

energy-producing state not included in these three regional groups is California. Two states alone produced almost half the total national energy production in 1971; more than 81 percent came from only nine states.

The geographic pattern of energy *consumption* (see figure 11.7) in the United States differs from that of energy *production.* As one might expect, consumption is highest in the most densely populated states (see table 11.3). However, there are significant differences in consumption per capita among the states, tending to be high where mining, processing, and the transportation of oil, gas, and coal form a substantial part of the state's economy, for these are energy-intensive activities. Per capita energy consumption for transportation, very high in the United States and Canada compared with the rest of the world, is related to population density and transport mode most used (see figure 11.8). The lowest per capita consumption is found in heavily urbanized states, where distances traveled are short and considerable

FIGURE 11.7
Per capita energy consumption in industry in the United States is higher in energy-producing states, such as Louisiana, Texas, Wyoming, West Virginia, than in more highly industrialized states, such as Pennsylvania, New Jersey, Ohio, Illinois.
(Based on data in United States Department of the Interior, 1973.)

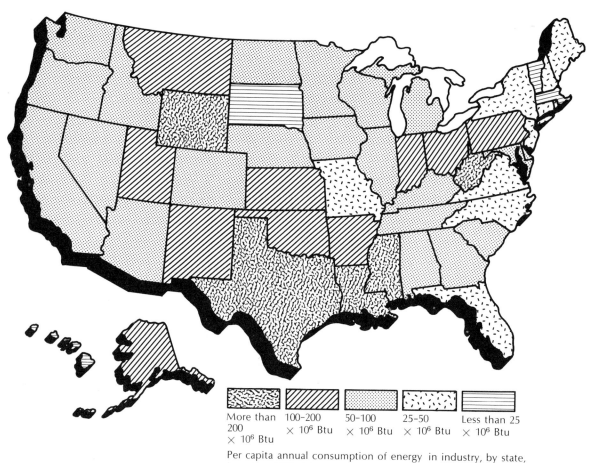

| More than 200 × 10⁶ Btu | 100–200 × 10⁶ Btu | 50–100 × 10⁶ Btu | 25–50 × 10⁶ Btu | Less than 25 × 10⁶ Btu |

Per capita annual consumption of energy in industry, by state, based on 1971 data

Table 11.3 Leading Energy Consuming States (1971)

Rank/State	Gross consumption* (10^{12} kcal)	Per capita gross consumption (10^6 kcal)
1. Texas	1,560	136
2. California	1,178	58
3. New York	989	54
4. Pennsylvania	953	80
5. Ohio	909	84
6. Illinois	872	78
7. Michigan	651	72
8. Louisiana	621	169
9. Indiana	558	106
10. Florida	453	64
11. New Jersey	441	60
U.S.A.	16,696	81

NOTE: *Gross consumption as given in the source has been reduced by an amount equivalent to the fictitious energy loss therein charged to hydroelectric power, as though it had been generated in a thermal power plant.

SOURCE: United States Department of the Interior (1973).

FIGURE 11.8
A plot of population density (on a vertical logarithmic scale) against per capita energy consumption in 1971 for transportation for 43 states reveals the effects of distances and methods of travel on energy use. The densely populated states where travel distances are short use the least energy per person in transportation. Rural states use more, and states where much or most travel is by air use the most.
(Based on data in United States Department of the Interior, 1973.)

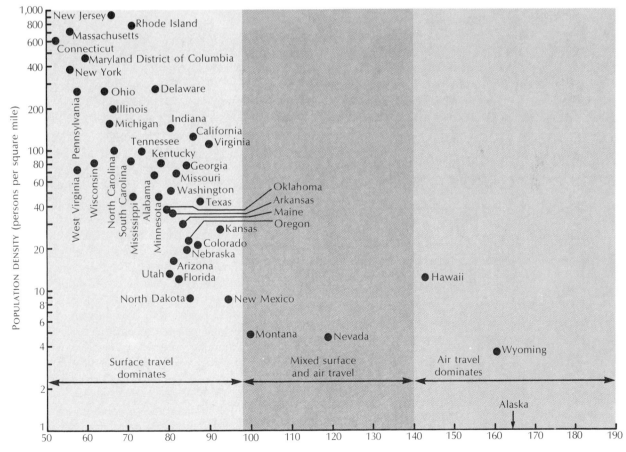

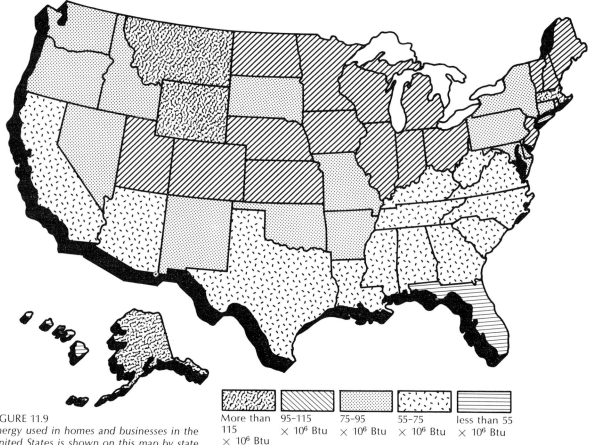

FIGURE 11.9
Energy used in homes and businesses in the United States is shown on this map by state, on a per capita basis. It does not appear to be related strongly to population density, although the two highest states are Wyoming and Alaska.

More than 115 $\times 10^6$ Btu	95–115 $\times 10^6$ Btu	75–95 $\times 10^6$ Btu	55–75 $\times 10^6$ Btu	less than 55 $\times 10^6$ Btu

Per capita annual consumption for household and commercial purposes, based on 1971 data

use is made of mass transit facilities. As the ratio of rural to urban population rises and as the population density falls, we find the per capita use of energy for transportation rising. Below a density of about 8 persons per square mile, the per capita transport use of energy increases abruptly, representing a substitution of the airplane for the car, as well as long travel distances. The relation of per capita energy use in home and commerce to climate becomes obvious when one plots the state averages on a map (see figure 11.9). The relation of energy consumption to the average annual temperature of each state, weighted for uneven population distribution, is only rudely linear, but would be more so if air conditioning did not exist (see figure 11.10). Except for Massachusetts and Wyoming, most of the states that consume considerably more than the average for their temperature zone use relatively large amounts of electricity for summertime air conditioning.

Only thirteen of the fifty states produced more energy than they consumed in 1971. These are the energy-exporting states (see table 11.4); together they produced more than 90 percent of the national

Energy and the United States

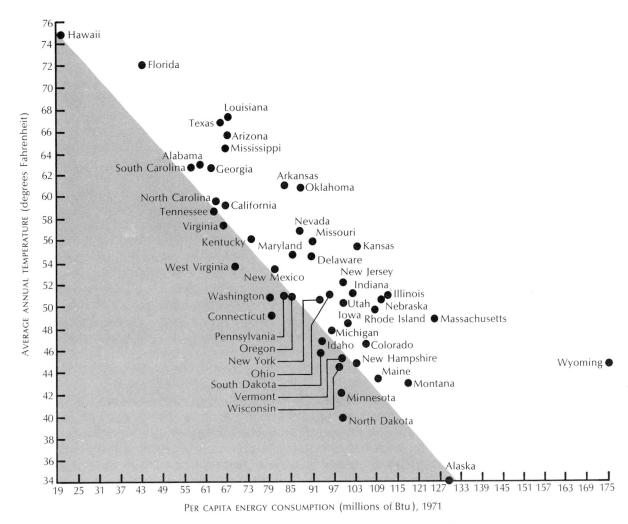

Table 11.4 The Energy Exporting States (1971)

Rank/State	Surplus (10^{12} kcal)	Sufficiency
1. Louisiana	2,942	573%
2. Texas	2,716	274
3. New Mexico	1,185	940
4. Wyoming	798	1,178
5. Oklahoma	556	333
6. West Virginia	536	315
7. Kentucky	467	256
8. Kansas	140	158
9. Colorado	117	168
10. Alaska	103	329
11. Utah	88	198
12. Montana	42	163
13. North Dakota	13	119

FIGURE 11.10
Most of the energy used in homes and businesses is for space or comfort heating. Consequently, there is a good correlation between per capita energy consumption in homes and commerce and the average annual temperature in each of the states. (Based on data in United States Department of the Interior, 1973.)

Table 11.5 Some Energy Importing States (1971)

State	Production (10^{12} kcal)	Consumption (10^{12} kcal)	Deficit (10^{12} kcal)	Sufficiency
California	716	1,178	462	61%
Pennsylvania	525	953	428	55
Illinois	410	872	462	47
Ohio	346	909	563	38
New York	24	989	966	2
New England (six states)	4	697	693	0.5
New Jersey	0	441	441	0

energy production and, after satisfying their own internal energy demand, the remaining surplus was equivalent to 70 percent of the gross energy consumption of the other thirty-seven states; the remaining 30 percent being made up of the importing states' own production and by foreign imports of energy. Some of the energy-importing states are listed in table 11.5; the amount of energy imported by the eight north-eastern states (New York, New Jersey, and those of New England) from the energy-producing states and from foreign countries in 1971 was equal to 301 million metric tons of coal equivalent and exceeded the entire fossil-fuel and electricity consumption of Africa (122) and South America (163) combined.

CONTEMPORARY ENERGY USE IN THE UNITED STATES

The most complete statistics on energy use in the United States are compiled by the United States Bureau of Mines. Although fundamental to an analysis of national energy flow, they leave something to be desired. They do not identify consumption in specific industries, such as iron and steel; they give no indication of the efficiency of energy use, except in the conversion of fossil fuel and uranium energy to electricity. Non-energy uses of the fossil fuels such as of coal in nylon manufacture and of natural gas in ammonia-fertilizer production are included in energy "consumption" totals. Totals for hydroelectricity are too large by about three times owing to the Bureau's practice of calculating energy consumed in hydropower production as if fossil fuel had been used in its generation. Other useful information on the use of electricity and natural gas is available from the Edison Electric Institute, which is supported by private utilities, from the Federal Power Commission, and from the American Gas Association, which represents distributors of natural gas (pipeline companies and private utilities).

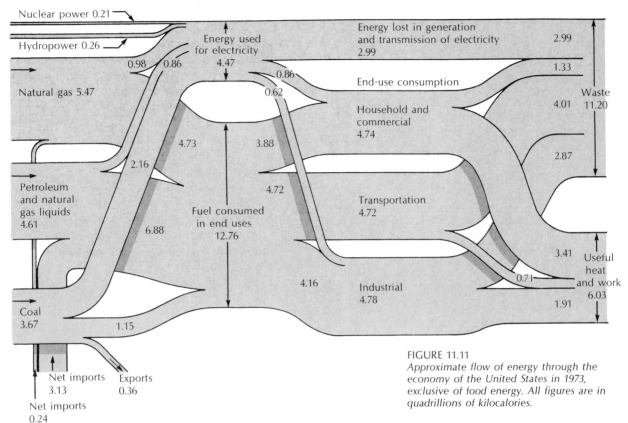

FIGURE 11.11
Approximate flow of energy through the economy of the United States in 1973, exclusive of food energy. All figures are in quadrillions of kilocalories.

Information on the efficiency of energy use has been hard to come by, requiring a search for data from both private and public sources. Recent interest in the subject, however, has inspired a number of good studies and it is no longer as difficult as it once was to get a fairly satisfactory picture of energy flow through the economy (see figure 11.11).

An energy balance sheet for the United States (see table 11.6) shows that in 1972, industry used 37.6 percent of the gross energy input, the household and commercial sector used 35.2 percent, and transportation took 27.3 percent. Fossil-fuel imports, mainly of crude oil and petroleum products, contained energy equivalent to about 14 percent of the total national consumption, imports being important because they accounted for 31 percent of the total petroleum consumption (including non-energy uses).

Table 11.6 Energy Balance Sheet, United States, 1972 (kcal $\times 10^{15}$)

Supply		Consumption	
FOSSIL FUELS		INDUSTRIAL	
Domestic production		Direct use of fossil fuels	3.387
Natural gas	6.321	Converted to electricity	2.400
Petroleum	4.908	Total	5.787
Coal	3.638	HOUSEHOLD AND COMMERCIAL	
Total	14.867	Direct use of fossil fuels	4.455
Imports	2.257	Converted to electricity	1.712
Stock change	0.013	Total	6.167
Total	17.137	TRANSPORTATION	
HYDROELECTRICITY	0.241	Direct use of fossil fuels	4.511
URANIUM (Net)*	0.152	Converted to electricity	0.013
TOTAL PRODUCED	17.530	Total	4.524
TO NON-ENERGY USES	1.052		
TOTAL AVAILABLE†	**16.478**	TOTAL CONSUMED	**16.478**

SOURCE: United States Bureau of Mines press release, March 13, 1974.
*For energetic use

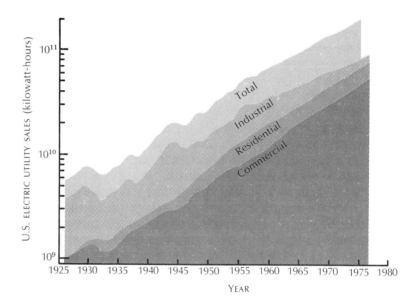

FIGURE 11.12
Use of electricity in the United States, 1926–1973, by major sectors. Note that the vertical scale is logarithmic.
(Based on data published by the Edison Electric Institute in its annual Statistical Year Book.)

Electricity generation diverted 25.1 percent of the gross energy input in 1972. Since 1926 electricity consumption has increased by 25 times and per capita use has increased 11-fold (see figure 11.12). Except during the struggle to climb out of the Great Depression and during World War II, there has been a steady decline in the proportion of electricity used by industry and a complementary rise in the proportions used in residences and commercial establishments. In the decade 1960–1970, industry's percentage of electricity use fell from 53.6 to 44.5 while the residential use rose from 29.6 to 33.0 percent.

Energy and the United States

The greatest proportional increase, however, was in commercial use, from 16.8 to 22.5 percent. The extraordinarily rapid increase in the air conditioning of homes, office buildings, and stores during this decade is an important factor in the rise of electricity consumption. Many spaces being cooled electrically are of poor thermal design; some, such as the open coolers and freezers in supermarkets, are designed for customers' convenience instead of economic use of energy.

Another way of breaking down the gross energy input of the economy is by end use (see table 11.7). Work accounts for the largest share and most of that is in the transportation sector. Process heating is next in importance, and space heating is third. These proportions are based on gross energy inputs; the outputs are not proportional to the inputs, because conversion and application efficiencies differ greatly. For example, most of the work performed by inanimate energy in homes, business, and industry is done by electrical motors and appliances; the conversion efficiency of fuel energy to work in these sectors is about 30 percent. The work done in transportation is mainly accomplished by internal combustion engines; the machines they power range in efficiency from 4 to about 30 percent, the average being 11 or 12 percent. Process heating efficiency ranges from 5 to 35 percent, the average being about 20 percent. The efficiency of air conditioning is 10 percent or less. Water heating and cooking use the chemical energy in fuel between 50 and 60 percent efficiency in direct heating and at about 30 percent for electric heating. Space heating is the most efficient end use, with an efficiency range of between 50 and 80 percent and an average of about 65 percent. The overall average of the national energy system can be calculated, from these estimates, to be about 35 percent.

Table 11.7 Electricity Generation, United States, 1972 (xcal $\times 10^{15}$)

Source	Input	Power output	Waste heat
Coal	1.895	0.605	1.290
Natural gas	1.039	0.332	0.707
Petroleum	0.802	0.256	0.546
Water power	0.254	0.242	0.013
Uranium	0.152	0.048	0.103
Totals	4.142	1.483	2.659

COMPARATIVE ENERGY COSTS IN THE UNITED STATES

Although the costs of energy in current dollars tend to change rapidly in times of transition from abundance to scarcity, a comparison of energy costs in a high energy economy such as that of the United States can yield some information about the way in which the natural energy subsidy is allocated within the society. Because the human labor costs and capital-cost energy equivalents required to exploit fossil fuels are a small part of the usable energy in the fuels themselves, the costs of fossil fuel in dollars and cents are low, compared with the cost of energy from alternative sources or with the value of products and services provided by the use of fossil energy. Distribution costs

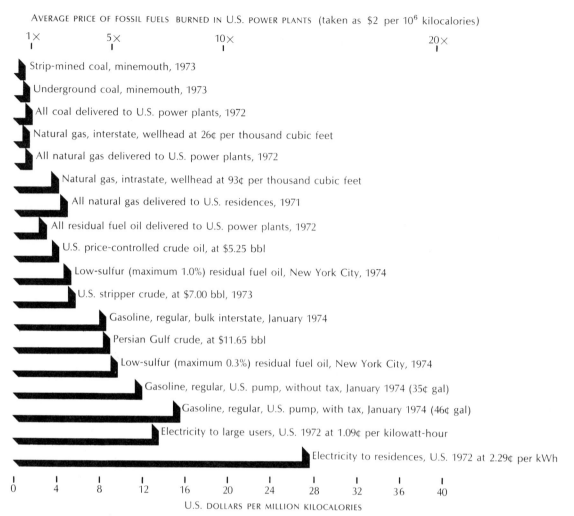

AVERAGE PRICE OF FOSSIL FUELS BURNED IN U.S. POWER PLANTS (taken as $2 per 10^6 kilocalories)

1×   5×   10×   20×

Strip-mined coal, minemouth, 1973

Underground coal, minemouth, 1973

All coal delivered to U.S. power plants, 1972

Natural gas, interstate, wellhead at 26¢ per thousand cubic feet

All natural gas delivered to U.S. power plants, 1972

Natural gas, intrastate, wellhead at 93¢ per thousand cubic feet

All natural gas delivered to U.S. residences, 1971

All residual fuel oil delivered to U.S. power plants, 1972

U.S. price-controlled crude oil, at $5.25 bbl

Low-sulfur (maximum 1.0%) residual fuel oil, New York City, 1974

U.S. stripper crude, at $7.00 bbl, 1973

Gasoline, regular, bulk interstate, January 1974

Persian Gulf crude, at $11.65 bbl

Low-sulfur (maximum 0.3%) residual fuel oil, New York City, 1974

Gasoline, regular, U.S. pump, without tax, January 1974 (35¢ gal)

Gasoline, regular, U.S. pump, with tax, January 1974 (46¢ gal)

Electricity to large users, U.S. 1972 at 1.09¢ per kilowatt-hour

Electricity to residences, U.S. 1972 at 2.29¢ per kWh

0   4   8   12   16   20   24   28   32   36   40

U.S. DOLLARS PER MILLION KILOCALORIES

FIGURE 11.13
Comparison of prices of energy derived from fossil fuels and prices of electricity in the United States. The scale at the top may be used to see that electricity delivered to homes in 1972 cost more than 13 times the average cost of fossil fuels burned in power plants.

for the fossil fuels commonly exceed the combined costs of finding, extracting, and processing.

Energy in the form of electricity is more expensive than energy in the form of fuel because more than two-thirds of the fuel energy is lost in the conversion to electricity and transmission to the consumer (see figure 11.13). Energy in the form of plant food, even at the farm gate, costs more than energy in the form either of fossil fuel or electricity, because the energy profit or natural subsidy from plant culture is much less than from the exploitation of fossil fuel (see figure 11.14). The most profitable feed crop, in energy terms, probably is corn; Heichel (1973) and others have shown that about three units of energy are produced for each unit invested. The energy return from coal mining or oil production is more like one hundred to one.

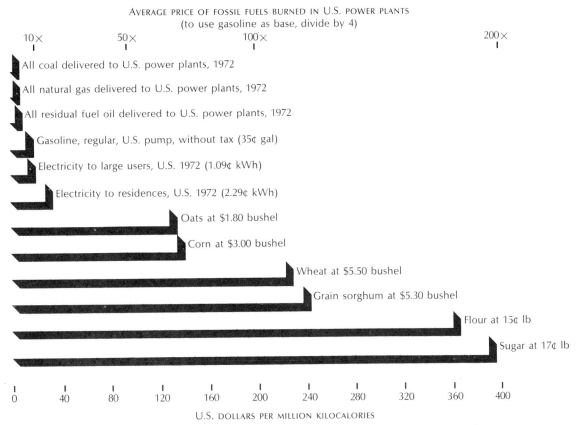

10×   50×   100×   200×

All coal delivered to U.S. power plants, 1972

All natural gas delivered to U.S. power plants, 1972

All residual fuel oil delivered to U.S. power plants, 1972

Gasoline, regular, U.S. pump, without tax (35¢ gal)

Electricity to large users, U.S. 1972 (1.09¢ kWh)

Electricity to residences, U.S. 1972 (2.29¢ kWh)

Oats at $1.80 bushel

Corn at $3.00 bushel

Wheat at $5.50 bushel

Grain sorghum at $5.30 bushel

Flour at 15¢ lb

Sugar at 17¢ lb

0   40   80   120   160   200   240   280   320   360   400

U.S. DOLLARS PER MILLION KILOCALORIES

FIGURE 11.14
Comparison of prices of energy derived from fossil fuels and electricity with those derived from common food plants. The upper horizontal scale may be used to see that energy in sugar at 17 cents a pound costs almost 20 times more than energy in electricity at 2.29 cents a kilowatt-hour.

Meat-energy costs, in turn, are higher than plant-food energy costs because of the inefficiencies of animal conversion of feed to meat (see figure 11.15). As we shall see, the present meat-production system supplies one unit of meat energy at an input cost of 17.6 units of plant energy, about half of which is supplied from pasturage, a cheap form of feed. The approximate 10-fold increase from plant (feed) energy cost to meat-energy cost is in accord, then, with the physical economy of meat production. Human labor is the most expensive form of energy in a high energy society.

For more than fifty years, the price of inanimate energy in constant dollars declined; in recent years, the price of all forms has been rising. The largest part of the increase has been due to the steep rise in the price of crude oil effected by the Organization of Petroleum Exporting Countries (OPEC), which in turn caused the price of other fuels, insofar as they were not regulated by governmental controls, to rise. But not all of the increase can be laid at the door of OPEC. Concern for environmental quality already had imposed substantial

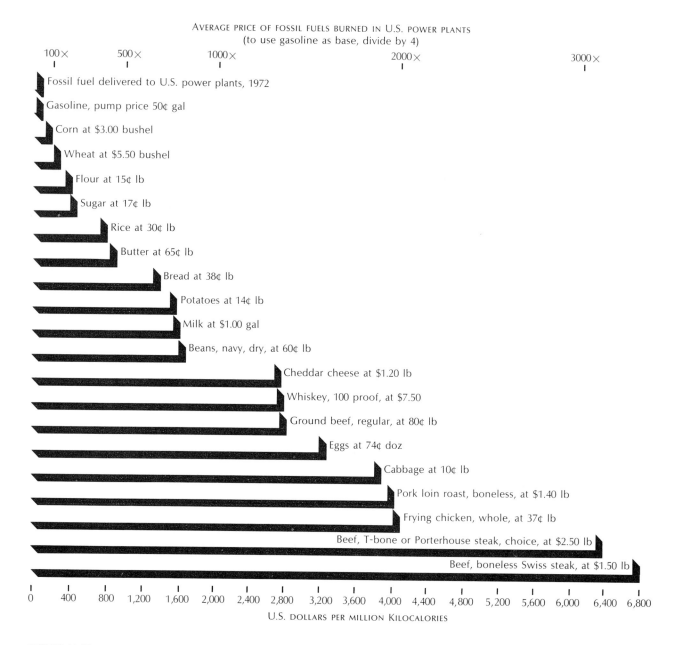

FIGURE 11.15
Comparison of prices of energy derived from fossil fuels with those derived from products of agriculture. By using the upper horizontal scale, one can find that the energy in choice steak at $2.50 a pound costs about 35 times more than the energy in the corn at $3.00 a bushel on which the steer that produced the meat may have been fattened.

new costs on the production, transport, and conversion of coal and crude oil. An impending and quite predictable shortage of natural gas was producing steep increases in the prices of gas not subject to price controls, while rising costs of finding and producing gas from less favorable geologic and geographic environments were forcing the Federal Power Commission to allow higher prices on production for

Energy and the United States

the interstate market. The capital costs of nuclear power plants, even allowing for price inflation, have been rising steadily, because of more expensive design and delays in authorization and construction, largely resulting from increased public concern about the hazards of nuclear power. At the same time, the price of nuclear fuel has risen in anticipation of the depletion of the relatively high-grade uranium ores.

The sudden upsurge in the prices of fuels and electricity was accompanied or somewhat preceded by similar increases in the prices of grains and soybeans. Between July 1973 and July 1975 the average price of crude oil processed by American refineries tripled, although the pump price of gasoline increased only 80 percent. Between January 1972 and January 1974, however, the market price of wheat almost quadrupled, while the price of soybeans, corn, and grain sorghum more than doubled. The price rise in fuels and the price rise in grains had different causes. The former was triggered by an increase in the price of rapidly rising imports of oil, the latter by a massive sale of grain to the Soviet Union that eliminated a surplus that had long overhung the market. In both cases, the soaring *prices* of the basic commodities and of the products made from them, such as gasoline, electricity, fuel oil, bread, and meat, tended to obscure the fact that the basic *costs* of the products also were rising, and for somewhat similar reasons.

These rapid increases in the cost of energy stimulated increased exploration for oil and gas, increased production of feeds, government-assisted exploitation of oil shale, a new surge of investment in Canada's tar sands, accelerated research and development of substitute fuels to be derived from coal, and expanded interest in the potentials of geothermal, solar, and fusion power. At the same time, ways to economize on energy use were given serious attention and in 1974 the consumption of both gasoline and meat dropped for the first time in many years.

ENERGY USE IN THE FOOD SYSTEM

Agriculture uses about 3 percent of the petroleum, 3 percent of the electricity, and 2 percent of the natural gas consumed in the United States. However, the total food production and delivery system, as Eric Hirst has demonstrated (1973, p. 21–22), requires more than five times this amount, because of the energy costs of transport, processing, and distribution. The efficiency of the system is low; it takes about 10 calories of fossil-fuel energy (Hirst's calculation for 1973 is 8.1 but he neglects the energy cost of machinery and buildings) to produce each calorie of food on the table. Only a nation in which the fossil fuels can be obtained at a relatively small energy cost can afford this degree of subsidization of the food system.

The overall system efficiency of the food-energy system is almost unbelievably low (see figure 11.16). Probably about 1 percent of the solar energy falling on pastures and planted fields in the United States is converted by photosynthesis to living plant tissue (of feed and food crops); of the amount of such plant tissue grown, perhaps 10 percent is actually eaten as feed by animals or as food by humans. In the United States, as we saw in chapter 9, it takes more than 16,000 calories of plant food and feed to provide the average daily per capita food supply of about 3,300 kcal; but to the loss in wasted materials and in animal maintenance, we must add the input of inanimate energy into the food-production system, about 33,000 kcal per person per day (at 10:1). It thus takes about 49,000 kcal of energy input to put 3,300 kcal on the table, an efficiency of about 6.7 percent. But we must also make an estimate of how much of our daily food supply we do not digest—discarded bacon grease, discarded fat on steaks and ham, food purchased but not eaten because of spoilage, surfeit, dislike, or feeding pets. In the United States that figure is at least 10 percent and

FIGURE 11.16
Energy flow through the food-supply system of the United States. Note that the drawing is not to scale. Conversion of solar energy to plant energy available for food and feed is less than 1 percent efficient and could not be shown to scale.

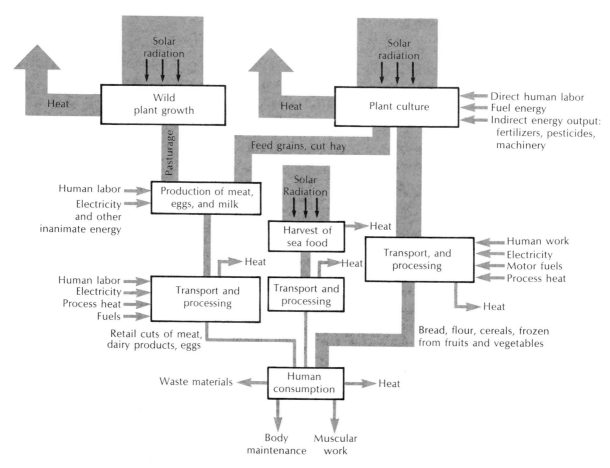

probably much higher. The efficiency of the food system, including photosynthesis, is the combination of all the step efficiencies:

$$(0.01) \ (0.10) \ (0.067) \ (0.90) = 0.00006 \text{ or } 0.006 \text{ percent.}$$

In early 1974, fossil energy as gasoline, fuel oil, and electricity was delivered to farms at an average price of about $12 per million kcal. This energy was used to produce work, useful heat, and light at an efficiency of about 20 percent. The useful output thus cost about $60 per million kcal. By contrast, farm labor at $3 an hour, producing perhaps 125 kcal of work an hour, cost $24,000 per million kcal. The price of inanimate energy will have to rise very greatly, or the price of labor decrease drastically, or both, before it would be worthwhile replacing machines with human beings on the American farm. This is true of the entire food system. At 1975 retail prices the food energy on the average table cost about $2,100 per million kcal; the inanimate energy that went into producing and preparing that food cost about $18 a million kcal, and even at an input-output ratio of 10:1, the inanimate energy represented only about 9 percent of the cost of the food on the table.

The concept of *feed-use efficiency* is important in analyzing food-energy systems. The pig's being three times more efficient than a steer in converting plant energy to meat energy is insufficient reason to argue that we should abandon beef in favor of pork. Range cattle get between 71 and 76 percent—these figures are for the period between 1940 and 1969 and are derived from data published by the United States Department of Agriculture (1970)—of their food from pasture, most of which is not fit to grow food crops on. In contrast, between 71 percent and 82 percent of a pig's diet is grain, and between 13 percent and 16 percent is other concentrated feed that is either acceptable food for man or is grown on land that could grow food. In other words, the range cow does not compete with man for the use of arable land; the pig does. If the human population of the United States started to press on available food supplies, we might give up pork rather than beef, despite their relative conversion efficiencies—or we would start raising hogs on waste materials. The grain-fattened beef animal, however, is another matter. Between 1965 and 1969, 94.1 percent of its food was grain, hay, and concentrates (United States Department of Agriculture, 1970). Man gets from the meat of grain-fattened cattle only 5 percent of the food energy he could get from eating directly the grain they are fed. In a food pinch, the grain-fattened steer would disappear while his range cousin would remain a valuable part of the national food system.

Under certain conditions even sheep and goats, creatures whose low plant-to-meat conversion efficiency is in part explained by the work they expend in seeking food, are net benefits to the human food system, especially when one adds their milk into the balance. However, if sheep did not produce wool, they would be displaced from most of their habitat by cattle. Feed-use efficiency can be calculated in the following way. A sheep is fed 16 percent of its food; it yields in meat energy 2.7 percent of the feed energy it takes in; its feed-use efficiency is

$$\frac{2.7}{16} (100) = 17\%.$$

A range steer is fed 27 percent of its food; it yields 4.6 percent in meat energy of the feed energy it takes in; its feed-use efficiency is

$$\frac{4.6}{27} (100) = 17\%.$$

A pig is fed 86 percent of its food; it yields 14.3 percent in meat energy of the feed energy it takes in; its feed-use efficiency is

$$\frac{14.3}{86} (100) = 17\%.$$

The feed-use efficiency of the average milk cow is 23 percent; of the average grain fattened steer, 5 percent. (All figures in these calculations are taken from or based on national statistics compiled by the United States Department of Agriculture [1970].)

In the forty years from 1930 through 1970, the demand on agriculture to provide food for domestic consumption has increased substantially, because of changes in the nation's meat eating and because of increased population. In 1930 (see table 11.9), 70 percent of the meat calories on the American table were pork and 26.5 percent were

Table 11.8 United States Energy Consumption by Use (1969) (kcal × 10¹⁵)

Sector	Space heating	Air conditioning	Water heating and cooking	Work	Process heating	Totals	Percentages
Residential	1.75	0.13	0.63	0.63		3.14	20.9
Commercial	1.08	0.32	0.21	0.53		2.14	14.3
Industrial:				1.17	4.43	5.60	37.3
Iron and steel					(0.87)	(0.87)	(5.8)
Aluminum					(0.19)	(0.19)	(1.3)
Cement				(0.02)	(0.12)	(0.14)	(1.0)
Transportation:				4.14		4.14	27.5
Automotive				(1.95)		(1.95)	(13.0)
Bus				(0.03)		(0.03)	(0.2)
Truck				(0.86)		(0.86)	(5.7)
Rail and subway				(0.16)		(0.16)	(1.0)
Air				(0.52)		(0.52)	(3.5)
Ship				(0.23)		(0.23)	(1.5)
Pipeline				(0.39)		(0.39)	(2.6)
Totals	2.83	0.45	0.84	6.47	4.43	15.02	100.0
Percentages	18.8	3.0	5.6	43.1	29.5	100.0	

NOTE: Figures with parentheses are included in the subtotals for the Industrial and Transportation sectors and therefore are not added to these subtotals to obtain the figures on the next-to-bottom line.
SOURCE: Associated Universities, Inc., "Reference Systems and Resource Data for Use in the Assessment of Energy Technologies" (a report made to the United States Office of Science and Technology, April 1972), table II-3, p. 12.

BOX 11.1 Is the Pig the Best Meat Producer?

Experiments as well as calculations made from feed-consumption and meat-production data show that the pig is the most efficient converter of plant energy into meat among the animals whose meat man eats in quantity. Under ideal conditions, the pig can convert more than 22 percent of the energy in its feed to edible meat, far and away better than any other common meat animal can do. Does this make the pig man's best meat producer? Should people whose population is pressing on their food resources abandon lamb and beef as protein sources? If Americans wish to export more grain, either to feed hungry nations or to pay for foreign oil, should they switch from beef to pork? Such questions are not as simple to answer as it may seem.

As we have seen from calculations in this chapter, the *feed-use efficiencies* of the cow, the sheep, and the pig as meat producers in the United States are almost identical—despite energy-conversion efficiencies ranging from 2.7 to 14.0 percent (on the basis of edible energy at the carcass stage of meat processing). This means that the same amount of raised feed goes into producing a pound of lamb or beef as goes into the production of a pound of pork. When it is considered that more of the energy in the pork carcass is discarded as unwanted fat (including bacon drippings) than is discarded from the carcasses of the other animals, a case can be made for the pig's being the *least* efficient of America's common meat animals.

In China there can be no question that the pig, both physiologically and socially, is the most efficient meat animal. It takes little room to raise, is fed entirely from waste materials, and requires minimal supervision.

The lesson here is that the physiological conversion efficiency of an animal becomes important to man only when a scarcity of food for both men and animals develops and where non-arable grazing land is also scarce.

---

beef; meat calories formed 15 percent of the diet. In 1970, pork had slipped to 55.5 percent, beef had risen to 42.8 percent, and meat formed over 17 percent of the diet.

In 1952 the average American consumed equal amounts of beef and pork, about 70 pounds of each. In 1972, twenty years later, per capita consumption of beef was up to 118 pounds but pork was only 67. It was not until 1968, however, that beef passed pork in energy content. These changes may appear small, but when calculated on the basis of plant-to-meat multipliers, it can be seen that the 20 percent increase in daily consumption of meat calories and the switch from pork to beef have caused a 31 percent increase in the plant-food requirement to support the new pattern of meat eating. Because the population increased 65.7 percent in those forty years, the total increase in plant food required to sustain the meat supply was (1.657) (1.31) = 2.17 times the 1930 level. Moreover, during this period there was a large increase in the percentage of the total food eaten by cattle and chickens that was supplied as grain and other prepared feed, so that the increase in the demand on arable land to produce feed for animals was even greater than these calculations indicate.

An energy input-output analysis for animal agriculture (see table 11.10) shows that the overall energy conversion efficiency dropped from 9.7 percent in 1955 to 7.6 percent in 1972; during the same period feed-use efficiency fell from 15.0 percent to 11.8 percent.

It is useful to distinguish between *cultural* energy, the flow of which is directed by man (muscle power, fossil energy, and so on), and *geophysical* energy (solar radiation, windpower, and the like) (see figure 11.17). Man could not exist without solar radiation and photosynthesis, neither of which is under his direction. How he uses cultural energy in agriculture has a great deal to do with the productivity and efficiency of his food systems and with the life-styles of his societies.

In the mechanized agriculture of the United States, cultural energy is used most efficiently in the production of hay, sorghum, and corn silage, all of which are animal feeds; each of these products yields between 5 and 6 units of digestible energy for every unit of cultural-energy input. One could speak of 500 or 600 percent efficiency, but only if one remembers the enormous geophysical energy input that has not been included in the calculation. The grasses (corn and other cereals) and the legumes (especially soybeans) are the great heat pumps of the food-energy system. Corn, for example, returns 2.8 to 4.4 times the cultural-energy input. Not all grasses and legumes are that profitable: sugarcane and rice, both grasses, return less than 1.4 units of digestible energy for each unit of cultural energy, while the leguminous peanut does no better. Legumes, however, more than make up for their

Table 11.9 Changes in U.S. Meat Eating Pattern (1930–1970)

Year	Meat on the table (kcal per capita daily)					Plant-food energy equivalent
	Beef	Pork	Lamb	Offal	Total	
1930	116.4	307.7	16.4	15.8	456.5	8,522
1935	129.4	221.9	17.8	14.4	383.5	8,084
1940	131.3	337.3	16.2	17.2	502.0	9,362
1945	148.6	305.3	17.8	22.4	494.1	9,860
1950	150.9	317.6	9.9	18.0	496.4	9,491
1955	190.9	306.2	11.2	19.6	527.9	10,913
1960	185.9	297.9	13.2	18.0	515.0	10,743
1965	209.6	269.2	10.1	17.8	506.7	11,098
1970	234.9	304.8	9.0	19.0	567.7	12,313

Sources: Based on data from United States Department of Agriculture (1965, 1968).

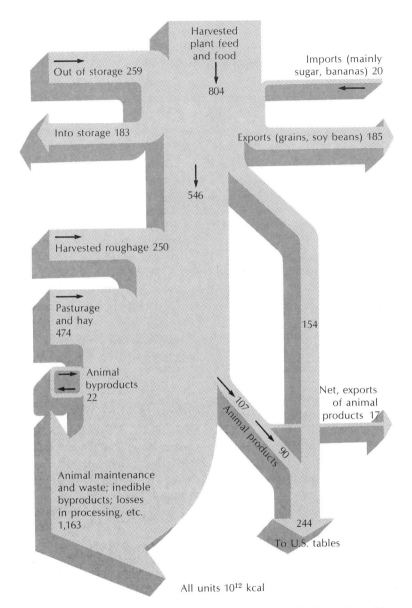

Out of storage 259

Harvested
plant feed
and food

804

Imports (mainly
sugar, bananas) 20

Into storage 183

Exports (grains, soy beans) 185

546

Harvested roughage 250

154

Pasturage
and hay
474

Animal
byproducts
22

107

Net, exports
of animal
products 17

Animal products

90

Animal maintenance
and waste; inedible
byproducts; losses
in processing, etc.
1,163

244

To U.S. tables

All units 10^{12} kcal

FIGURE 11.17
Energy flow through agriculture in the
United States in 1970, exclusive of cultural
energy.

relatively low energy return by their bounteous yield of protein and by
the fixation of nitrogen in the soil that accompanies their growth. The
energy in the food eaten by humans and food-producing animals in
1970 totalled 1.25×10^{15} kcal. The conversion efficiency, if one
neglects the unused energy in discarded vegetable material such as
stalks, shucks, and cobs of corn, was 20.3 percent.

The average daily diet contains about 2,350 kcal of plant food and
950 kcal of meat and animal products. Each calorie of edible animal
product required (in 1972) 13.2 calories of plant feed to produce, of

Energy Use in the Food System

which 4.7 were supplied from pasture and 8.5 by raised feed (some fish and animal protein was included in this portion). The total plant feed and food produced in 1970 was 804×10^{12} kcal (see figure 11.17), net exports were 165×10^{12} kcal, and a net of 76×10^{12} kcal was drawn from storage. A total of 700×10^{12} kcal of plant energy was used to produce food for tables in the United States. Only 22 percent of this was consumed directly by humans; the remaining 78 percent was fed to animals. Similar calculations for subsequent years can be made from data published by the Department of Agriculture in its annual publication, *Agricultural Statistics.*

Striking trends in agriculture in the United States during the past fifty years have been mechanization and increased productivity. Tractors and self-propelled machines have replaced horses, mules, and human labor. Farm output has increased as a result of higher-yielding crops and intensive use of fertilizers, herbicides, insecticides, and irrigation. Farm output per man-hour has increased very rapidly, because of the additional effect of mechanization.

ENERGY USE IN INDUSTRY

In 1974, 37.6 percent of all inanimate energy consumed in the United States went into industry. The primary metals industries used a fifth of this amount, principally in iron and steel making and aluminum manufacture. The chemical industry consumed another fifth of the industrial total. Petroleum refining and related industries required over 5 percent of the total, as did paper and paper products (see table 11.11). At least half the energy used was devoted to the extraction and processing of nonrenewable resources.

The energy required in 1971 to add one 1967 dollar of value to the product are given, from the most energy-intensive to the least, in table 11.12. The figures show that those industries that exploit and process natural resources require much more energy per dollar of product than other industries. Also significant, but not shown in the table, is the fact that, whereas other industries show a progressive historical decline in energy use per constant dollar of product, energy use in the energy-intensive industries recently has been increasing. The largest increase is in petroleum and coal products, which in 1971 used 34 percent more energy per constant dollar of value added than in 1954; this group has shown a sustained increase in energy intensiveness for twenty years. Four other groups used more energy in 1971 per dollar of added value than they did in 1967: food, up 8 percent; primary metals, up 7 percent; paper, up 3 percent; and stone, clay, and glass, up 2 percent. In striking contrast to the trend in chemicals and all other manufacturing, which has shown a steady improvement in the economic efficiency of energy use (decline in energy used per dollar of

Energy and the United States

Table 11.10 Input-Output Energy for U.S. Animal Agriculture

Form of energy	1955	1965	1970	1972
Input (10¹² kcal)				
Concentrates	425.6	556.8	617.6	656.0
Harvested roughage	211.2	240.0	249.6	259.2
Pasture	339.2	470.4	473.6	508.8
Totals	976.0	1,267.2	1,340.8	1,424.0
Output (10¹² kcal)				
Beef	16.83	23.22	26.85	27.76
Veal	1.07	0.69	0.40	0.31
Pork	20.07	20.34	24.52	24.89
Lard	10.88	8.36	7.65	6.24
Lamb	0.91	0.78	0.66	0.65
Chickens	1.37	2.54	3.31	3.48
Turkeys	0.59	1.10	1.27	1.40
Eggs	5.99	6.08	6.34	6.43
Milk	36.35	36.63	34.50	35.37
Edible offal	1.17	1.20	1.30	1.52
Totals	95.23	100.94	106.80	108.05
Input-output ratio	10.3	12.6	12.6	13.2
Conversion efficiency	9.7%	7.9%	8.0%	7.6%
Feed-use efficiency	15.0%	12.7%	12.3%	11.8%

SOURCE: United States Department of Agriculture, *Agricultural Statistics* (Washington, D.C.: United States Government Printing Office, 1956, 1966, 1971, 1973).

Table 11.11 Industrial Fuel Consumption by Major Users (1969) ($\times 10^{15}$ kcal)

Industry	Coal	Natural Gas	Petroleum	Electricity	Total	Percentages
Primary metal industries	710	235	88	129	1,162	20.1
Chemicals and allied products	167	332	409	162	1,070	18.5
Petroleum refining and related industries	0	276	456	22	754	13.1
Food and kindred products	66	162	39	34	301	5.2
Paper and allied products	117	93	61	28	299	5.2
Stone, glass, clay, concrete	102	122	25	28	277	4.8
All mineral and other manufacturing industries	244	1,303	207	156	1,910	33.1
Total	1,406	2,523	1,285	559	5,773	100.0

SOURCE: United States Bureau of Mines (1971), p. 26.

value added), the raw-materials processing and manufacturing industries, in the 1967–1971 period, declined in economic efficiency.

Although the industrial sector of the economy consumes more energy than any of the other three sectors, its energy consumption has been growing more slowly than that of any of the others. There are two reasons for this. Industry produces goods. As an industrial nation matures, its internal market tends to become saturated. Then, as other industrialized nations mature, the external markets for all tend to become saturated. But the internal and external markets for services—rather than goods—commonly increase as the demand for goods decreases; thus the economy of a mature industrial nation tends to produce proportionately more services and fewer goods. The shift becomes apparent in the distribution of energy consumption: relatively

Energy Use in Industry

more goes into the commercial and residential sectors than into industry. The second reason is that competition and the ability to effect energy economies of scale keep the use of energy in industry improving in efficiency; consequently, the energy consumed in production grows at a lesser rate than production itself.

Especially where energy cost is a substantial portion of the cost of the product, industry has moved to improve the efficiency of energy use. The electric-power industry is the outstanding example. The thermal efficiency of fuel-burning power plants was only 3.6 percent in 1900; it reached 32.5 percent in 1970, and some of the newest plants are operating at 40 percent. The increase was achieved by raising the temperature and pressure of steam leaving the boiler, by replacing piston-driven engines with well-designed steam turbines, by lowering the condenser temperature and pressure, by feedwater heating, and by increasing the size of generating units.

The aluminum-reduction industry also has paid attention to efficiency. To produce metallic aluminum from the parent material (aluminum oxide) requires a good deal of electrical energy. Because aluminum made its markets by taking them away from other metals, there being no unique use for aluminum, the matter of production cost has always been an important competitive factor in the market strategy of aluminum producers, who have managed steadily to reduce the amount of energy required to produce a pound of aluminum metal, to the point that in 1974 it took half as much electricity as it did forty years before.

Iron and steel makers have been moving in the same direction. From 1959 through 1969 the amount of energy required per ton of pig iron from blast furnaces decreased 19 percent. In addition, a partial change-over from open-hearth steel furnaces to basic-oxygen furnaces strikingly reduced the energy cost of making steel. A study at the Lackawanna plant of Bethlehem Steel indicated that the open-hearth process required 3.05×10^6 Btu per ton of steel, whereas the basic-oxygen process consumed only 0.38×10^6 Btu per ton. In 1972, 53 percent of steel-making furnaces were of the new type. Even when the energy cost of producing the oxygen required in the new furnaces is added in, they represent a very significant energy savings.

ENERGY USE IN TRANSPORTATION

The distribution of the various quantities of energy used by the different modes of transportation in the United States in 1970 is shown in table 11.13. From 1950 through 1970, the consumption of energy by automobiles increased 171 percent; the energy cost per passenger-mile increased from 1,200 to 1,350 kcal. During the same period there was a 179 percent increase in energy use by trucks, and airplanes' share

Table 11.12 The Economic Efficiency of Energy Use in U.S. Industries (1971)

Industry	Energy required (kcal $\times 10^3$) per $1.00 (1967)
Petroleum and coal products	73.0
Petroleum refining	49.1
Primary metal industries	44.4
Stone, clay, and glass products	41.1
Paper and allied products	36.1
Blast furnaces and steel mills	20.9
Food and kindred products	11.2
All other manufacturing	5.2

SOURCE: The Conference Board (1974), p. 21.

of transport energy rose from 2 to 11 percent. At the same time, there was a general decrease in the efficiency of energy use in passenger transport, marked by a large increase in the energy cost of moving people in airplanes, as jet planes replaced propeller-driven aircraft. An even larger *decrease* in the energy-cost of moving people by railroad, representing the wholesale cancellation of lightly used intercity trains and the advent of the diesel locomotive, had little effect on the overall system efficiency because of the continued switch of passengers from trains to less efficient automobiles and planes.

The importance of the automobile is emphasized by the fact that most workers drive to their jobs. Despite the costs and parking problems, most drive alone. In 1973 there were, according to the Motor Vehicle Manufacturers Association (1974), 101,237,000 passenger cars and 23,241,000 trucks and buses registered in the United States, representing 46 percent of the world's automobiles and 40 percent of its trucks and buses. There were 797,000 persons employed in manufacturing cars, and 2,731,000 employed in selling and servicing them. There were also 568,000 people working on state, country, and local roads. Automobile manufacturing in 1973 consumed 33 percent of all zinc, 20 percent of all steel, 9 percent of all aluminum, 8 percent of all copper, and more than 60 percent of all rubber consumed that year. In 1972 motor vehicles were involved in 56,600 traffic deaths and they consumed 105 billion gallons of gasoline and diesel fuel (Federal Highway Administration, 1974, p. 5), 47 percent of the total consumption of petroleum products. About half of this amount, 52 billion barrels, is believed to have been used for trips of less than three miles. Elimination of one out of every three of those short trips would have saved 17 billion gallons of gasoline, an amount equal to more than half of all the oil imported that year. If no driver had exceeded 55 miles per hour in 1972, an additional 7 billion gallons of gasoline would have been saved.

A summary of motor-vehicle travel in the United States in 1972 is given in table 11.14.

The United States has about five hundred automobiles for each 1,000 inhabitants, or for each 640 inhabitants between the ages of fourteen and sixty-five. This is far and away the highest such figure in the world. Towns and cities in the United States for the most part have been designed to accommodate the car; suburbia and the automobile are symbiotic. Furthermore, the distances between cities and between cities and their suburban or satellitic centers are greater than in Europe or Japan, so that mass transit is more expensive here than there. Also there is less precedent for public transport in the United States; it is considered demeaning to use the subway or a city bus. Finally, there is the entrenched political power of the automobile, oil, and highway interests that tends to reduce consideration of alternative modes of transport.

Table 11.13 Distribution of Energy Used in U.S. Transportation (1971)

Type of Vehicle	Energy (kcal × 10^{12})	Percentage of total
Automobiles	2,180	48.2
Trucks	868	19.2
Pipelines	401	8.9
Military	363	8.0
Aircraft (civilian)	358	7.9
Waterways	176	3.9
Railroads	137	3.0
Buses	31	0.7
Motorcycles	9	0.2
Total	4,523	100.0

SOURCE: Mutch (1973), p. 15.

ENERGY USE IN HOME AND COMMERCE

The energy used in the residential sector is difficult to distinguish from that used in the commercial sector; the United States Bureau of Mines does not do so in its annual summaries of energy use. But Stanford Research Institute has estimated the respective shares for major end uses in 1968 (see table 11.15). Note that three times as much energy is used in the home for heating water than is used for cooking. The useful heat difference is even greater, because water-heating systems are more efficient than cooking systems. On a per capita basis, energy consumed for household refrigeration increased 66 percent from 1960 through 1968 (Stanford Research Institute, 1972); for television, 91 percent; for clothes drying, 97 percent; for food freezers, 135 percent, for air conditioning, 190 percent; for dishwashers, 299 percent; for cooking—not at all. For the same period in the commercial sector, marked increases were shown in energy consumption for air conditioning (93 percent) and for lighting and electric drive (about 700 percent for the two combined).

Electric lighting has had a revolutionary impact on human life, but so have long-distance communication of sounds and images, and labor-saving devices in the home. The first is illustrated by television, the second by clothes dryers. The automobile as well as the organization of industrial society have tended to take the family out of the home and to fragment it. Television is a centripetal force acting to pull the family back into the home. Restrictions on the consumption of energy for unproductive transport may be accepted much more readily by a society that can substitute television for a day in the country than by one that cannot. Television can be a labor-saving device; countless mothers use it for "baby-sitting," a mechanism to tie down and entertain the children and free the mother for other tasks. Those other tasks involve more and more the operation and monitoring of powered machines such as clothes dryers. Electric and gas-fueled appliances supposedly reduce the time required of the housewife to maintain her home and supply her family's needs. In fact, it is doubtful that they do. What they do is substitute for the

Table 11.14 Motor Vehicle Travel in the U.S. (1972)

Type of vehicle	Share of mileage	Share of fuel	Average miles per vehicle	Average miles per gallon
Passenger cars	77.8%	69.6%	10,184	13.49
Motorcycles	1.3	0.3	4,500	50.00
School buses	0.2	0.3	7,414	7.37
Commercial buses	0.2	0.5	30,968	4.39
Trucks	20.5	29.2	12,229	8.46

SOURCE: Federal Highway Administration (1974), p. 52.

Table 11.15 Estimated Distribution of Residential and Commercial Shares of Energy Consumption (1968)

End use	Residential shares	Commercial shares	Combined shares
Space heating	57.5%	53.7%	56.0%
Water heating	14.9	8.4	12.3
Air conditioning	3.7	14.3	7.9
Refrigeration and freezing	7.5	8.6	7.9
Lighting	3.6	7.7	5.2
Electric drive	2.5	5.4	3.7
Cooking	5.5	1.8	4.0
Television and clothes drying	4.8	0.1	3.0
Total	100.0	100.0	100.0

SOURCE: Stanford Research Institute (1972), p. 36, 67. Data include allowance for waste heat generated in electricity production, but exclude non-energy uses of fuels, such as of asphalt and road oil for roofing and paving.

help the housewife once had from daughters, hired help, and a widowed mother or maiden aunt. In addition, they make the time available that the modern housewife needs, because of the demise of home-delivery services, to do her shopping. The amount of energy required by such appliances depends upon the efficiency of the process involved and the amount of use made of the appliance. Approximate percentages of energy used in the average home for heating, cooling, cooking, lighting, television, and various appliances are shown in table 11.16.

Both the residential and commercial portions of the nation's energy consumption have increased over the last few decades because of the increasing use of electricity in these sectors and the relative inefficiency of electrical systems. The increase in the commercial sector is due mainly to increased air conditioning, increased lighting, and increased electrical power for elevators, escalators, and office machinery.

Home refrigeration has increased much faster in recent years than has commercial refrigeration. In 1960, 44 percent more energy was used for commercial refrigeration than for home refrigeration. By 1967, the two were almost equal and, the following year, energy consumed by home refrigerators for the first time exceeded that used in commercial establishments.

MILITARY USE OF ENERGY IN THE UNITED STATES

The petroleum products procured by the United States Department of Defense during the twenty-two years from 1952 through 1973 grew from about 2.5 percent of the gross energy consumption of the nation in 1952 to 4.2 percent in 1968, principally because of a great

Military fuel procurement in the United States between 1952 and 1973. It can be seen from this diagram that the Air Force takes the most fuel, the Navy comes next, and the Army runs on what is left. (Based on data supplied by the United States Department of Defense.)

increase in the use of jet fuels for aircraft (see figure 11.18). One must remember that such figures do not include the energy expended in manufacturing military hardware such as combat aircraft, tanks, battleships, and supplies such as uniforms, ammunition, and food; they do not include energy used by civilian contractors in constructing and repairing military installations and in transporting military personnel, supplies and equipment; they do not include the energy used by the civilian employees of the Department of Defense in maintaining themselves and their families and in traveling to and from their places of employment; and they do not include purchases of electricity, natural gas, and coal by the military. If all these were added in, it would be seen that perhaps 10 percent of the national energy consumption is used for military purposes. Between 1956 and 1974, the military services obtained more than 35 percent of their petroleum needs each year from foreign sources. During this period foreign oil was cheaper than domestic oil. When this price differential abruptly reversed in 1973–1974, direct purchases by the military from foreign sources dropped quickly, ending in late 1974.

THE OIL COMPANIES

The operations of the major international oil companies, most of which are based in the United States, were discussed in chapter 9. Because of the widespread interest in the domestic operations of the

Energy and the United States

oil companies, their importance to the national interest, and persistent charges of monopoly and conspiracy, a brief description of the "oil companies" may be useful as an introduction to a sketch of the way energy is managed in the United States.

In 1870, the Standard Oil Company (Ohio) was incorporated, John D. Rockefeller being one of the founding partners. The company grew rapidly, and in 1882 the Standard Oil Trust was created in order to limit production, fix prices, and allocate markets. The Standard Oil Trust, first of its kind, succeeded in cornering the petroleum market and driving most of its competitors out of business. The spread of the trust form of organization to iron and steel, sugar-refining, and other industries and their market control led to wide-spread public concern and protest; two of the consequences were enactment of the Sherman Antitrust Act in 1890 and a suit brought by the federal government under that act against Standard Oil in 1906, which ultimately resulted in the forced divestiture of thirty-three subsidiary companies. Some of these subsidiaries retained their original names, several of which began with "Standard Oil Company of" followed by the name of the state in which they were incorporated, and thus a persistent belief grew up that these companies were still controlled by a central group, when in fact, one person could not be a director or hold a position of management in any two of the divested companies. Three of the Standard Oil companies were, at the start of 1974, among the four largest oil companies in terms of market value.

After the Standard Oil Trust was dismantled, the market for petroleum products in the United States mushroomed and new fields were discovered; other companies emerged as potent factors in production, refining, and marketing. Texaco, Mobil, and Gulf ultimately became major international companies, along with Standard Oil Company (New Jersey), Standard Oil Company of California, and two others that are not American, British Petroleum and Royal Dutch-Shell.

After World War II, some of the smaller oil companies in the United States began to expand into international oil production, refining, and marketing, among them, Standard Oil Company (Indiana), Phillips, Atlantic Richfield, Continental, Sun, Marathon, Union Oil of California, and Getty. The top thirty companies in the United States, measured by the value of their stock on January 1, 1975, included nine oil companies (table 11.17). Although the profits of these oil companies jumped in 1974, their average returns on equity during the previous five years were lower than those of most of the other companies in the top thirty, and even their 1974 profits, termed "obscene" by Senator Henry Jackson, fell below those of Coca Cola, Merck, American Home Products, Eli Lilly, and Dow Chemical. The 1974 profits proved ephemeral. In 1975 the profits of the large oil companies fell again because an oversupply of high-priced crude oil

Table 11.16 Energy Used in an Average U.S. Household

System or device	Share of total energy consumed	Share of electric energy consumed
Heating system	56.9%	11.5%
Water heater	14.7	15.5
Refrigerator	5.9	17.4
Stove and oven	5.4	6.6
Air conditioner	3.7	10.8
Lighting system	3.5	10.5
Television	3.0	9.0
Food freezer	1.9	5.6
Clothes dryer	1.8	3.6
Irons	0.4	1.3
Washing machine	0.4	1.0
Electric skillet	0.3	1.0
Dishwasher	0.3	0.9
Coffeemaker	0.3	0.9
Radios	0.3	0.8
Electric blankets	0.2	0.6
Fans	0.2	0.5
Portable heaters	0.2	0.5
Vacuum cleaner	0.1	0.4
Toaster	0.1	0.3
Other*	0.4	1.3
Totals	100.0	100.0

NOTE: *Includes dehumidifier, broiler, hot plate, clocks, humidifier, mixer, shavers, food disposers, hair dryer, etc.
SOURCE: Stanford Research Institute (1972), p. 33–62.

had resulted in weakened retail demand and because a large part of their 1974 gains had come from oil "in the pipeline" revalued upward by the OPEC price hikes.

Investments of American oil companies in foreign exploration for oil and in domestic reserves of coal and uranium have been dictated by a desire to stay in business as reserves of oil and gas in the United States become depleted. By 1970, oil companies had purchased coal companies whose combined output represented 35 percent of the coal market in the United States. At that point, the federal government started to contest further acquisitions of active coal companies by oil companies, but did not attempt to prevent the purchase of coal reserves. Among the companies reported to hold extensive coal acreage are Exxon (formerly Standard Oil of New Jersey) and Atlantic Richfield. Oil companies produce quite a bit of the natural gas of the United States, some of which is associated with oil in natural reservoirs. They also own some of the pipelines through which the gas is transported. In addition, however, there are many large and small natural-gas producers and pipeline companies that are not oil companies and are not subsidiaries of oil companies. Some of the large companies whose main business is natural gas production, gathering, and transmission are Texas Eastern Gas Transmission, El Paso

Natural Gas, and Coastal States Gas Producing Company. In all the major phases of the oil and gas business—producing, refining, transport, and marketing—there is strong concentration of assets and facilities in the hands of fewer than a dozen companies.

Apart from the big oil companies, there are hundreds of medium and small enterprises, collectively known as "independent" producers. The increasing cost of staying in a business in which the chances of success continually decrease has caused a great attrition among the independents over the past twenty years.

The large petroleum companies that are based in the United States and operate also in other countries are of great economic importance to this country. They have returned a great deal of money to the United States as profit from their foreign operations and this money has helped the nation's balance of payments. The balance of payments consists of trade (exports less imports), services (inflows less outflows for royalties, management fees, and so on), and capital investment

Table 11.17 Thirty Top U.S. Corporations by Market Value (January 1, 1975)

Company	Assets	Market value on 1-1-75 (in millions)	Earnings	
			Five-year average return on equity	Latest 12-months (as of 1-1-75)
American Telephone and Telegraph	$74,047	$24,979	9.7%	10.7%
International Business Machines	14,027	24,908	19.7	21.1
Exxon	42,532	14,453	16.3	22.4
Eastman Kodak	4,695	10,145	20.9	19.2
General Motors	20,381	8,807	15.3	7.8
Sears Roebuck	15,216	7,615	14.0	12.4
Procter and Gamble	3,071	6,714	18.8	18.1
Standard Oil (Indiana)	8,915	6,393	13.2	22.1
General Electric	11,977	6,096	17.2	18.0
Texaco	17,176	5,672	16.6	23.6
Minnesota Mining and Manufacturing	2,841	5,267	20.4	20.8
American Home Products	1,242	5,218	29.5	31.0
Dow Chemical	5,114	5,083	18.7	34.1
Merck	1,243	5,003	28.2	28.7
Eli Lilly	1,265	4,694	23.8	24.3
Johnson and Johnson	1,406	4,673	19.2	19.5
E. I. Dupont	5,969	4,428	14.9	14.8
Atlantic Richfield	6,152	4,267	9.1	15.1
Xerox	4,090	4,081	24.3	22.0
Schlumberger	1,328	3,963	17.3	22.6
Standard Oil (California)	11,634	3,775	13.7	20.0
Mobil Oil	14,074	3,667	14.3	20.5
First National City Corporation	57,839	3,503	14.5	16.4
Weyerhaeuser	3,122	3,470	18.1	23.4
Gulf Oil	12,503	3,431	11.9	18.8
Phillips Petroleum	4,028	3,291	11.2	21.0
Coca Cola	1,536	3,172	27.0	24.6
Shell Oil	6,129	3,158	11.2	17.5
Ford Motor	14,153	3,110	12.0	6.5
Burroughs	2,046	2,964	13.9	13.3

SOURCE: *Forbes Magazine*, January 1, 1975 and May 15, 1975. Data represent annual compilations of financial information on several hundred American companies.

(outflows for investment less inflow of profits). Until 1973, when grain exports reversed the trend, a growing trade deficit had to be offset by a net flow in the two other components, and this was obtained mainly from oil companies' profits and service revenues. Tanzer explains (1969, p. 44–48) the history of the close cooperation between their home governments and the major oil companies as due to the repatriation of profits, which he holds has made "the difference between a shaky solvency and bankruptcy" for the United Kingdom since World War II and has been of increasing importance to the United States. In the absence of direct government intervention in negotiations with oil-exporting countries, the oil companies are the only national instrument for obtaining foreign petroleum for use in the United States.

MANAGEMENT OF THE NATIONAL ENERGY SYSTEM

The national energy system is managed by a combination of profit-seeking enterprise and governmental control. Private enterprise is responsible for most of the exploration and all of the extraction and processing of energy raw materials, as well as for a large part of the conversion and distribution of energy to the ultimate consumer. It also sponsors much of the research related to energy.

The mixture of private and public involvement is different for different resources and forms of energy. Oil, coal, natural gas, and uranium are found and produced by private industry, with some useful exploration and production information being generated and provided by the government through agencies such as the United States Geological Survey, the United States Bureau of Mines, and the Energy Research and Development Administration. The government intervenes in exploration and extraction as the landlord of the public domain, protector of the physical environment, and guardian of the health and safety of workers. In the cases of crude oil and natural gas, but of neither coal nor uranium, the government has gone two paces further; it has regulated production rates for the maximum ultimate recovery (a form of conservation) and it has imposed price controls at the wellhead (on natural gas destined to cross a state boundary and on a certain portion of the nation's oil production whether it is to cross a state line or not) in order to protect the consumer from gouging by monopolies.

The processing and wholesale transport of oil, coal, and natural gas are in the hands of private enterprise. Although the milling of uranium ore and extraction of the metal are done by private companies, enrichment, because it is done by a secret process and because the investment required has not been justified by a commercial market, is a government monopoly—as has been until recent years the reprocessing of spent nuclear fuel elements to recover fissionable residues. The transport and distribution of natural gas, although mainly in the hands of private companies, is commonly subject to regulation by state and local governments.

The government may protect and encourage as well as restrain. The uranium industry in the United States was stimulated in its youth by high price levels guaranteed by the federal government. Later, research, development, and insurance subsidies similarly encouraged growth of the nuclear-power industry. For almost fourteen years, 1959 through 1972, the domestic petroleum industry was given market protection by an import-control program designed to maintain prices for domestic crude oil higher than those for foreign oil. The rises in the price of foreign oil that began in 1971 and sharply accelerated in 1973 rendered the program superfluous. By 1971 it was clear that the ostensible aim of the program, to maintain a surplus domestic production capacity, was no longer achievable, since all the domestic fields were then producing as much as they could without risking loss of ultimately recoverable oil.

Government regulation of the coal industry has focused on the health and safety of underground miners and the protection or restoration of surface values, especially in strip mining. There are no conservation regulations or production limitations such as those for petroleum. The market for coal is dominated by large users rather than small as in petroleum and the product price is determined mainly by negotiated long-term contracts.

Natural-gas production from fields where it is associated with crude oil is governed by the conservation regulations imposed on petroleum production by state conservation commissions. The Federal Power Commission exercises price control at the wellhead over all natural gas destined for shipment out of the state of origin and monitors the reporting of interstate pipeline companies. In addition, state regulatory commissions set allowable prices and grant franchises to local distributors of natural gas. In other words, natural gas, mainly because it is distributed to millions of residential and commercial users on a noncompetitive basis, is regulated as a utility is, from wellhead to consumer. Even within those states, such as Texas, where there is essentially a free market in natural gas, the state government may intervene to assure essential deliveries or to determine priority of distribution under conditions of scarcity. When the energy in coal, fuel oil, uranium, or falling water is converted into electricity, the conversion devices (power plants) and transmission facilities (high-voltage power lines) may be owned and operated by private or by public entities. Because most large dams could not have been justified for power alone, but can produce electricity incidental to other functions such as flood control, the federal government today owns most of the hydroelectricity generating capacity of the nation. Interestingly, however, a considerable portion of the power generated at such dams is transported and sold by private utility companies.

Permission to build electricity generating plants is under the supervision of the Federal Power Commission if they are to be on navigable streams or of the Nuclear Regulatory Commission if they are to be nuclear. Interstate transmission of electricity, no matter how generated, is regulated by the Federal Power Commission. State utilities commissions regulate distribution to consumers. Both state and federal agencies regulate the emission of deleterious substances from power plants.

Utilities are regulated for the protection of the consumer. The emphasis on low prices is understandable but has some drawbacks. First, it encourages a utility company to build its power plant close to the "load center"; and thus we find coal-fired power plants inside cities polluting the urban air, and nuclear plants being built on the outskirts of cities, with no consideration given to placing them underground. Second, this type of regulation fits a nonrenewable, abundant source of energy better than it fits a nonrenewable or scarce source. Present tensions between those who claim that costs of extraction and transport should govern the price of natural gas to the consumer and those who maintain that its value as an increasingly scarce resource that needs to be conserved should determine the price the consumer pays illustrate the problem of the regulatory agencies as well as the growing conflict of interest between the major consuming states and the major producing states. The trend of energy regulation, however, is clearly in the direction of extending the utility regulatory mode to complete energy-delivery systems.

Until recently, in peacetime, little attempt has been made to regulate *end uses* of energy. Indirectly the Clean Air Act of 1968 had a great impact on end use, causing high-sulfur coal to be replaced by low-sulfur coal and low-sulfur fuel oil and putting a great stress on available natural-gas supplies. In 1973, however, authority to allocate crude oil supplies and petroleum products began to be exercised by the Federal Energy Office and a gasoline-rationing system was devised, although not implemented. At the same time, end-use restrictions on the use of natural gas were imposed in some states and discussed at the national level. The expanded federal role in the management of the energy system produced a federally mandated national speed limit of 55 miles per hour and yearlong daylight-saving time.

IMPORTS AND EXPORTS OF ENERGY

The energy imports of the United States are mainly in the form of petroleum; the energy exports in the form of coal and agricultural products, mainly grain and soybeans. In 1974, 1,269 million barrels of crude oil and 967 million barrels of refined petroleum products were imported (United States Bureau of Mines, 1975). The total of 2,236 million barrels was 37 percent of the petroleum consumed in the United States. The net petroleum energy imports amounted to 3.11×10^{15} kcal, or about 18 percent of the total national energy consumption exclusive of food energy. Petroleum imports increased slowly in the years before domestic production peaked in 1970. Because foreign oil was considerably cheaper in those years, imports would have been much larger had import quotas not been imposed. When it became apparent that American oil no longer could supply the demand, the import restrictions were removed. In recent years, the volume of imported oil has grown much more rapidly than before 1970, in order to make up for the diminishing domestic supply and to take care of increasing demand. Imported oil, which used to come almost entirely from Canada and Venezuela, now originates to an increasing extent in Africa, Indonesia, and the Middle East. The flow of oil and natural gas from Canada, for a long time determined only by the market, has been restricted by actions of the government of Canada, moving to assume an adequate future supply for Canada in the face of declining discovery rates. Venezuela, a member of OPEC, has not been able to increase its oil production to keep pace with the rising needs of the United States; the latter, once the world's greatest exporter of oil, now finds itself being drawn into increasing dependency on imports of oil from distant countries whose political and economic interests differ so much from those of the United States that the difficulties of maintaining an independent foreign policy and an affluent society may be greatly intensified. Projected imports of natural gas in liquid form (LNG) will not alleviate the growing problem of energy dependence, for they would come from OPEC countries such as Algeria, Indonesia, and Saudi Arabia, and from the Soviet Union.

Although exceeded in energy content by exports of coal, the most valuable exports of energy are in agricultural products. In 1972, for example, exports of coal contained 0.38×10^{15} kcal of energy, while net exports of energy in farm and food products amounted to only 0.18×10^{15} kcal. But the agricultural exports carried prices equivalent to about $85 per million kcal while the coal was valued at about $1 per million kcal. The great difference in value of energy as food and energy as fossil fuel is of unusual importance to the balance of trade between the United States and other countries. Since 1971 energy in the primary agricultural products that are exported in large amounts, such as grains and soybeans, has been valued at forty times higher than energy in crude oil. This means that a 1 million kilocalories of exported wheat can pay for 40 million kilocalories of imported oil. This relation was not altered by the steep price rise of oil in 1973–1974 because the prices of exported agricultural products rose correspondingly.

As American industry long has had the capacity to produce more goods than Americans need or can consume usefully, so has American agriculture had the capacity to produce food surpluses. For both, the export market has helped absorb what Americans could neither use nor eat. From 1961 to 1973, for example, more than half the wheat grown was exported (United States Department of Agriculture, 1974, p. 1). Grain exports are much less predictable than oil imports because they depend to a great extent on good weather at home and bad weather abroad. In 1968, only 33.4 percent of the wheat produced was exported, but in 1972, 75.3 percent went to foreign buyers. The United States is the world's major exporter of rice; from 1961 to 1973 slightly less than half of all rice grown was exported (United States Department of Agriculture, 1974, p. 23). Likewise, about half of the soybean crop moves out of the country—as beans, oil, cake, and meal. In recent years, exports of corn (maize) and sorghum grain have equaled almost a quarter of the total raised. The ability to produce large surpluses of food may prove to be, in the long run, more important to the United States than Saudi Arabia's oil will be to it.

NATIONAL ENERGY PROBLEMS

The energy problems faced by the nation differ according to the time scale of resource depletion. Immediate problems are those of moderate scarcity and allocation of shortages stemming from bottlenecks or stoppages in the supply systems. We have seen sharp and for some people painful increases in the prices of basic foods and fuels. These result from increasing demand, restrictions of supply (both physical and political), and the workings of a market economy. The United States produces a surplus of food. For many years that surplus was stored, sold, even given away; it supported an ever less-efficient (in energy conversion terms) animal agriculture. In 1973, the remaining surplus melted away under the pressure of dramatically increased export sales. As for oil, production in the United States was falling and imports rising when the major oil-exporting nations seized command of international pricing (by having control of a large part of world production) and raised the price by five or sixfold within two years. The problems of high food prices, high petroleum prices, and the slowing down of a dynamic economy are interrelated. We could have lower food prices by sacrificing export earnings on farm products that could pay for most of the foreign oil we want. We may continue to produce our own oil more cheaply than we can buy foreign oil, but only at the cost of hastening the exhaustion of our own reserves and delayed but greater ultimate dependence on foreign oil. If we do not decide to trade what we can still produce cheaply (compared with the rest of the world)—food—for that which we can no longer produce cheaply—oil—our economy will be slowed, if not reversed, first by a heavy balance-of-payments deficit and then by persistent and debilitating fuel shortages.

In the intermediate term (between twenty and fifty years), we face the problem of replacing natural gas and domestic petroleum as sources of more than half of our energy consumption. Some combination of substitute fuels from domestic coal and oil shale, of increased nuclear power, and of foreign hydrocarbon imports will be required. These should not be much more costly than petroleum and coal are today; the need is to avoid incurring great social costs, from adverse environmental impact at home or hostile actions abroad, while building an efficient replacement system. In the distant term (more than fifty years) the problem is to replace the existing mined energy economy with one based on renewable resources or resources that are so abundant they are almost inexhaustible. Overlaying this sequential frame is a continued concern for environmental health and quality and as well for social hazards that may be engendered by eagerness to find solutions or by desperation under conditions of severe shortages.

These problems appear sequential, because the consequences of failure to deal adequately with them will become severe at different times in the future, and because solutions will take longer to implement for those problems termed intermediate and distant. As problems, however, they all exist right now. We cannot deal with them sequentially if we are to deal with them satisfactorily. If we concentrate on present shortages we will not even be in the right frame of mind to deal with the more distant problems. There is a human tendency to believe that bad tomorrows never come. The biggest problem of all may be to create a decision mechanism whereby short, intermediate, and distant problems can all be given consideration at the same time.

The energy system of the United States reflects an economy of abundance and a large natural subsidy that has contributed greatly to the economic development of the country. It is consumer-oriented and growing more so. It is a wasteful and dirty system, but some parts of it are more wasteful or dirty than others. The system provides the average citizen with a great deal of choice, the hallmark of a free society. The system appears to be undergoing a major change, from being one of rapid growth, abundance, and waste, to one of lessened growth, limited scarcity, and economy. The change is being caused by increasing costs of obtaining fuel and other mineral resources and of maintaining an acceptable environment.

SUPPLEMENTAL READING

Conference Board, 1974. *Energy Consumption in Manufacturing.* Cambridge, Mass.: Ballinger.
Much information about energy use in manufacturing in the United States, most of it in economic, not energetic, form.

Congressional Research Service (Library of Congress), 1973. *Energy Facts.* U.S., Congress, House, Committee on Science and Astronautics, 93rd Cong., 1st sess., serial H.
Useful potpourri of statistics on energy in the United States and foreign countries from fifty sources.

Dewhurst, J. F., 1955. *America's Needs and Resources.* New York: Twentieth Century Fund.
1955 was a good year for energy books. Cottrell's *Energy and Society* appeared, Palmer Putnam's classic was but two years old, and this study by Dewhurst and his associates came out. In his treatment of energy, Dewhurst emphasized conversion efficiencies and included calculations of the contributions of draft animals and human musclepower to the national work output.

Jones, Howard Mumford, 1971. *The Age of Energy.* New York: Viking.
Although he states rather casually the physical basis for the energetic expansion of the American people in the fifty years from 1865 through 1915, Jones describes in cogent detail the web of attitudes that both guided and rationalized the "development" of a continent.

Landsberg, H. H., and S. H. Schurr, 1968. *Energy in the United States: Sources, Uses, and Policy Issues.* New York: Random House
Mainly condensed from *Energy in the American Economy;* contains good accounts of the depletion allowance (p. 197–202), natural-gas price regulation (p. 203–208), and the issue of private versus public power (p. 208–218).

Perry, Harry, 1971. *Energy—The Ultimate Resource.* Committee print, serial J, Task Force on Energy of the Subcommittee on Science, Research, and Development of the Committee on Science and Astronautics, 92nd Cong., 1st sess.
Entirely composed of tables and charts from other sources, this is a valuable fact book.

Pierce, John R., 1975. "The Fuel Consumption of Automobiles." *Scientific American,* v. 232, n. 1, p. 34–44.
Reasons for the poor fuel economy of American cars and how their efficiency could be increased 40 percent within a few years.

Rice, Richard A., 1974. "Toward More Transportation with Less Energy." *Technology Review,* v. 76, n. 4, p. 44–53.
An energetic comparison of what is with what could be in transportation in the United States.

Schurr, S. H. and B. C. Netschert, 1960. *Energy in the American Economy, 1850–1975, Its History and Prospects.* Baltimore: Johns Hopkins.
A basic work; its bias is econometric, its projections useless, its use of "resource base" debatable, but it contains a wealth of historical data.

Stanford Research Institute, 1972. *Patterns of Energy Consumption in the United States*. Washington, D.C.: United States Government Printing Office. + appendixes.

Good example of a straightforward analysis of energy consumption in the United States for two selected years, 1960 and 1968.

Vansant, Carl, 1971. *Strategic Energy Supply and National Security*. New York: Praeger.

Brief but valuable in its analysis of national options related to security in a broader sense than military.

Yager, Joseph A. and Eleanor B. Steinberg, 1974. *Energy and U.S. Foreign Policy*. Cambridge, Mass.: Ballinger.

A concise world geography of energy rather than a treatise on foreign policy, this book has the strong economic bias of Energy Policy Project publications.

William Tenney

Chapter 12   DECISIONS ABOUT ENERGY

Most with whom you endeavor to talk soon come to a stand against some institution in which they appear to hold stock.

—Henry David Thoreau, 1863

I wish to God there were some more wise men in the world, I do find it so lonesome.

—Mark Twain, 1907

A BACKWARD LOOK

Looking back can be useful if we know what we are looking for. In pondering present problems related to energy, it might help us to consider how they may have been made in the past. The Chinese used coal and natural gas for comfort and process heat at least as early as the British used coal, but the Chinese did not move toward an industrial society, whereas the British did. The reasons for the difference have been discussed by many, and no attempt will be made here to settle the question. It must be obvious, however, that in the broad sense, the two societies arrived at different *decisions,* since neither appears to have faced any insuperable resource or environmental barrier to industrialization. Adam Smith, writing as the Industrial Revolution in Britain was gathering momentum (1850, p. 32–33), chose Chinese society as an example of the "stationary state," which he deemed stagnant and unhealthy for its members, in contrast to the "progressive state," which he thought the best for mankind and in whose prime example he lived and taught. In this judgment Smith was speaking for the decision makers of his social environment as well as for himself. In Great Britain the merchant descendants of the Vikings became the real kings of society, while invested kings became merchants in order to live beyond the means Parliament provided them. The monarchy and the military were used as instruments to extend markets, obtain materials, and make the world safe for the capitalism of a market economy. No national referendum, no government plan, ever set forth a proposal of industrialization to be considered and voted upon; nonetheless, over the years, and through thousands of individual decisions, both private and public, a national commitment to what Adam Smith called "the progressive state," what we would call "the growth state," was made. A decision, with no date, no signatures, and no witnesses, was arrived at. In the United States, such formless but nevertheless real decisions govern the evolution of society and the economy. The so-called silent majority has a great deal of power.

In China the merchant was low in the social order. Invaders on horseback instead of in ships came seeking a softer, more secure life, rather than loot and they seem to have been absorbed into a culture that attained a remarkable ability to survive and a resilient resistance to shock through shared values and a teleology inculcated through a hierarchical structure of government and society. More important than dress and language, the manners and aspirations of the Chinese differed from those of the British. The British looked outward for material opportunity; the Chinese looked inward for the elements of a good life. In both cases, there was an inchoate, stable belief in what it was that was considered as worthwhile for the individual and the nation to do; the strategies could be argued about, even fought over, but the goals were agreed upon.

We tend to think of a decision as the expression of the will of individual participants in some act involving judgment, whereas it may be as much or more the expression of historical social forces that lead us and others to a restricted view of the options available and of what it is good or right to do. These forces, moreover, have fashioned attitudes both in our minds and in our societies that favor certain decisions over others. Christianity and western industrial society have emphasized the worth of the individual over the worth of the state. We see this ideal in almost grotesque form today in pressures against discrimination that threaten the continuance of capitalistic meritocracy and in decisions that require expensive modifications in the design of public buildings and vehicles to accommodate the physically handicapped. It skews our views of acceptable public strategies for protecting the majority against the obtrusive or disruptive actions of minorities, for the preservation of common resources and amenities against the selfish appetite of individuals, and even for the maintenance of public utilities and services against the reluctance of the individual to shoulder a full share of a common burden.

Four centuries ago, the decision to limit the cutting of French forests for firewood and that to limit the burning of coal in London could be made more readily than they could today, because the decision structure (a monarchy with advisers who were representing a limited number of interest groups) and the decision mechanism (a royal decree) were simple, the costs (soil erosion from deforestation and ill health from air pollution) and benefits (continued hunting and grazing in the forests, relief from the stench and dirt of coal smoke) of the proposed action could be weighed easily, and the acquiescence of society outside the small group of decision makers could be expected with some degree of confidence. It is not correct to assume that the monarchs were acting solely in the selfish interests of the nobility. Although French forests were hunting preserves for the nobility, the impetus for their preservation came from early concern for the bad effects of deforestation and subsequent erosion on agricultural production. In London no citizen was eager to burn the weathered coal that came from the early shallow pits because its smoke was thick and choking and neither homes nor stoves had been designed to accommodate it. Early use of it was largely by industry and it was industry forced out of the city, not citizens made to freeze, that was the result of the ban.

As the Industrial Revolution developed, however, demand for charcoal, ships' timbers, and food put great pressure on the remaining forests, while the high cost of charcoal and fuel wood forced the use of coal as a substitute. The earlier decisions were reconsidered. The forest, especially in England, diminished further, and coal—which became less vile as the pits got deeper and heating systems were designed to use it—rapidly became the main fuel of British and European cities. From the lifting of the royal ban on coal until modern times, there have been few public or governmental decisions in western society directly affecting the use of energy. During the Industrial Revolution private decisions determined where mines, mills, railroads, and factories were built. Private decisions determined discharges of waste into streams and skies, as well as the choice and design of technological processes for mining and burning coal and oil. Private decisions for the most part determined working and living conditions, wages and rents, hours and prices, despite the emergence of unions and the entrance of government to curb the worst abuses of the factory system. This was the period of laissez faire, of the unregulated market economy.

But for several reasons, even in the market economy, public or governmental decisions about energy resources and energy use have had to be made in the present century. The inefficiencies and potential for monopoly involved in private utility competition prompted the franchise system for companies providing gas or electricity (or both) to a city and, in most states, a certain amount of state control. The desire to use the resources of the public domain under conditions of competitive equity led to the establishment of a federal leasing system for oil and natural gas prospecting and exploitation on federal lands. The effectiveness of the Standard Oil Trust provoked antitrust legislation and the breaking up of the trust into several competing companies. The critical importance of petroleum in warfare was the reason for creation of the Naval Petroleum Reserves. Later, the potential impact of an atomic explosive encouraged the government to subsidize an intensive program of uranium exploration and atomic-energy development. More recently, the need to subsidize high capacity in the United States for the production of oil—for security purposes—was the ostensible reason for a national quota system for oil imports. The potential for monopoly pricing brought the imposition of federal control over interstate movements of electricity and natural gas, and of the price of the latter when designated for interstate sale. The structure for decisions about energy is complex and reflects differences in the types of resource, the levels of government involved, the ownership of the resource or the transport system, whether or not the transport system crosses a state line, and the context of the decision (utility regulation, taxation, environmental protection, antitrust or fair trade objectives, resource conservation, resource development, national security, or occupational health and safety). In most other countries, the decision structure is more simple, largely because—even in the capitalist countries—the national government plays a greater role in energy decision making and the individual citizen a lesser role than in the United States.

THE ROLE OF ENERGY IN THE DECISIONS OF DEPENDENT NATIONS

Some of the most important decisions made by nations dependent on external sources for all or most of their supplies of inanimate energy are dictated by that dependence. They range from propitiatory political actions pleasing to their suppliers to the imposition of harsh restrictions on domestic consumers.

The reluctance of Israel to consider withdrawing from the Sinai territory it won in 1967, a precondition for peace in the Middle East according to the Arab nations, stemmed directly from the vital importance of the Sinai oil fields to Israel's survival. After they were captured, these fields supplied Israel with about 90 percent of its oil consumption, 60 percent of its total energy needs. Although the Secretary of State, Mr. Kissinger, offered to see that the country was supplied with enough oil to compensate for the loss of the Sinai fields, Israel doubted that Congress would confirm the offer, that any major supplier would want to supply Israel, and that the added burden on the Israeli trade deficit could be supported, and thus the Israeli-Egyptian stalemate continued for some time.

Yugoslavia, a relatively poor nation running a substantial energy deficit in its determined effort to industrialize and raise the living level of its people, has maintained a stubborn independence in the face of Soviet power in central and eastern Europe. As the price of oil imported by Yugoslavia has increased more than the price of Soviet oil piped to the satellite countries, its independence has become more costly. Should the Soviet Union offer a particularly good long-term deal for oil and gas, Yugoslavia's will might weaken.

The two examples will suffice to suggest that a country will decide against what it would like to do in favor of what it feels compelled to do for its own survival and that one of the strongest forces in such a decision can be its energy needs. The rest of this chapter will pertain mainly to decisions on energy within the United States.

THE ROLE OF THE INDIVIDUAL
IN ENERGY DECISIONS

The role of the individual citizen in making decisions is restricted but decisive. It is restricted because he or she commonly has a choice only among gas stations and of quantities consumed. Dwellings are generally not lived in by the people who designed them; most cities do not offer real transport options to many of their residents; energy prices tend to be either fixed or to range within narrow limits in any city or region. Yet the response of citizens in the aggregate to lowered speed limits, higher prices, pleas for conservation, and suggestions for lowered (or raised) environmental standards that are related to the use of energy will decide the success or failure of government efforts to achieve energy objectives.

Completely rational economic man would base his decisions on a rigorous analysis of the costs and benefits to him. He would conserve electricity or gasoline only if the resultant saving in money would purchase a greater benefit through an optional expenditure, say for food. Economists are relying on the responses of completely rational economic man when they advise government to reduce consumption of gasoline by raising its price to the consumer. There is abundant evidence, however, that completely rational economic man does not exist. Advertising and marketing people know this very well. They know that men and women make purchasing decisions—not just occasionally, but frequently—on factors other than economic, such as taste preferences, product "loyalty," desire for status, friendship, convenience, and so on. Often individuals will make decisions contrary to their own economic interest for the benefit of family, religious, or ethnic groups, nation, or even mankind. Were it not so all states would be police states.

Granted that people act on impulse or "emotion," that they often do not seek information that would appear to be relevant to the decision, there is still one element that appears common to all conscious decisions—an excess of expected benefit over perceived cost. As mentioned, the benefit may not be to the individual but to some person or concept he or she cherishes. It is the wide range of expected benefits and perceived costs that causes individuals to vary so much in their decisions, given the same set of alternatives. Some individuals demand a great deal of information about the problem; others do not, preferring to substitute their own bias or someone else's recommendation. For the former, decisions are difficult; for the latter, they may be easy. Decisions by groups of private individuals are more apt to be made on a basis of economic interest and relevant information than are decisions by individuals. Included here are labor unions and professional organizations, as well as companies hoping for profit. Governmental decisions in a representative democracy, as might be expected since government attempts to represent and lead both irrational, non-economic individuals and rational economic groups of men and women, tend to be a mixed bag, attempting as they do to satisfy both the impulses of individuals and the perceived needs of groups within the society.

PUBLIC ENERGY DECISIONS IN THE UNITED STATES

Decisions about the production and use of energy resources and about the allocation of the social benefits and costs of energy use are of vital importance to everyone in the world. There are many problems involved in reaching and implementing such decisions: insufficient knowledge of the physical existence of oil, natural gas, and uranium; difficulties in forecasting the progress of energy technology; conflicts of interest between energy-exporting states and energy-importing states; the necessity for compromises between environmental quality and continued or increased energy use on the one hand, and between living standards and energy conservation on the other; social, ethical, and political stresses inherent in the use of food and oil as instruments for increasing national income or advancing other national goals; and the great difficulties involved in adjusting energy policy to population policy.

Within the United States, public decisions about energy involve not only the several levels of government but also, at the national level, are made in fragments by a number of federal agencies. In 1974, the Federal Energy Administration was created, to advise the president on national energy policy, to assess the national energy-resource base and to develop plans for dealing with energy shortages. In 1975, the Energy Research and Development Administration (ERDA) was established, taking over the existing fossil-fuel research and development activities of the Department of the Interior and the Atomic Energy Commission's civilian-reactor and nuclear-fuels research programs. Also activated in 1975 was the Nuclear Regulatory Commission, which carries out the licensing and regulatory functions formerly vested in the Atomic Energy Commission.

The federal government regulates natural gas moving in, or destined for, interstate commerce; the health and safety of underground coal miners; the interstate transmission of electricity; the siting of hydroelectric dams on any navigable stream in the United States;[1] the safety, design, and construction of nuclear facilities of all types; and interstate compacts related to energy. It implements any import and export restrictions that may be placed on energy materials; it taxes the income of those engaged in energy production and delivery and administers tax provisions, such as the depletion allowance, that may be of singular importance to the energy industries; it is a large consumer of energy, with extraordinary powers to requisition, allocate, and control production and prices under conditions of scarcity; it is the owner and lessor of lands, both dry and submerged, containing energy resources; and it is the promulgator and principal enforcer of laws relating to the adverse effects of energy use on the environment.

State governments regulate the activities of organizations designated as utilities, commonly including those that distribute electricity and natural gas but not those that distribute gasoline, fuel oil, coal, firewood, or bottled gas, implement interstate compacts through state oil and gas conservation commissions, and lease state lands, including those offshore but within the territorial waters of the United States, for energy-resource exploration and exploitation. They enforce air and water quality laws, may enact laws to control or ban strip mining or to prevent polluting facilities such as refineries from being constructed in their coastal zones. They may tax the transport of energy materials across their lands. They may collect severance taxes from energy-resource producers.

In general, the role of American cities in energy decision making is greater than that of counties. Cities can tax and zone and can prohibit or curtail activities deemed inimical to municipal welfare. The role of cities varies with the powers granted them by their home states.

PRIVATE ENERGY DECISIONS
IN THE UNITED STATES

Private decisions are made at two levels: that of corporate management and that of the individual. The individual's choice is limited to the retailers he prefers for gasoline and fuel oil (for natural gas and electricity he has no choice) and to the quantity he consumes. Most

---

[1]The legal definition of a navigable stream covers all streams, as some wag said, "that would float a toothpick in a rainstorm."

individuals have no choice of alternative forms of energy (electricity or natural gas) or much influence on the efficiency of their energy-support system, except in the matter of personal transport, because they rent or buy their dwellings and have not designed them. Because he can vary the amount of energy he consumes only within fairly narrow limits and because he has a very limited range of alternatives, the average consumer has no bargaining power over price and must rely upon the government for protection against gouging.

For oil and gas, corporate management decides where to explore, where to drill, how much to offer for leases in the United States or foreign concessions, where and when to build refineries and lay pipelines, what size tankers to buy or lease, where to buy crude oil or petroleum products if they do not produce as much as it can sell, where to sell and at what prices, what the product mix of its refineries will be, what portion of its profits should be reinvested and in what, and from whom to borrow money in what quantities at what rates. In Japan such corporate decisions are made in close cooperation and agreement with the national government, which is even an overt co-sponsor of some of the consequent actions. The British government owns a large share of a major oil company (British Petroleum) and thus is directly and continuously interested in such decisions. But in the United States, where the government's attempts to help the big oil companies in some of their troubles with producer nations are regarded by many citizens as wrong and where the government sometimes refuses the proffered aid of industry experts, corporate decisions that may vitally affect the nation's welfare are made without benefit of representation in the public interest.

The role of corporate management in the coal industry is similar but somewhat simpler and considerably less hazardous politically because there are few small consumers of coal to fuss about "windfall profits" and long lines at the coal bin. Except for those "captive" coal mines that feed steel companies the raw material for coke, the coal industry is not vertically integrated as is the oil industry. Most coal is sold on long-term contract to large consumers. Unlike the oil industry, coal production is mainly a domestic, not an international, business.

Management decisions in the electricity and natural-gas portions of the energy system are constrained by the franchise requirements of regulated public and private utilities. The utility manager's role is to forecast demand within his market area and to select from among alternate ways of meeting that demand the way that will best meet the constraints of the utility commission on prices and on profit. The siting of power plants and the choice of conversion method and fuel to be used, although subject to approval by public regulatory bodies, are initiated by utility management. In recent years, uncertainties about the availability of fuel and escalation in prices have made power plant selection more difficult as more stringent environmental regulations have made the approval of sites more drawn-out, expensive, and uncertain. The great number of approvals necessary in some cases before construction of a power plant can begin and the costs of years of delay have brought proposals from government agencies and the Congress for streamlining and telescoping this portion of the decision process.

DOGMA AND DOCTRINE OF THE GROWTH STATE

The available mechanisms and structures for making decisions, as well as the perception of problems and alternative solutions, are very much fashioned and conditioned by prevailing dogma and doctrine within a society or its ruling class. Thus national decisions are products of the historical "memory" as well as of present volition. Doctrine and dogma, in most societies, are promulgated by a priesthood as well as by government. In the United States this priesthood is composed mainly of economists, who are inclined to reason in terms of dollars, economic growth rates, and percentages of unemployed laborers rather than in kilocalories, depletion rates, and supplies of amenities. The federal government leans heavily on the advice and forecasts of economists. For several hundred years the Roman government relied on the advice of the College of Augurs, whose members read portents in the sky and whose interpretations could not be questioned by lay observers of the same phenomena. The Council of Economic Advisers fills a similar role in the United States today.

"The flow of energy," said Frederick Soddy (1933, p. 56), "should be the primary concern of economics." Soddy was a distinguished British chemist who tried in vain during the Great Depression to get his neighbors in the industrialized world to recognize the egregious stupidity recorded in an economic system that allowed millions to be in want and hunger while labor, natural resources, and manufacturing plants existed in underutilized abundance. Now, more than forty years later, although the availability and costs of energy have become a concern of almost everyone in industrialized countries, the flow of energy still is not a major concern of economics. The importance of the second law of thermodynamics and the concept of entropy to the study of economic processes, as cogently put forward by Nicholas Georgescu-Roegen (1971), Herman Daly (1973), and treated seriously by both Kenneth Boulding (1966) and Charles Madden (1972, p. 25-28), has been rejected or ignored by "mainstream" economists. The suggestion that the nonrenewable resources have energy-profit exploitation limits that cannot be deferred indefinitely by advances in technology is derided by economists who point to the history of increasing resources over the past two hundred years as refutation of the shallow judgments of those who have not read this record or have not been able to interpret it correctly.

Mainstream economists are direct descendants of Adam Smith. Their vestments of priesthood sparkle with econometric sequins; their rational edifice is supported by the same two basic assumptions that underlay Smith's "political economy." The first assumption is that the sum of a vast number of selfish acts will, by the action of a benevolent *invisible hand,* turn out to be a social good; this is the philosophic justification of the free market or laissez faire economy. From this assumption flows the concepts of *the doctrine of positive externalities* (that pervasive social benefits will accrue from public economic stimulation of productive enterprise), the idea that justified giving away the national patrimony through the General Mining Law of 1872, the Homestead Act of 1862, and the Reclamation Act of 1902, that rationalized grants to pioneer railroads and utilitarian (land-grant) universities, that now supports public investment in water-management projects, new highways, and nuclear-power development. Faith in the "invisible hand" also undergirds *the doctrine of unlimited ingenuity,* which holds that man forever will be able, through technology, to outwit the demon of scarcity, will be able, in other words, to develop useful knowledge and resources as fast as he needs them. To this doctrine is added the caveat that man, in order to stay ahead of scarcity, must accept the gospel of Smith (or Marx—the eastern brethren differ mainly in replacing the combination of selfish acts and invisible hands by central planning). The second basic assumption is that of the existence of *economic man:* that each person attempts to maximize his or her personal gain by rational choice among perceived options with gain defined in terms of wealth rather than social status or religious grace. If a society can be assumed to be made up entirely, or even mainly, of such beings, the behavior of that society under economic stimuli becomes predictable.

The theory of the "invisible hand," challenged recently by such arguments as Garrett Hardin expressed in the "Tragedy of the Commons" (1968), for more than a century has seemed to thoughtful students of environmental degradation and resource depletion to be increasingly inconsistent with observable processes. The assumption of economic man has been challenged by many anthropologists, among whom is Sahlins (1972), who shows that if economic man exists he must be a result of rather recent cultural imprinting because primitive man behaved according to social goals and standards that were not economic. Although some have argued that economic man exists in that he tries to minimize his own efforts—that economic man is lazy man—there appears to be considerable evidence that even modern man has a stubborn tendency to do things the hard way rather than the work-efficient way.

The idea that private ownership of land is the basis of many of our decision problems appears to be held by many people who might regard communal ownership as preferable. But one finds that a system in which the state owns the land and productive facilities, as in the Soviet Union and China, is rationalized by economic concepts similar to those in private-enterprise countries, notable among which is the Marxist theory that population, resources, and living standards will not constitute problems if all production is in the hands of the state, which represents all the people. Both private and state enterprise stress economic efficiency and productivity. Both measure their achievements mainly in rates of production of *goods* (one of many economic words with a theological overtone). On the distribution of the benefits of production the centrally planned or command economies differ from the market economies. Central planning reflects a conviction that the social benefits of production can be maximized only if the state controls the production and distribution processes completely, whereas a market economy demonstrates a belief that the maximum social good is achieved when each person is free to fight, on the basis of ability and desire, for a share in the benefits of production. In both socioeconomic systems, a fixation on the means of increasing the supply of goods has had the effect of obscuring perception of longer-range social costs in degraded environments and depleted resources. Where such problems have evoked concern, conventional wisdom has appealed to a faith in technology both to clean up the mess it has been generating and to replace the resources it has been consuming.

Faith in technology is rooted in the achievements of the industrial and technological revolution of the past two hundred years. During this brief time, one portion of the human population shot from a condition of relative scarcity of goods and of frailty in the face of disease to a condition of abundance and strength and appears to have achieved a much wider distribution of available benefits in the new condition than in the former. Persons born into such a condition of acceleration find difficulty in applying the principles of Newtonian mechanics to an analysis of future human conditions. Instead they champion the "unlimited ingenuity of man" to overcome all physical problems. At base, their faith is in the proposition that useful knowledge always will be obtained by man at a rate equal to or exceeding his need for it.

A decision structure that encourages the doer and producer reflects also the work ethic, which was a significant part of the philosophical luggage of most of the European emigrants to North America. The work ethic and a strong desire for material security underlay their rapid spread westward across the continent, their avid exploitation of lands, forests, and minerals, and the mushroom growth of an integrated, efficient industrial economy. A decision structure created to facilitate and encourage increases in production, either through private enterprise or government planning, tends to magnify immediate benefits, to discount future costs, and to discourage alternatives that do not promote development. In the absence of central planning and direction, it evolves as a fragmented and poorly coordinated structure, representing private decisions by individuals and groups of widely differing economic and social strength overlain by government regulatory bodies, each of which was created in response to a perceived social cost arising from activities of a single private structure, such as the Standard Oil Trust, within the exploitation system for a single commodity, for example natural gas producers and distributors, or affecting a single common resource, as does air pollution.

INCENTIVES TO GROWTH-ORIENTED DECISIONS

William Ophuls remarks,

> It is just not rational for a producer to make an indestructible, easily repairable, low maintenance-cost car. If consumers were perfectly rational creatures, they could no doubt oblige producers to turn out nothing else, but we know that consumers are not completely rational (about cars least of all) and that producers do everything in their power to encourage them in their irrationality. By comparison, the incentives to satisfy needs with minimum inputs of energy and materials are quite weak. [In press]

Depletion allowances favor the use of primary raw materials rather than recycled material and thus work against both energetic efficiency and the conservation of resources. Regulatory limits on their profit margins encourage the utility companies to expand their sales and production of energy because it is the only legal way to increase total profits. Consequently, utilities advertise and offer lower rates to large consumers. Competition among oil companies for shares of the gasoline market has been fierce and wasteful of energy resources. Venture capital moves toward the promise of high return, which in a market society comes from high rates of use of materials and energy. Investment decisions force growth.

When sales of automobiles slacken, it becomes almost patriotic to buy a new car whether one needs it or not. Appeals are made to get America rolling again and no attention is paid to either the real need for the objects to be produced or the real costs of depleting energy and material resources to produce them.

The national income tax law favors married people over single and gives allowances—if not incentives—for dependent children without limit. The depreciation provision in the income tax law, not all that different in principle from the depletion provision, is designed to help businessmen and industrialists replace productive equipment more often than they would otherwise; it subsidizes the replacement of old equipment so that more capital can be invested in more equipment for increased production. When it is desired to stimulate the economy—or the growth of one sector of it—accelerated depreciation allowances are granted. Those who have regarded the depletion allowance as inimical to the nation's welfare should take a hard look at depreciation—although they might not want to because they themselves benefit substantially from that tax loophole.

Subsidies are another means of stimulating economic growth and resource consumption. There would be no commercial nuclear-power industry today without the large subsidies that support research and development, fuel enrichment, and insurance in the nuclear-power field. The development of new energy-delivery systems to replace natural gas and petroleum will require heavy government subsidy.

MANAGEMENT OF THE COMMONS

The emphasis on augmenting supply rather than on diminishing demand and the persistent belief that the development of resources always will yield an excess of social benefits over social costs have a great influence on the management of common resources, indeed on the very perception of a common resource.

In the early years of the public domain, land owned by the federal government but not dedicated to a government function (such as land reserved for military reservations or post offices) was the subject of controversy between those who felt that such land should remain in the hands of the government and those who argued that the welfare of the nation would be served best by allowing the land to be purchased or appropriated by those who wished to develop its resources. The latter view prevailed and became formalized in the Homestead Act of 1862, the General Mining Law of 1872, and in laws governing the acquisition of timber on the public domain. Later modifications of this appropriation doctrine resulted in lease arrangements for the production of certain types of mineral resources, including oil and gas, as well as for the use of the public domain for the grazing of livestock.

The appropriation doctrine reflected the view that a common resource was to be divided up for private gain whenever conditions were such that private gain could be realized from the resource. Quite a different concept of a common resource has since emerged, one that is in fundamental conflict with the older view. Today, to many people, a common resource is any resource whose misuse can affect adversely a great number of people. This definition goes far beyond those resources such as national parks (owned by everybody) and international fisheries (owned by nobody). The atmosphere today is generally recognized as common resource. Water, while in principle accorded the status of a common resource, is in some situations closely allied to private-property rights—in arid country, water holds the value, not the land on which it is used. Privately owned land, once regarded as a possession to be used freely by the owner, is fast acquiring the status of at least a quasi-common resource. Energy resources, in the public view, are in a similar state of transition. The change may have started in the 1930s when the private appropriation and exploitation of water power to produce electricity started to give way to government enterprise; municipal, federal, and public utility districts built dams and power plants and marketed electricity. In most countries other than the United States mineral rights *are* vested in the state, which sets the conditions under which concessions or exploitation rights may be granted; underground energy resources are both legally and in the vernacular view part of the national patrimony. A private company may be granted a use right (concession or lease) but only because that is considered an appropriate way to develop the resource; should the company, because of rise in the price of the resource or extreme good fortune in striking a prolific deposit, start taking more profit out of the ground than is deemed necessary to keep it in business, it may be faced with forced adjustment of the concession terms, revocation of the concession, or expropriation of its assets, because the state finds the private company to be diverting too much of the national *common* to its own use. In the United States both the doctrine of positive externalities (in this case, holding that the creation of a new supply of energy for the nation will entail social benefits greater than the social cost of having a few individuals get a large reward) and the doctrine of *the sanctity of contract* would prevent revocation solely on the ground that windfall profits were being taken from the national patrimony. Far more likely would be a move by taxation or by price controls to reduce such profits and give the windfall to the consumers of the energy being produced—another way of distributing a national common resource stock, a way being followed in the United States in the case of interstate natural gas.

The provision insisted upon by Congress in the strip-mining bill pocket-vetoed early in 1975 by President Ford (that would have given the owner of the surface rights to any land on which the federal government retained the mineral rights the power to prevent strip mining on his land) was a sign of a value structure that keeps the farmer ahead of the miner as an instrument of social policy and that also incorporates a clear dislike for the separation of surface and subterranean rights. In the United States these rights are separable under the law and for many tracts of land the owner of the surface does not hold the mineral rights. In North Dakota and Montana many lending institutions foreclosed on property loans during the Depression. When they resold a property it was common practice to retain the mineral rights and subsequent oil strikes and recent coal exploitation in those states prove that the managers of the lending institutions were farsighted. The owners of the surface lands can expect nothing but fair market value for damage or destruction to the land incurred in mineral exploitation. The federal government holds the mineral rights to some land that has been appropriated by or sold to private owners; these were the rights involved in the strip-mining bill.

Wellhead price control of natural gas was imposed in order to protect the small consumer against the power of the producer under conditions of uncompetitive distribution. This type of regulation, when it works as it was designed to work, has a strong bias in favor of the consumer, leading regulatory bodies to suspect pleas of resource scarcity and to be slow to allow price increases on that basis.

The history of the management of common resources that were easily recognized and well accepted as such—grazing commons such as pastures in Great Britain and national forests in the United States, fisheries commons such as those of the blue whale, the Peruvian anchovy, and the North Pacific salmon; game commons such as the buffalo and the passenger pigeon; energy commons such as oil fields with many owners—is not reassuring to one who looks on the fossil fuels as an international common. Not only is there a wide range of estimates of how much fossil energy remains to be exploited by man, but there is the viewpoint that it does not really matter because man will have found an adequate substitute energy system for the fossil fuels before they run out, and he needs to consume fossil energy at an ever-faster pace in order to maintain the dynamic society, the growth state, and the ability to form capital that are necessary to achieve successful replacement. In such a view of human progress, prudence and conservation appear counterproductive.

Whales were a common resource for mankind when whale oil fueled the lamps of the world. Whales grew scarcer, but no attempt was made to conserve the resource, except to let the price of whale oil rise. Then kerosine replaced whale oil, and there was no longer a great need to kill whales. Did the pressure on the whales relax? Was there then created an effective mechanism to limit the kill rate? Not at all. They continued to be hunted and slaughtered to the point where one or more species will almost surely become extinct, and all this long while, about a century, since kerosine took over lighting, the whale has provided a miniscule portion of human support.

What is treated as a common resource in one culture may not be so treated by another. It is contended that the American Plains Indian did not slay bison wantonly, as those Eskimos who live off reindeer do not waste their prey—even though the numbers of bison must have seemed inexhaustible. As a result they were a perpetuated common resource of the Indians. The white man killed bison for meat occasionally, for hides more often, for sport in great numbers, and finally almost to extinction, "to solve the Indian problem," in the words of General Phil Sheridan.

Texas for years has found it impossible to enact an effective oil-field unitization law. Oil-field ownership may be fragmented among a number of persons or companies. Because oil and gas migrate in the direction of lowered pressure, uncoordinated drilling not only can cause oil and gas to move laterally underground, but also often results in leaving more of the oil in the ground than would have been the case if a comprehensive planned sequence of carefully spaced and managed wells had been used. Such a scheme, however, requires that all owners agree to surrender their management prerogatives to a single manager who operates the field for the benefit of all. Any effective secondary-recovery operation likewise virtually requires such a "unitizing" agreement. It is not unlawful in Texas to make such an agreement. But one owner can block it. In fields with several owners, voluntary unitization has been so rare that several attempts have been made in the state legislature to pass a law that would force unitization if the owners of three-fourths of the field agree; this would force the other fourth into the pact, in a manner similar to that by which a reluctant property owner is forced to pay his share of street paving if enough other property owners agree to pay their shares. The conservation value of such a law, in the sense of maximizing total production from a field, can be demonstrated conclusively. Yet it fails to be enacted because some operators fear that an operator who owns perhaps a larger portion of the unitized field would manage the field in such a way that the interests of the former would not be protected by the latter—or because the former could not take advantage of the latter?

Posterity does not vote. The interests of future citizens in a common heritage are rarely taken into account in public decisions, though they may be in private ones. The flaring of quadrillions of cubic feet of natural gas, first in North America and now in the Middle East, is deemed necessary by those whose interest lies in the associated crude oil; nevertheless, this flaring is a social cost that will be paid by all those who might have been able to use gas had exploitation been delayed until it was possible to ship the gas through pipelines to places of use, to liquefy it for tanker shipment, or to use it for the production of nitrogenous fertilizer. The helium now being wasted in the United

States is another example of a common resource with great potential value being wasted because the present generation is unwilling to pay the costs of separating it from natural gas and storing it for future use. Helium has some uses, notably in cryogenic technology, for which there appear to be no good substitutes. The largest concentrations of helium have been found in natural gas from wells in Kansas, Oklahoma, and north Texas. A helium conservation program was financed by the federal government for a number of years but has been scrapped because estimates project storage costs that are very high compared with the present value of helium. What the future value of helium will be is the unknown, but it is apt to be very high if cryogenic transmission of electricity becomes advantageous to conserve energy and if, in the meantime, most of the world's original stock of helium has been allowed to dissipate into the atmosphere with the burning of natural gas.

A MODERN COMMON MANAGED FOR PERMANENCE

Although the history of many resource commons is tragic, there is a small spot in southern Louisiana that has a different history (Cook, 1974). Avery Island has been owned for more than a century by a single family. About two miles across, the circular island rises not from

FIGURE 12.1
The tranquil appearance of Avery Island, Louisiana, whose post office is shown in this photograph, gives little hint of its production of oil, gas, salt, and Tabasco Sauce.

the sea but from coastal marshland; it is the surface expression of a salt dome. A resource matrix that includes a salt mine, fields of tabasco peppers, a bird sanctuary and botanical garden, about one hundred oil and gas wells, and marshland used for cattle raising, rice farming, hunting and trapping is managed through a family corporation for a primary goal of permanence (stability of the man-environment system) and a secondary goal of optimization of profit and environment quality.

Most of the island is covered by grass and woodland, much of it reclaimed from earlier sugarcane fields, now grazed and browsed by horses and deer. There are several lakes with abundant fish and bird life. The island's resources are held as a family common, with limited use rights granted to individuals and groups both within and outside the family. Shares in the family corporation may be owned only by blood members of the family and their spouses; it is governed by a board of directors elected by the shareholders from among themselves; the board in turn selects the managers of the family enterprise, who have always been blood members of the family. The management philosophy is illustrated in the different attitudes toward salt and petroleum income. Salt, although certainly not renewable, is regarded as relatively inexhaustible in terms of present and probable rates of production and the known and inferred reserves; consequently, salt royalties are channeled through the family corporation and some is invested in the management of the surface resources. Oil and gas are regarded as ephemeral and royalties therefrom go directly to family members. In this way is avoided the temptation to regard a depleting capital resource as an income resource.

If the goal is permanence, the strategy of management is to try to develop and maintain a balanced ecosystem, in which the resources are not depleted and the environment is not degraded. Soil-retention and soil-building measures are employed. Surface disturbances due to salt, oil, and other operations are held to a minimum. The environment is maintained as biologically diverse and as humanly pleasing as possible. This management philosophy strongly reflects a sense of living in a dynamic natural system, an obligation to family posterity, and dedication to the use of resources without wasting them. Although family members express deep interest in nature and its forms and processes, there seems to be little or no mysticism in their attitude. They value pleasant surroundings, but have not insisted that oil well-head structures or storage tanks be hidden or camouflaged. They place a high value on natural systems, but do not define natural as untouched by man; the only plot of ground on the island and its environs that the family closed to oil drilling was the family graveyard. They do insist that the integrity of their ecosystem be not permanently impaired. The most serious existing threat to that integrity lies in the extensive system of canals cut by the oil company to facilitate access to drilling sites in marshland. Saline waters introduced by the canals destroy plant life in the marsh and eat away the land. To meet this problem the family corporation has established a fund to clean up and block off these canals when they no longer are needed, in order to allow the marsh to restore itself.

Avery Island, unlike the outer world, has a simplicity of decision structure and a homogeneity of decision makers. The island by no means represents a closed system, or even a fully balanced one; nonrenewable resources are being depleted and surplus population is being exported. Were the constraint of supporting a steadily increasing population to be placed on the managers of the island, they would be faced with a choice of using capital resources as income until exhausted, or maintaining environmental integrity at the cost of a steadily decreasing standard of living. Only if population were to be perceived as a management variable could permanence as a goal be retained.

It is just here, in this matter of population as a variable, that Avery Island may be most relevant to the broader world. In four generations of the family since 1830 thirty-five members lived to be adults; they had a total of seventy-three children, an average of 2.1 per adult scion (blood member of the family), close to a zero-growth mode. In the fourth generation, which is only now completing its siring cycle, an interesting division appears: among those who spent much of their childhood on the island, the average number of children is only 1.7, whereas among those who spent most of their youth elsewhere the average is 2.7. About half of the thirty-nine members of the fifth, present, generation are now married and they already have an average of almost 3 children each. From this record, it would appear that the reproductive restraint of the past may reflect a perception of limits that was clearer to those raised on the island. A sharp increase in birthrate among the present generation more or less parallels the increase in family affluence since 1942 (when oil and gas were found at the island). Although the family saw some danger in expanding its corporate activities on the capital-resource base of oil and chose not to do so, it seems to see no danger in expanding its own numbers on the same base. Perhaps the lesson of Avery Island is that the effectiveness of Malthus's "moral restraint" depends on the perception of consequences and on the degree to which individuals feel obliged to protect a family or communal heritage.

PROBLEMS OF PERSPECTIVE AND POSTERITY

A perspective of time is difficult to incorporate into democratic decision processes. Political representatives are under pressure to produce immediate solutions to problems as perceived by their constituents. Although the problem may have been forming for years as the cumulative consequence of the interaction of social, political, and physical trends affecting the availability of a resource, the wellbeing of a region, or the sanity of the society, so that only incremental adjustments are possible when something goes wrong, the demand for a quick fix by the nation's political plumbers can prevent or deter ameliorative actions based on a careful analysis of the situation and on a comparison of the benefits and costs of alternative actions.

Gasoline and fuel oil shortages have elicited calls for quick solutions ranging from nationalizing the oil and gas industry to freeing the fuels market from all controls, from the opening of the naval petroleum reserves to the invasion of oil-producing nations. Solutions that are perceived as being of somewhat longer range, in that they could not be expected to solve the problem within the next two or three years, invoke provident technology for fusion, solar, geothermal, wind, or garbage power. To find and develop new oil fields, to build pipelines, tankers and refineries; to test ways of making substitute fuels and of getting energy out of tar sands and oil shale; to train coal miners and engineers; to amass venture capital and allocate it; to lay the foundations of new types of government assistance, monitoring, and regulation for new kinds of energy-delivery systems all take time as well as money. Private decision processes handle time better than do public decision processes; the perspective in the former may be foreshortened by a time rate of discount, but it is there and is dealt with in terms of balancing present sacrifices against future returns; it seems much more difficult to do this in public decisions where present sacrifices are exaggerated and the future extends a short way past tomorrow.

Another barrier to perspective is what might be called "mission vision." Decision structures (agencies) are apt to be set up to meet an existing public need in a certain way; these agencies then continue what they were set up to do (and may get very good at) even when the original need no longer exists. Even if the need does not change, there may be new ways of meeting it that are not within the mission and orientation of the organization. This difficulty was publicized in 1972 when the Secretary of the Interior, Mr. Rogers Morton, objected strongly (and in vain) to a court decision ordering the Department of the Interior to include in an environmental impact statement pertaining to a sale of oil and gas lease off the coast of Louisiana a discussion of alternative sources of supply even though some of those alternatives were not within the responsibility or expertise of his department.

NOISE, PREJUDICE, AND CONFLICTS OF INTEREST

Both public and private decision making may suffer from myopia and tunnel vision, but only public decisions are constrained and distorted by the passion for identifying scapegoats and conspirators that flares up in times of social stress, whether that stress involves military threat, social malfunction, or an energy shortage.

The role of the news media in public decision making is important. In the United States and other representative democracies, elected representatives attempt to respond quickly to surges of public discontent. Such surges tend to be channeled, both by human inclination and the dynamics of a free press, into charges of conspiracy, malfeasance, betrayal, subversion, and gouging by monopolies. The average citizen knows little about the origin, distribution, physical characteristics, and political and economic control of energy resources. His is a foxhole view of energy systems and conflicts. Energy is of little concern to him as long as it is available at prices he considers reasonable. He becomes concerned when confronted by a restriction on the availability of gasoline, fuel oil, or electricity, by a sharp rise in the price of energy, or both. Then he seeks information from newspapers and television. His informants in the communications media are in most cases no better informed than he; furthermore, they are imprinted with a working bias in favor of controversy, because their business is to sell goods and services by attracting attention to the entertainment in which advertisements, the *raison d'être* of newspapers and television, are embedded. That entertainment is called news, but its information content is incidental to its capacity to evoke or incite emotions. Dispassionate objectivity does not sell, because it is a taste acquired by very few people in any society.

The news media highlight charges and countercharges as if they were covering a war. They are careless with facts, because facts are not of prime importance to them. As a result, the public has a variety of views and purported facts from which to choose and it generally chooses the simplest view that absolves the viewer of responsibility for either the situation or the solution. When this view is communicated with sufficient intensity to the government representatives, it either paralyzes government responses to the problem or causes precipitate action that makes the situation worse instead of better. The history of the response of the United States government to energy shortages in 1973 and 1974 illustrates the sequence.

Ideological bias easily enters into judgments made in an adversary mode. Liberals have fought the depletion allowance because it caused "unearned" wealth and financed ultraconservative political campaigns. The intellectual is against business control and in favor of governmental control of scarce resources. The businessman and many economists are in favor of a freer market and of letting high prices allocate scarce resources. The underdog is against any identifiable topdog, and demagogues flourish by appealing to the underdog's prejudices.

A persistent conflict of interest in private decision making is the conflict between the self-interests of corporate longevity and maximum profit rate, on the one hand, and public or national interests on the other. Corporate decisions often are made on a longer-range, less emotional base than are public decisions, but they are decisions for the corporation's welfare and for that alone. They will coincide with the national interest only insofar as national goals of economic growth and security are furthered by corporate strategies. Executives of large corporations may confuse corporate objectives with national interests, but the nation agrees at its peril. For many years, it could be argued that the practices and international operations of the major oil companies based in the United States were more or less parallel to the national interest in maintaining a favorable balance of payments and assured access to cheap energy. In the past few years, as these corporations have lost control of the world market and have become weak in places where they were strong, the same argument has become less tenable, and calls for direct intervention by the government in negotiations with exporting nations have grown louder. The system of energy exploitation and delivery that we have relied upon, while it has greatly contributed to the swift development of an affluent, high energy society and an extraordinarily productive physical plant, also has hastened the depletion of domestic oil and gas resources and has moved great amounts of capital into development of foreign oil and gas without laying the basis for a secure alternative energy system at home.

A conflict of interest may arise within a government agency. A classic case was the Atomic Energy Commission's conflicting missions of promoting the use of nuclear energy and protecting atomic workers and the public from the ill effects of that very promotion. It was almost inevitable that when the promotional function collided with the regulatory function, the latter would give way and expediency, either political and economic, would govern the decision.

EXPEDIENCY, SUBCULTURE BIAS, AND TRIBALISM

It was expedient for the Atomic Energy Commission to disclaim regulatory authority for the conditions under which uranium miners on the Colorado plateau worked in the 1940s and 1950s that led to the early deaths of hundreds of them. It was expedient for the Commission to allow millers of uranium ore to do their own monitoring of radioactive "contributions" to the streams along which the mills piled their tailings and some of which became contaminated well above maximum levels set by the Public Health Service; it was expedient to allow mill tailings to be used as foundation material for homes, schools, and office buildings, some of which had to be evacuated years later when hazardous levels of radon gas were discovered in them, seeping up from the radioactive material below. It was expedient to store thousands of gallons of highly radioactive liquid wastes in tanks in Washington, Idaho, and South Carolina while trying to develop a safe and *economical* means of permanent storage or disposal. The tanks are still there, some of them more than thirty years old. It was expedient to bury radioactive wastes in shallow pits over one of the nation's great freshwater aquifers in southern Idaho. And it was expedient to select the abandoned Kansas salt mine in which waste-storage experiments had been carried out for the first national permanent depository for high-level radioactive wastes—until it was demonstrated that the site had hazards that should have thrown it out of consideration as a permanent depository.

There is also the obfuscatory expediency of the elected official who tries to please two disputing parties by appearing to decide in the favor of one while actually doing nothing to hurt the interests of the other.

It has often been noted that regulatory agencies have a tendency to work rather sympathetically with the firms they were set up to regulate, and the statement that regulators become the captives of the regulated is more common than demonstrations of the truth of the statement. A more balanced view might give some weight to the proposition that the enabling legislation as well as the philosophy of the administration of which the regulatory agency is a part may clearly tell the regulator not to regulate so harshly that he will hurt the economy of the city, state, or nation. Another reason for sympathetic interaction that may not involve subornation is that regulator and regulated may belong to the same subculture, may have attended the same schools, may have similar views about political matters and the proper use of resources, and indeed may have experience as individuals on both sides of the regulatory fence. It is hard to see how this problem—if indeed it is a problem—can be avoided in regulation that demands a high level of technical or professional education.

It is when subcultural bias becomes so hardened that values no longer can be transmuted into interest for negotiated decisions that any public-decision mechanism breaks down. Communal dropouts from the mechanized world, fanatical religious fundamentalists, Archie Bunkers of the boardroom—a small knot of any of these can form a destructive embolus in the body politic if they refuse to "compromise" their rigid value structure in order to come to a workable agreement with the rest of society. Such intransigeance has been called "tribalism" and may be a sympton of serious illness in a democratic society (Crowe, 1969). Both tribalism and a growing tendency to try to bypass the representative structure of government to deal directly with the "enemy" were visible in the controversy over the oil spilled in the Santa Barbara channel.

A TALE OF TWO CITIES

Attempts to resolve environmental and resource issues are hampered by conflicts among local, regional, and national interests. A city in Texas may want a more or less permanent variance from state air-quality standards because it values the income from a smelter more than the associated benefits to health and the environment that would result from closing down the smelter. Oil-producing states may view their interests in a national energy plan very differently from the way New England sees the same plan. Two coastal cities in the United States found that their attempts to solve the same local, energy-related, problem ran into conflict of interest at higher levels, and the results are useful to a student of such decision problems.

Corpus Christi Bay is a prominent cuspate indentation of the Texas coastline, somewhat more than a hundred square miles in extent, separated from the Gulf of Mexico by a low barrier island. Corpus Christi is unique among Gulf Coast cities in that the major residential part of the city is built on a bluff that rises about forty feet above sea level and gives a view of the broad expanse of bay from any point along about eight miles of bluff. Not only does the city owe much of its charm, beauty, and recreational opportunities to the bay, but also a considerable part of its economy depends on commercial and sports fishing in the bay, on tourism related to the bay, on vessels that use the bay (Corpus Christi is the third largest port in Texas), and on oil and gas production from the bay. In 1950 when the city annexed most of the bay, there were dozens of producing oil and gas wells in the bay; all of them, however, were far away from the city and nearly invisible

FIGURE 12.2
Corpus Christi, Texas, wanted to keep its oil and gas in its backyard (note the storage tanks and refinery complex in the photograph) but the discovery of a gas field in Corpus Christi Bay (see map) within 4 miles of the marina, posed a threat to its prized waterfront. (Photograph courtesy Corpus Christi Caller-Times.)

from the municipal shoreline. In the early 1960s, however, exploration moved closer to the city and citizens who valued their shoreline became concerned. When in 1965 it was announced that a new gas field had been discovered two miles offshore from downtown Corpus Christi and a request was made to the city council for erection in the bay of a production platform to separate water and natural-gas liquids from the wet gas of the new field, a public outcry developed against the proposal.

Until 1965, local lawyers, petroleum engineers, and geologists had played a key role in the municipal regulation of oil and gas development within the city limits. They were called upon to draft relevant ordinances and to interpret them after they were enacted. The community does not seem to have been concerned about the conflict of interest inherent in the drafting by these people of legislation to

FIGURE 12.2 *(Continued)*

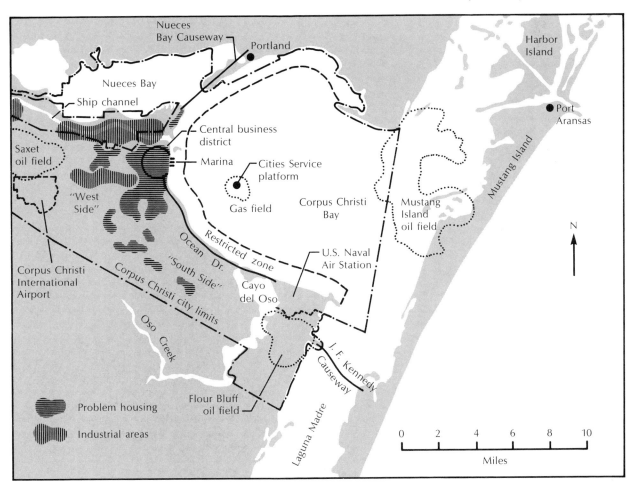

regulate activities in which they were professionally and economically associated. In 1965, the local experts suddenly were rejected as advisors. To replace them, an *ad hoc* committee was appointed by the mayor. The committee was designed to represent group interests and community values and no member was chosen for professional expertise. The nine members included representatives of the three major oil companies that had large interests in production from the bay, three members of the city council, and three citizens. The committee labored diligently for more than a year, not always in harmony, and finally brought forth a new ordinance that was adopted by the city council late in 1966. Subsequent amendments to the ordinance have been minor; in other words, the decision has been accepted by all interested parties. The new ordinance established a permanent committee advisory to the city council to be known as the Bay Drilling Committee, created a Division of Petroleum Inspection within the city government, set up a drilling sanctuary extending approximately one mile offshore, required the city council to prescribe approved exposed silhouettes of well structures within two miles of shore, the silhouettes to be as low as practicable and all equipment to be submerged that could feasibly be submerged, required cluster drilling of wells, banned new production platforms, required that well structures above the water be painted a conspicuous color and production platforms an inconspicuous color, banned signs and lights from such structures except for working lights and signs forbidding smoking and trespassing, required drillstem tests (which can be very noisy) to be started only during daylight hours, provided that blowout preventers and other safety features be used, and required insurance and a bond against accident and noncompliance. The decisions incorporated in the Bay Drilling Ordinance of Corpus Christi were made entirely at the municipal level. The state of Texas, which owns the mineral rights beneath the bay, took no action; its Land Commissioner, however, took a public stance *against* depriving the state of any of its production royalties. The county took no stand. The United States Corps of Engineers was asked by individual citizens to deny approval of further drilling and did delay the approval of a pending request.

In Santa Barbara, events took a different course. Offshore oil production there had not been directly off the municipal shoreline. Santa Barbara, like Corpus Christi, derives most of its attractiveness from its marine view and associated recreational opportunities; like the Texas city, it has a large number of wealthy long-established inhabitants, a thriving tourist trade, and some ugly reminders nearby of irresponsible oil exploitation. In 1966, the first federal lease was sold in the Santa Barbara Channel beyond the three-mile limit of state sovereignty. As Carol and John Steinhart say, in their excellent report on the Santa Barbara oil spill (1972), "a storm of protest broke loose (p. 29)." By that time, the state had established a drilling sanctuary

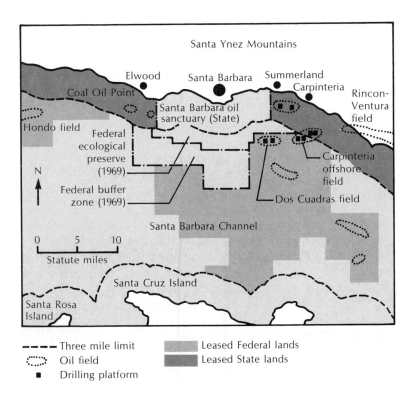

Santa Ynez Mountains

Elwood Santa Barbara Summerland

Coal Oil Point Carpinteria

Rincon-Ventura field

Santa Barbara oil sanctuary (State)

Hondo field

Federal ecological preserve (1969)

N

Federal buffer zone (1969)

Carpinteria offshore field

Dos Cuadras field

Santa Barbara Channel

0   5   10
Statute miles

Santa Cruz Island

Santa Rosa Island

FIGURE 12.3
On this map of the Santa Barbara Channel are shown: the 3-mile wide zone of state offshore sovereignty, within which is the Santa Barbara Oil Sanctuary; the federal ecological preserve and federal buffer zone; the Dos Cuadras Offshore Oil Field, where the great blowout occurred; Carpinteria, where early oil wells a few hundred yards offshore left unsightly relics; and the lands leased for oil, which extend across the channel to the 3-mile state limit off Santa Cruz Island.

---- Three mile limit   Leased Federal lands
⊂⊃ Oil field   Leased State lands
▪ Drilling platform

within the three-mile limit for sixteen miles along the shoreline of Montecito, Santa Barbara, and Goleta. The strong opposition in these communities to offshore drilling won for them this concession: a federal lease sale scheduled for 1967 was delayed until February 1968, and a two-mile buffer zone was established off the state sanctuary, effectively prohibiting drilling within five miles of shore. On January 29, 1969, a well being drilled on one of the leases awarded in that sale blew out and set in motion what is still referred to as a disaster, although no human lives were lost, the beaches were cleaned up, and permanent ecological damage appears to have been negligible. Leakage continued for almost a year and litigation for much longer.

The federal government, in whose jurisdiction the blowout occurred, made most of the decisions in what followed. The secretary of the interior, the director of the United States Geological Survey, and the president of the United States were under great pressure from citizens of Santa Barbara and from environmentalists across the nation to suspend drilling in the Santa Barbara Channel and rescind the existing leases. Some of the ultimate decisions were: the two-mile buffer zone was turned into a permanent ecological preserve, further leasing in the channel was indefinitely suspended, oil companies were made liable for pollution damages stemming from operations on their leases, whether accidental or not, regulations governing drilling and production on the outer continental shelf were strengthened.

The state of California, the city and county of Santa Barbara and the city of Carpinteria jointly submitted a claim to the secretary of the interior on February 18, 1969, for $500 million. The same group also filed suit against the Union Oil Company and its partners, the drilling contractor, and assorted federal agencies, for $560 million. Another suit, on behalf of the citizens of Santa Barbara, against Walter Hickel, William Pecora, and other federal officials and the Union Oil Company and its partners, asked for a total of $1.3 billion in damages. Since both the cleanup expense and property loss were paid for by Union Oil and its partners and as the other economic loss to local business was a minor fraction of the total claimed, the Steinharts conclude that "much of that astonishing figure of more than two billion dollars must then be laid to the cost in human anger and grief, if not to a desire to punish those who had wronged Santa Barbara." (Ibid., p. 108). A plainly biased but more explicit discussion of the legal, administrative, and social equity problems raised by the Santa Barbara event by Nash, Mann, and Olsen (1972) contains this statement: "In the best of all possible judicial worlds local feeling should quite possibly outweigh developmental interests (p. 104)."

Certainly, the plaintiffs in the cases cited above would agree. But in that case, how would the national interest in energy supply be served? And how does one place a monetary value on "local feeling"? Furthermore, would not the same reasoning require that a community that is willing to accept pollution in return for jobs be allowed a variance from national standards? The vocal citizens of Santa Barbara alleged that the federal government "betrayed" them in allowing oil drilling in "their" channel. The question about resources and environment that comes to the mind of a remote observer, and is a vexing one, is how much of the channel belongs to Santa Barbara and how much to the rest of the nation? Perhaps the most significant difference in the tales of the two cities is that Corpus Christi had the mechanism available to arrive at a local decision that satisfied local interest groups, whereas Santa Barbara did not. Many recent controversies suggest that people who feel that a social cost is being *imposed* upon them without their having had a reasonable share in the decision are apt to object strongly to a cost they might have accepted had they felt themselves *participants* in the decision.

THE CHANGING ROLE OF THE EXPERT

In both controversies just described there was an obvious rejection of the traditional role of technical experts in decision making. In managing public resources, in planning public facilities, and in protecting the public welfare, technical and legal experts are required. They define resources, interpret statutes, identify hazards, do research, calculate benefits and costs, and design or approve technological and managerial means of meeting public needs. In all these activities, their role is supposed to be that of informing or carrying out political decisions. As experts, they are not supposed to make political decisions or moral judgments. But expert decisions often do entrain moral, political, or esthetic judgments.

In Corpus Christi the group of petroleum engineers and lawyers who advised the city council before September 1965 were firm in their belief that they acted not only from an unchallengeable basis of technical expertise but also in the "best interests" of the citizens of the city. Their recommendations, incorporated into law, involved implicit compromises between economic health and environmental quality. When in 1965 this advisory group recommended that production platforms for natural-gas processing be allowed on each of some one hundred state lease tracts in Corpus Christi Bay, "the sludge hit the fan." Growing sensitivity about the environment, focused on the bay as the city's outstanding amenity, exploded in a demand that the question be taken out of the hands of "experts" and placed squarely in the political arena, which it was. To this day, no further use has been made of the old Oil and Gas Wells Commission; it was neither discharged nor thanked for its earlier work; it was simply ignored—because the social values implicit in its technical judgments no longer accorded with the values of the community.

As the technical experts in government have tended to make political judgments along with technical ones, so do their severe critics tend to make technical judgments along with their political ones. Critics of oil drilling in the Santa Barbara Channel rejected the technical judgment of government experts that continued pumping of oil would relieve geological pressures and abate the continuing leakage; one of those critics, an assistant professor of geography in the local university, stated that the pumping itself caused the pressure that kept the ocean floor leaking, just the reverse of the technicians' consensus. The rejection of a technical judgment frequently is accompanied by an expressed or implied charge of venality laid on experts in government, the universities, and industry, charges that were rampant in the Santa Barbara affair.

A SHIFT OF THE BURDEN OF PROOF

Along with a rejection of expert judgment, at least insofar as that judgment may be applied to questions of values that cannot be determined in the market place, there is clear evidence, in the public attitude toward environmental controversies, of a desire to shift the burden of proof of a positive social benefit-cost ratio from the defender to the developer. Hitherto the initiative has been with the developer or exploiter of resources, and the burden of proof has been on the person who, objecting to the proposed alteration of the environment, has wanted to prevent or restrict it. In the political forum, the opponent has had to persuade decision makers that the cost to the community or to the nation of the proposed development would exceed its benefits. In the courts, he has had to demonstrate a clear and present physical or financial threat of himself in order to deter development. Now both rules are changing. Extending the new philosophy that drugs, to be approved, must be demonstrably beneficial, to the environment, many people now argue that the burden should be on the developer to prove that his proposal has a clearly positive benefit-cost ratio for the nation, in terms of environmental and ecological impact as well as market values, before he should be permitted to alter the environment by building a dam, mining for coal, spraying pesticides, or drilling for oil. Further, the courts have recognized the right of a citizen to maintain a suit on the grounds of the esthetic deterioration of the environment, and the National Environmental Protection Act of 1969 has opened the door to suits questioning the validity of the basis for public decisions affecting the environment.

It is easy to change the rules of a game before the game starts. It is exceptionally difficult to change them once the game is well under way. The game of resource development, played with a strategy of profit maximization for the private investor and a goal of material well-being for the public, has been under way in this country for a long time. Individuals, corporations, and governments play the game, profit from it, and bet on it. Any fundamental change in the rules will require wholesale shifts in vested interests. But as resources diminish and the environment decays, formidable constraints will be placed on the initiative and choices of the resource exploiter. It is unlikely that his initiative will be entirely taken away, for that would destroy the function of private enterprise in resource development and make all development decisions political. If the role of private initiative in resource development is to be retained, the new constraints will need to be predictable and not imposed after the fact. We can expect to see a more comprehensive zoning and permit system for the private sector as its actions affect the environment, for in the present system there are unacceptable discords between what may be done on or with adjacent tracts of land (and bodies of air), and there is too little consideration of long term needs and consequences in the planning and approval process. For public decisions we can look for a continued demand for new decision mechanisms that will allow the public and its representatives to compare alternative plans, in terms of all the costs and benefits both tangible and intangible, will allow non-development as an alternative, will regard the preservation of choice for the future as a public benefit and the maintenance of environmental quality as a practical goal, and will allow the decision to be made according to the best alternatives available.

SOME NEEDS AND SUGGESTIONS

In the United States the system for making private decisions about energy is relatively simple and, except for regulatory hurdles, swift. These regulatory hurdles, however, have markedly stretched out the time required to implement private decisions to build a nuclear power plant or to open a strip mine. Many public decisions, slow enough under clear grants of discretionary authority, now are similarly delayed in implementation or overturned after lengthy judicial challenges. Other public decisions, still to be made by the Congress, are delayed by conflicts of interest, by political infighting, and by lack of knowledge of the problem for which solutions are being offered.

For good decisions about energy, the following are needed.

1. Information on the availability of energy resources—reserves, both proven and potential, their location and the degree of uncertainty of their availability, which depends upon both physical and political factors.
2. Information about the appropriate technology for exploitation, both available and potential.
3. Analysis of the system efficiencies of alternative energy-delivery systems.
4. Analysis of the social, environmental, and economic impact of each alternative system, and an assessment of the social acceptance of each system and alternative ways of achieving acceptance.
5. A decision structure that is competent to analyze alternative solutions, to preserve them during the analysis, to present them for political decision and that is accessible to any individual or interest group and is yet able to resolve conflicts and produce a reasonably prompt decision.

Available information on which to base decisions for the nation's future welfare will always be woefully incomplete, perhaps even inconsistent or contradictory. The problems of estimating amounts of undiscovered oil and gas will be discussed is chapter 13. Expert estimates range widely and there can be no certainty. Yet decisions must be made that require a choice of one expert or one method over others.

In 1975, the *cost* of getting a barrel of crude oil out of Arabian ground and into a refinery in the United States was about $2, but the *price* was more than $12. Expert estimates of the long-term ability of the cartel (OPEC) to maintain such a differential range as widely as do those of the experts on undiscovered oil and gas. Yet, in the face of this kind of uncertainty also, the nation must make a choice. A token effort to conserve while hoping for the collapse of the cartel? A determined effort to increase the domestic supply of energy, a choice certain to entail large environmental and social risks related to coal mining and nuclear power? A mixture of conservation, development of additional domestic supply, and encouragement of the producing countries to invest in the United States—or, short of that, negotiation with them of trade agreements designed to assure the inflow of oil at a stable price?

Another type of decision fraught with uncertainty is a decision to invest in research applied to energy needs. The long-term future of the United States may be substantially affected by present public investments in research and development of fusion, solar, or geothermal power. We have no way of knowing whether any investment in any of these three will pay off as handsomely as we would like. Yet such investment decisions must be made and to have them made by Congressional action based mainly on hope and the testimony of parties directly interested in the appropriations leaves a great deal to be desired.

Even when the information available is abundant and the degree of uncertainty low, as in the dilemma of the United States as to how to use its ability to produce large agricultural surpluses, policy may be difficult to formulate because of conflicts among interest groups, the demagoguery of some politicians and labor leaders, and the reverberatory effects on public opinion of sensational reporting. If crops are larger than sales, the consumer enjoys direct benefits in lower food prices but suffers indirect costs (through taxation) to pay for governmental programs of help to farmers, for storage of surpluses, and for encouragement of exports. If, however, foreign sales absorb the surplus, food prices rise and the consumer suffers direct costs; although his indirect contribution—through taxes—to the cost of food production may have diminished, his tax return will not show it immediately. Consequently, the food consumer is against large export sales of farm products, for what he considers a very good reason. Because only the farmers (4 percent of the population), the Department of Agriculture, the Treasury Department, a few grain dealers, some railroad executives, and a smattering of economists are in favor of such sales, we very likely shall continue to have agricultural surpluses in storage even when they could be sold.

The development of substitute or alternative energy sources to the point of substantial availability will require a great deal of capital and years of time. Whether made by subsidies and guarantees (for example, of a profitable market for shale oil) to industry or by the government itself, the investment can be returned only through production and consumption. It will be costly to hold alternative technologies and substitute facilities out of production and it will be impossible if the development method selected has included a guaranteed domestic market for the substitute energy. A nation with a high material level of living confronted with an increasing shortage of energy may face no more important set of questions than the following:

1. How can wise choices be made among alternative investments in research and development designed to produce economical substitutes for domestic oil and gas?
2. Are there effective ways of developing substitute sources to the point of significant producing capacity *without being forced to use them* (in case, for example, foreign oil again becomes cheaply available)?
3. How should the costs of developing substitute capacity be allocated?
4. How does a nation weigh the costs of developing substitutes against the risks of dependence upon imports?

In the perception of problems related to energy, the pressure of immediate economic, social, and environmental concerns may obscure one or more issues that overshadow all the rest. Such may be the case with problems involving the hazards of a greatly expanded nuclear-power industry.

Industrialized society has become highly susceptible to interdiction, sabotage, and blackmail, whether by Arab oil suppliers, disgruntled garbage collectors, or hijackers of airplanes. So far, responses to such actions have ranged from acquiescence and payoff to the imposition of an expensive search of all airline passengers. In an expanded nuclear-power industry, there will be irresistible pressure to concentrate power generation and fuel-recycling facilities at relatively few sites, in order to minimize the direct environmental and social hazards and to take advantage of economies of scale. When to the hazards of critical failure of an energy system are added the blackmail and sabotage potentials of clandestinely diverted materials such as plutonium and uranium 233, the necessity of a permanent search-and-guard protective system, at least for the nuclear-power portion of the economy, becomes obvious. Although perpetually perfect surveillance is a logical absurdity, the short-term aims of a well-designed protective system may be achieved. But at what social cost? Perhaps at the cost of extensions of the system, especially if it appears to work or succeed, as many say the airline search procedure "works," to other critical parts of high-energy society. The danger here is not so much the direct threat to individual liberty but the risk that the critical elements of society, thus guarded from illegal disruption, may become relatively easier for a small group to take over legally or quasi-legally, thereby creating a garrison state. Small may or may not be beautiful, but applied to energy systems it carries an implication of social security that deserves serious consideration. But in the existing decision system, there is almost no way in which it can be given serious consideration. The greatest decision facing the United States in 1976 is whether or not to proceed with the development of commercial breeder-reactor power plants. A national referendum on the issue would alert the public to the issues and uncertainties involved. But the odds are against the Congress's considering such a referendum. Such decisions are made a little at a time, not in one great exercise.

The national decision structure does appear inadequate for the tasks it faces in the energy field, tending to ignore long-term costs and benefits in favor of perceptions of immediate needs and interests. This orientation is perpetuated by the practice of having most public decision makers directly responsible to the public, by deficiencies in the ability to forecast consequences, and by a pervasive faith in technology or Providence to save us from the adverse consequences of decisions based on immediate desires. It is perhaps instructive to note that the less the managers of large industrial enterprises are in fact responsible to their stockholders, the greater their ability to make decisions for the long-term welfare of the enterprise; in other words, as they grow less constrained by the perceived interests of the individual shareholder, the more they can make responsible decisions in terms of healthy corporate longevity.

There seems to be no place in democratic governments for relatively apolitical, dispassionate analysis of alternative national futures, of the needs and resource constraints of ten, twenty, or fifty years from now, and of the measures that might be taken to assure large, far-term benefits from small, near-term sacrifices. More's the pity, for there is a great need for some such analytical arena in which to plan for the optimization of energy use, environmental quality, and our future standard of living. Revolutionary changes in ethics and goals that might reduce energy demand can hardly receive useful consideration in the adversary arenas we seem to prefer for social decision making. In the United States, the universities, the National Academy of Sciences, and philanthropic foundations can play an important role in formulating and presenting analyses to government, but can be effective only insofar as government heeds their counsel.

The question of man's fitness today hinges on his ability to make and maintain decisions that require individuals to make sacrifices now for the benefit of the species in the future. Contemporary human decisions are not reassuring to one who wants to believe in the continued progress of his species. Social conscience has yet to advance to the species level; the teachings of the world's great religious leaders about the brotherhood of man have been honored more in the breach than in the performance. He will sacrifice to gain advantages for himself or his family, and sometimes for his nation or his cultural group, but such advantages must generally be realized fairly soon. The weakening of the family, the decline of patriotism and religion, and the stress of individual worth both in Christianity and in market democracies are barriers to community-preserving, not to mention species-preserving, decisions. In contemporary market democracies we have no mechanisms for long-range, species-oriented decision making. Our politicians are under intense pressure to provide quick and cheap solutions to perceived problems. It is a paradox that an affluent society, possessing more resources for study and experimentation than a poor society, appears to be prevented by that very affluence from making decisions that require any sacrifice. Faith moves *people*, not mountains. Faith—in a market economy, in provident technology, or in "built-in" population control—is a hazardous substitute for the tough decisions that need to be made if mankind is to adapt to environmental stress brought about by resource depletion, environmental perturbation, and population growth. For the first time, man possesses sufficient information to forecast the probable long-term consequences of his immediate decisions. His cultural evolution no longer needs the assumption of divine guidance in cultural evolution (implicit in the Western philosophy of "progress"), as he has not needed, for almost two hundred years, the assumption of metaphysical natural law to explain the sensed world about him. Whether he will now use his knowledge to guide his own future and to avoid the predictable catastrophic risks of letting the cultural machine run on momentum and faith is still an open question.

SUPPLEMENTAL READING

Baldwin, Malcolm F., 1970. "The Santa Barbara Oil Spill." In *Law and the Environment*, edited by M. F. Baldwin and J. K. Page, p. 5-46. New York: Walker.

Written from the legal point of view, this paper is an excellent introduction to the jurisdictional and regulatory problems involved in offshore oil drilling.

Berry, R. Stephem, and Margaret F. Fels, 1973. "The Energy Cost of Automobiles." *Bulletin of the Atomic Scientists* v. 29 (December), p. 11-17, 58-60.

Automobiles, could be made to last three times as long with only a 5 percent increase in energy cost; the authors propose a policy of thermodynamic thrift.

Bezdek, Roger and Bruce Hannon, 1974. "Energy, Manpower, and the Highway Trust Fund." *Science*, v. 185, p. 669-675.

If construction monies were shifted from highways to railroads, the energy required for construction would be reduced by about 62 percent and employment would increase by 3.2 percent; if a transport shift to railroads then occurred, the energy cost of transporting passengers and freight would be much reduced.

Crowe, B. L., 1969. "The Tragedy of the Commons Revisited." *Science*, v. 166, p. 1103-1107.

Pertinent exposition of the role of "tribalism" in contemporary public controversies—the refusal to let commensurable interests stand in the place of values, and the resulting breakdown of democratic negotiation.

Bohi, Douglas R., and Milton Russell. 1975. *U.S. Energy Policy: Alternatives for Security.* Baltimore: Johns Hopkins University Press.

Contains an illuminating discussion of the probable dynamics of decision within the Organization of Petroleum Exporting Countries. Outlines alternative strategies for secure supplies of energy. Weak on the question of the elasticity of domestic crude-oil supply.

Daly, Herman E., 1974. "Steady-State Economics versus Growthmania: A Critique of the Orthodox conceptions of Growth, Wants, Scarcity, and Efficiency." *Policy Sciences* v. 5, p. 149-167.

An economist argues that absolute scarcity and relative wants are more realistic concepts than the orthodox doctrines of relative scarcity and absolute wants.

Fabricant, Neil, and Robert M. Hallman, 1971. *Toward a Rational Power Policy.* New York: Braziller.

A rarity—a report written under bureaucratic aegis by two lawyers that is clear, concise, vigorous, internally consistent, highly readable and well bolstered with pertinent information and citations.

Ford Foundation Energy Policy Project, 1974. *A Time to Choose—America's Energy Future.* Cambridge, Mass.: Ballinger.

Heavily larded with economists and sadly deficient in earth scientists, the project staff produced a report helpful in its exposition of alternatives, ridiculous in its treatment of domestic oil and gas resources.

Hardin, Garrett, 1968. "The Tragedy of the Commons." *Science* v. 162, p. 1243-1248.

By now a famous argument, but based on the myth of economic man, in which it is held that common resources are bound to be degraded by selfish overuse.

Linton, D. L., 1965. "The Geography of Energy." *Geography* v. 50, n. 228, p. 197-228.

Argues for immediate worldwide controls on energy distribution and conversion.

Lowe, William W., 1972. "Creating Power Plants: The Costs of Controlling Technology." *Technology Review* v. 74, n. 3, p. 22-30.

Slowing demand for energy would greatly ameliorate the planning and decision process.

Mesarović, Mihajlo, and Eduard Pestel, 1974. *Mankind at the Turning Point.* New York: Dutton.

If mankind chooses to control its own future, the result will be "a dawn, not a doom." Not very helpful on how the choice can be made.

Metzger, H. Peter, 1972. *The Atomic Establishment.* New York: Simon and Schuster.

Well-documented critical history of conflicts between the promotion and protection missions of the Atomic Energy Commission.

Nash, A. E. K., E. D. Mann, and P. G. Olsen, 1972. *Oil Pollution and the Public Interest—A Study of the Santa Barbara Oil Spill.* Berkeley: University of California, Institute of Governmental Studies.

A study of social equity in which the authors' values become entangled.

Ophuls, William, in press. *Ecology and the Politics of Scarcity.* San Francisco: W. H. Freeman and Company.

An excellent and thorough treatment of the problems of adjusting the institutions that influence decision in the United States to an economy of scarcity and of making the quality of life, not the quantity of material production, the aim of economics.

Polanyi, Karl, 1944. *The Great Transformation.* New York: Rinehart.

A historic diagnosis of the decline and fall of the market economy.

Ray, Dixy Lee, 1973. *The Nation's Energy Future.* Washington, D.C.: Atomic Energy Commission.

The title is misleading; it is the first comprehensive proposal for a national energy research and development program; the stated goal of the recommended investments is energy self-sufficiency.

Rose, David J., 1974. "Energy policy in the U.S." *Scientific American* v. 230, n. 1, p. 20-29. (Available as *Scientific American* Offprint 684.)

The attention given in national energy policy and decisions to energy supply far outweighs the attention given to utilization and conservation. Describes a useful technique for assessing energy research-and-development investments.

Steinhart, C. E., and J. S. Steinhart, 1972. *Blowout—A Case Study of the Santa Barbara Oil Spill.* North Scituate, Mass.: Duxbury Press.

One of the authors was a participant in the events related; rather awkwardly presented, but provides valuable testimony.

© William Carter 1971

Chapter 13   THE DEPLETION OF GEOLOGIC RESOURCES

Sight is a faculty; seeing, an art.
—George Perkins Marsh, 1864

ARE THERE RESOURCE LIMITS TO GROWTH?

Will shrinking resources limit population and economic growth? This question has been argued since the nineteenth century English economist Robert Malthus first challenged the optimistic doctrine of growth published in 1776 by Adam Smith. Now the debate gains strength from the need to define national policy and design national strategies to blunt the economic impact of a scarcity of resources, whether real or artificial. It grows heated when interests are threatened and it reveals differences in professional training and social philosophy. On the one hand are those who profess belief in a "natural" economic growth rate, in the market economy's ability to allocate resources for maximum social benefit, and in the perpetual development of new resources through technology; they argue for the disintegration of cartels, the freeing of the market, and increases in production as appropriate national strategies. On the other hand are those who believe that no exponential rate of increase of anything tangible can be sustained indefinitely, that the exploitation of natural resources is subject to the clear constraints of the law of diminishing returns, and that the ingenuity of man will not overcome the laws of physics; they argue for increased controls on the market, government allocation of scarce resources, and decreases in consumption. The concerned citizen wonders which view is correct or better to follow in making national decisions.

Deciding what to believe is difficult. The wide range of expert views, for example on the future availability of fossil fuels to the consumer

Much of this chapter appeared in the June 1975 issue of *Technology Review* (v. 77, n.7, p. 14–27) and is reproduced here by permission of the Alumni Association of the Massachusetts Institute of Technology.

in the United States, and the poverty of knowledge of many of the reporters, economists, and politicians who seek to interpret the information given them, allows ample opportunity to adopt a view that accords best with one's own interests, beliefs, and desires. It will be extraordinarily important to the nation to see the future wisely, if not clearly, and for this reason I shall attempt to give here an argument that, although not new, may be helpful to the perplexed. It is an argument based on my own bias in favor of the use of geological knowledge in the interpretation of geologic resources, and of history as a guide to the future.

This discussion will deal only with nonrenewable resources although a similar argument could be made for the renewable ones when they are depleted by use faster than they are replenished. *Depletion* is here defined as reduction in the total amount of a resource ultimately available for use by mankind.

THE IMPORTANCE OF DEPLETION

In previous chapters we have seen that there are physical limits to the economic exploitation of natural resources and that it is not possible for technological progress to extend indefinitely, either in time or rate, the use of a resource. Only if abundant, cheap energy were to become available would man be able to "burn" the rocks and "mine" the oceans of the world, and then only to an extent dictated by energetic or ecological limits or both. If we cannot extend particular resources indefinitely, the argument is often raised that substitute materials will be found almost ad infinitum as economic pressures rise and technology advances. Although substitute materials can extend the satisfaction of man's needs, there appear to be no good substitutes for several materials now in use, and for many other materials available or potentially available substitutes represent either a substantial loss in desired properties or a heavy investment in new processing and use systems, or both. Substitutes are nice, but there are the same sorts of limits on substitute materials as exist on the original materials, and so the day of reckoning is but delayed by the discovery of an adequate substitute.

If one accepts these conclusions or goes only part way, preferring to cherish the hope that adequate substitutes always will be found when needed badly enough, it becomes of obvious importance to be able to judge when a specific resource is approaching either a rate limit or a quantity limit of availability. With *renewable* natural resources it seems that we often learn the limit of sustained yield only after we have overharvested, sometimes sharply depleting or extinguishing the resource in the process. With *nonrenewable* natural resources there is no question of sustainable yield rates, but only the question of the ultimately recoverable quantity. This chapter will explore the problems of estimating or forecasting the depletion and ultimate exhaustion of nonrenewable resources.

THE NATURE OF GEOLOGIC RESOURCES

Geologic resources are geochemical concentrations of materials that can be recovered and used at a profit. The profit may be in the form of an energy surplus, as from the exploitation of deposits of fossil and nuclear fuels, or it may be in the form of an energy saving, as in the lesser expenditure of human energy (and time) required per unit of work (or utility) achieved when one uses steel in place of wood in implements and utensils. Defined thus in terms of direct or indirect energy profit, the concept of a geologic resource becomes one of energy economics and is dependent upon the ratio of the utility of the product (measured by energy income or savings) to the work cost of production, a ratio that must be greater than one. The work cost of production in turn depends upon the geochemical concentration, the degree to which the host or reservoir rock is refractory or recalcitrant, the costs incurred by a hostile environment, and the energy or work costs of transport and conversion to useful form. A fact that must not be lost sight of is that the reference currency, the hard money, of energy economics is the man-hour. Without subsidization and augmentation of human effort by the fossil fuels acting through powered machines, geologic resources would be meager indeed, because human effort is limited in power, in efficiency, and in convertibility to other forms of energy.

Useful elements in the earth's crust show not only a wide range of average abundance but a wide range in degree of concentration (see table 13.1). Geochemical concentrations of useful mineral substances are far from uniformly distributed throughout the world: almost 70 percent of the proved reserves of crude oil are in the Middle East; five countries produce more than 65 percent of the world's copper. The non-random distribution of geologic resources is real and does not result from differences in exploration effort. Geologists have long known that metallogenetic as well as petroleum provinces exist and that the world's coal deposits are strikingly concentrated in the temperate belt of the northern hemisphere. They were astonished when Herman Kahn once stated (Mitre Corporation, 1972, p. 97) that the geology of Latin America is the same throughout that "continent" as it is in the barren highlands where most of the mineral resources have been discovered and that little has been found in the other parts of Latin America because "people just have not looked . . . ." The search for geologic resources has been unremitting and has become increasingly sophisticated throughout the world. New discoveries have

Table 13.1 Ratios of Cutoff Grades to Crustal Abundance for Selected Elements

Element	Crustal abundance (ppm)	Cutoff grade* (ppm)	Ratio
Mercury	0.089	1,000	11,200:1
Tungsten	1.1	4,500	4,000:1
Lead	12	40,000	3,300:1
Chromium	110	230,000	2,100:1
Tin	1.7	3,500	2,000:1
Silver	0.075	100	1,330:1
Gold	0.035	3.5	1,000:1
Molybdenum	1.3	1,000	770:1
Zinc	94	35,000	370:1
Uranium	1.7	700	350:1
Carbon	320	100,000	310:1
Lithium	21	5,000	240:1
Manganese	1,300	250,000	190:1
Nickel	89	9,000	100:1
Cobalt	25	2,000	80:1
Phosphorus	1,200	88,000	70:1
Copper	63	2,000	35:1
Titanium	6,400	100,000	16:1
Iron	58,000	200,000	3.4:1
Aluminum	83,000	185,000	2.2:1

NOTE: Approximate lowest concentration economically recoverable in 1975.
SOURCES: Except for carbon, crustal-abundance figures are from Tan Lee and Chi-lung Yao, "Abundance of Chemical Elements in the Earth's Crust and Its Major Tectonic Elements" (*International Geology Review* v. 12, 1970), p. 778–786.
The figure for carbon is from Mason (1958). The cutoff grades are author's estimates.

been adding reserves of crude oil to the world's total faster than consumption has been increasing and are in part responsible for the present great surge in the world's copper-producing capacity; but in general, the golden age of discovery of large, cheaply exploitable deposits appears to be drawing to a close. The persistent dream of vast undiscovered resources, a dream unclouded by available geological knowledge, is told again by John C. Fisher in his recent book, *Energy Crises in Perspective* (1974): "I expect that as better data become available, the energy resources of the world will prove to be more or less uniformly distributed (p. 34)." Such hopeful ignoring of the facts of the earth's constitution and history might be harmless were it not used as the basis for recommended national energy and minerals policy.

A fact highly important to man is that the natural mechanisms of concentration of the elements operate most effectively on and near the earth's surface. Weathering, erosion, sorting during transport, and leaching by groundwater are effective only in the upper few hundred feet of the earth's crust. Conditions favoring the maintenance of open fractures and the deposition of mineral materials from rising emanations exist only in the upper few thousand feet of the crust. The formation, migration, and entrapment of oil and gas takes place within the sedimentary skin of the continents, at depths of fifty thousand feet or less. Certain conclusions emerge: many geochemical concentrations

The Depletion of Geologic Resources

have been destroyed by erosion, uplift and erosion have made many deposits of geologic resources more accessible than they otherwise would have been, geochemical concentrations (ore bodies, oil pools) are not likely to exist below levels that are almost within the range of current drilling technology.

Geologic resources display habits or modes of occurrence that affect the ability to extract them economically. Some, such as coal, salt, phosphate, and potash occur in relatively large, commonly tabular bodies and may be used directly with little or no processing other than breaking and cleaning, and the energy or work costs of production are related mainly to the depth and thickness of the beds and their distance from places of use. Although the ultimate limit of such a resource will be determined by geochemical boundaries, contemporary limits are clearly economic. Thin beds cannot be recovered economically, nor can thick beds if they are too deep. Different are those resources that tend to occur in small, sharply bounded, high-grade deposits of irregular shape. Mercury, silver, and crude oil are in this group. Once discovered, such a deposit usually can be mined to physical exhaustion, in other words, to its geochemical boundaries, no matter how deep or small the deposit. Material will be left behind only because of inefficiencies in extraction technology.

Then there are the geologic resources that, although they may be found in small, high-grade deposits, also occur in larger deposits of gradational character that grow in recoverable tonnage, at least for a time, as fast as or faster than the exploitable grade decreases. Commodities in this group include copper, iron, aluminum, uranium, and oil shale. Even these deposits, because of the lack of homogeneity in the host environment, generally show a marked break in the strength of metallization as the barren surrounding rock is approached. These geologic differences in mode of occurrence are clearly reflected in production histories, as we shall see. First, however, let us consider the matter of limits to geologic resources.

LIMITS OF GEOLOGIC RESOURCES

There are three kinds of limits to geologic resources. First is the *limit of comparative utility:* a resource is a resource only while it can be used to perform a function desired by man better or more cheaply than can another resource. If the cost of one resource rises to a level at which another resource can be substituted at a lower cost for comparable utility, substitution will take place and a limit will have been imposed on the first resource. If however, there is no substitute of comparable utility available, the limit to the utilization of a resource will be the point at which no one in the society is willing to pay the cost of production, because doing so would lower his level of living more than foregoing the use of the resource. The *living-level degradation limit* will be reached sooner for some resources, say diamonds, than for others, say food; in fact, forgoing the use of food entails such penalties that people will continue to produce or purchase it even though their level of living deteriorates thereby. The ultimate limit to the use of geologic resources will be set by the limits of the natural energy subsidy in fossil and nuclear fuels, in solar radiation and the hydrologic cycle. Wherever and whenever that natural subsidy—the excess of useful energy above the work required to obtain it—is used entirely in the supply of food and shelter—or in food, shelter, and leisure—there can be no non-energy geologic resources. Finally, the ultimate limit to energy resources is the *limit of net work profit;* when it takes more energy to find and recover the fossil fuels than can be obtained from them in useful form, there will be no more oil, gas, or coal resources—although there may be a considerable amount of each left in the ground.

The basic question of depletion is whether or not the work-profit limit will be reached, or demand will cease short of mining common rock and sea water. For some resources we can say with assurance that the work-profit limit will be reached long before ordinary rock is mined. The fossil fuels are the best and most important examples. The energy potential represented by the average concentration of carbon in the earth's crust is 2.6 kilowatt-hours per ton, not nearly enough to crush and grind the rock to liberate the carbon for use. (In one large modern copper-ore mill, grinding and classification alone require 7.87 kilowatt-hours per ton of ore milled.) We can thus be sure that the sharp physical boundaries that characterize coal and petroleum deposits are also economic boundaries. Because no recycling of the fossil fuels is possible, depletion is approximately coincident with production and ultimate exhaustion a certainty.

But in the case of uranium, there exist very large low-grade deposits in which the potential energy is much more than sufficient to break, transport, and pulverize the rock and then to recover the uranium. Even average crustal rock may contain enough uranium (1.7 ppm) to justify mining and processing into fuel elements for breeder reactors that will be able to produce 13,680 kilowatt-hours from a ton of such rock at a conversion efficiency of 60 percent. The unknown factor here is the energy requirement for the extremely fine grinding and multi-stage recovery that would be needed.

WAS C. K. LEITH WRONG?

Forty years ago, the geologist C. K. Leith called attention to the coming exhaustion of mineral resources in the United States, claiming that "despite a magnificient endowment [of metals and fuels], depletion is further advanced than even mining men generally realize (1935, p. 169)." At the time Leith wrote, proved reserves of crude oil, zinc, and lead in the United States were fifteen to twenty times larger than production, the Lake Superior iron ores appeared to have less than twenty years of measured supply remaining, and known copper reserves were about forty times the 1934 production. He went on:

> Further discovery and the use of lower grade resources will extend the life of most of these resources, but the range of possibilities is now pretty well understood, and with maximum allowance for such extension, the figures are sufficiently small, when compared with what we hope will be the life of the nation, as to be matters of public concern. . . . Discovery has not stopped, but the rate has been slowing. . . . Of 33 metal-mining districts that have yielded the greatest wealth to date only five have been discovered since 1900 and none at all since 1907. . . . The rate of discovery of oil and gas continues high, but . . . . the chances of finding another East Texas or Kettleman Hills are not promising. [Ibid., p. 169]

Was Leith right or wrong?

Since 1935, more crude oil (77.3 billion barrels) has been discovered in the United States than had been discovered from 1857 through 1934 (62.0 billion barrels). The rate of discovery, however, has been declining ominously in recent years. The sole exception to this trend occurred in 1968, when the Prudhoe Bay oil field on Alaska's North Slope was discovered; it was added to proved reserves in 1970. During the latest five years of record, domestic production has exceeded discovery by an order of magnitude. In the same period, 4.5 times as much natural gas has been produced as has been discovered. Neither technological improvements in exploration and recovery nor higher prices have been able to overcome these sagging discovery rates. At the end of 1974, the ratio of proved reserves of crude oil to annual consumption was less than eight, and would have been about five but for Prudhoe Bay.

Since 1935, when Leith wrote, the United States has produced more zinc than it did prior to that year. In 1968, the ratio of measured domestic reserves to primary consumption (defined as demand less secondary or scrap supply) stood at twenty-four, despite the fact that demand had soared and more than a third of the national supply was being imported as metallic zinc. Mine production of zinc in 1968 (529,000 tons) could have been maintained for some sixty-four years on the then-known reserves (33,730,000 tons).

Although lead production in the United States since 1935 does not equal the pre-1935 total, the ratio of measured reserves (35,300,000 tons) to primary consumption (898,000 tons) in 1968 was thirty-nine, and the 1968 mine production could have continued for ninety years without further discovery.

The Lake Superior iron ore that Leith discussed has been exhausted, but it has been largely replaced by taconite, a low-grade iron-bearing rock not considered to be ore in Leith's day. In 1975 the ratio of measured reserves to primary iron consumption was seventeen. At the 1968 rate of mine production, the reserves would last for more than thirty-five years.

Since 1935, more copper has been mined in the United States than in all the years prior to that time. Based on 1968 figures, the ratio of reserves to primary consumption was fifty-six, and mine production could have continued at the 1968 level for seventy-four years without new discoveries.

What has happened between 1935 and 1975 is that new discoveries have extended crude-oil, zinc, and lead reserves, new technology has "created" large new reserves of iron ore and copper, while recycling has decreased the primary demand for zinc, lead, iron, and copper. The percentages of the nation's consumption represented by secondary or scrap metal in 1968 were 20 for zinc, 38 for lead, 36 for iron, and 45 for copper. Increased prices have extended copper reserves substantially, iron ore moderately, lead and zinc modestly, and crude oil hardly at all. Because consumption rates for these geologic resources have risen substantially since 1935, the achievements of geologists, mining engineers, and metallurgists are all the more outstanding.

In the short term, at least, Professor Leith appears to have been wrong. The continuous-creation school of resource analysts would classify him as a doomsayer of the past whose forecasts went awry for the same reasons that those of present day Cassandras will miss the mark. The physical-limits school, on the other hand, would point to the very important role in prolonging domestic reserves of imports from countries where depletion is not as far advanced, and to the sharp downward trend of the reserves-to-production ratio for domestic oil and gas as a sign of the future for all geologic resources. The fact remains, however, that in regard to the major industrial earth resources of Leith's time, the United States is substantially worse off now only in oil and gas.

THE EFFICIENCY OF EXPLOITATION

A great deal of the expansion of the geologic resource base of the United States and the world has come about through increases in the efficiency of exploitation. If copper still had to be mined as it was a hundred years ago, the enormous low-grade "porphyry" copper mines that today produce much of the world's copper would not be possible; the energy or work costs of mining and milling the copper-bearing rock would be so high that only relatively rich ores, say 5 percent or more in copper content, could be mined. At least one copper mine today can mine at a profit rock containing only 4 pounds of copper per ton. At that grade, every ton of copper produced requires the breaking, transport, and milling of five hundred tons of rock and the removal of perhaps an equal amount of waste material. Modern copper mining represents a large-scale replacement of human musclepower and inefficient steam power by diesel engines and electricity. A great deal of energy (more than 20,000 kWh) is required to produce a ton of copper today, but the cost of that energy is low compared to the energy costs of supporting the equivalent in men and mules; furthermore, the efficiency of modern power shovels, trucks, and locomotives greatly surpasses that of animals or steam engines. In addition, the power and scale of modern mining allow high rates of production that, although expensive in energy units, are economic in financial terms, helping to overcome the high interest costs of a capital-intensive activity.

But now it appears that further increases in the efficiency of extracting geologic resources will be difficult to achieve. There is in view no more efficient transport device than the diesel-electric locomotive or diesel-powered truck. Draglines, power shovels, grinders, crushers, flotation machines, and air compressors are best powered by electricity; there is in view no economic way of increasing the efficiency of electricity generation in a thermal power plant much above the present 40 percent. Economies of scale in mining are still being achieved in surface operations, but will be limited ultimately by the size of individual deposits or by the lack of flexibility in the use of enormous machines.

It appears that a further great increase in the reserves of geologic resources through cumulative incremental improvements in exploitation efficiency is not likely. Only new technological breakthroughs, such as an economic method of leaching very low-grade copper deposits in situ or the development of a cheap fusion power reactor, offer hope for large additions to reserves of geological resources. Not only the efficiency of available technology but also the nature or mode of occurrence of geochemical concentrations of valuable natural materials has a great deal to do with the way depletion takes place. Let us now look at depletion in terms of types or modes of geologic occurrence.

ORE-BODY DEPLETION:
MERCURY, SILVER, CRUDE OIL

Mercury, silver, and crude oil illustrate the mechanics of what might be called ore-body depletion, where the economic limits to exploitation are coincident with physio-chemical boundaries in nature. Mercury and silver are more rare than uranium in the earth's crust, and geo-chemical concentrations (ore bodies) of both metals tend to be hundreds to thousands of times higher than the average crustal abundance and to have very sharp boundaries. We should expect, therefore, to find evidence of depletion in the production histories of both metals, as technological improvements and higher prices combined prove impotent against the steep geochemical gradients that bound the deposits. The United States has accounted for about 10 percent of the world's mercury production but is estimated to have only 2.4 percent of the remaining reserves. The relation of price to production (see figure 13.1) suggests three phases of exploitation: a waxing phase, during which price was a decreasing function of production (increases in production caused the price to fall), a mature phase, in which price and production were more or less in equilibrium, and a waning phase, during which production has been a decreasing function of price (successive surges in price have evoked progressively weaker responses in production). The pattern in silver is similar (see figure 13.2), with the flagging production responses to sharp price surges in the waning phase even more pronounced. The waxing phase represents a period of falling real cost because of new discoveries, improvements in technology, and economies of scale. The falling cost stimulates demand. Increased production hastens the exhaustion of high-grade, low-cost deposits and puts pressure on technology to counter increasingly adverse geologic and geographic conditions. When technology begins to lose the battle, which will be sooner for some materials than for others, real costs rise and the waning phase is entered. Prices rise with the marginal costs of production (faster, if producers can control the

FIGURE 13.1
U.S. mercury production and prices, 1850–1971. Successive surges in price (numbers in squares) brought weaker and weaker responses in production (numbers in circles).
(After figure 46, p. 404, in Bailey, Clark, and Smith, 1973.)

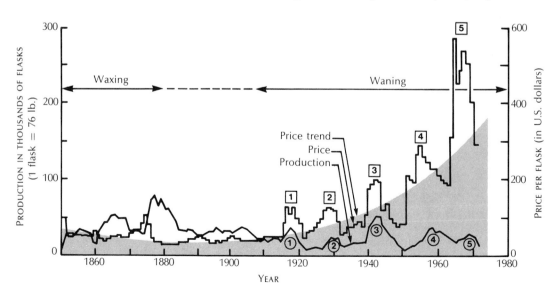

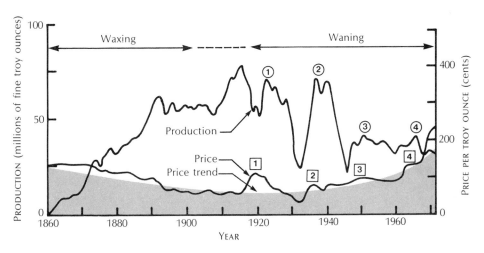

FIGURE 13.2
U.S. silver production and prices, 1860–1971. Successively higher price levels (numbers in squares) evoked weaker production surges (numbers in circles).
(After figure 66, p. 585, in Heyl and others, 1973.)

market, if hoarding takes place, or both), and demand will fall with the marginal utility of resource use. Sharply rising prices will stimulate the search for a substitute; if found, the production of the primary resource will cease at the point at which the cost of an additional increment produced exceeds the cost of an equally useful increment of the substitute.

These historical patterns reflect the depletion histories of individual mining districts. The production history of the Comstock Lode (see figure 13.3) illustrates a simple depletion pattern characteristic of high-grade, sharply bounded, and vertically limited mineral deposits. The Comstock vein system, a candelabra with tabular branches, is typical of fractures found and filled with ore minerals at shallow depths: rich and intricate near the surface; barren and simple a few thousand feet below. Its production history shows three distinct stages. In the first stage, during which by far the greatest part of the total value of the lode was extracted, high-grade ore was mined at a fast rate. In the second stage, it became possible to mine lower-grade ore bypassed in the first stage, because mining, transport, and milling costs had been lowered by technological improvements. In the final phase, waste material and some very low-grade ore were processed by a new technology but little was added to the value already produced.

The production history of crude oil in the United States (see figure 13.4) has not run its course as far as the mercury and silver histories, but it appears to be developing in similar fashion. During the waxing phase, the United States was both the largest producer and the largest consumer in the world. Several times new discoveries produced gluts of oil and drove the price down to or below the actual cost of production. The proliferation of automobiles and trucks was stimulated by cheap fuel and demand grew. Passage to the waning phase in 1971 was abrupt because of the wide difference in cost between domestic

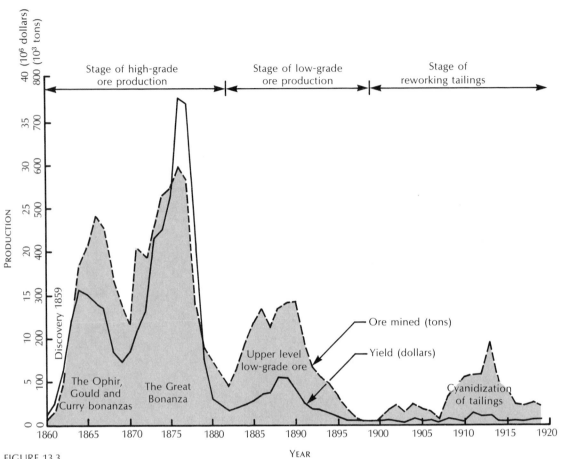

FIGURE 13.3
The history of silver production of the Comstock Lode in Nevada shows three stages of depletion.
(Sources of data: for 1860–1881, Lord, 1883; for 1882–1920, Smith, 1943.)

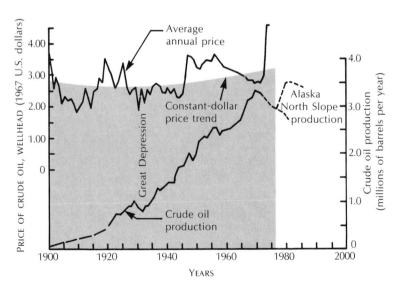

FIGURE 13.4
Crude oil production and prices in the United States in this century.

and foreign crude oil at a time of strongly rising demand, a difference that forced a rapid shift from domestic production to imports. The price rises of 1973–1974 have encouraged exploration for new domestic reserves and have stimulated new efforts to recover more than the 31 percent so far recovered of the crude oil discovered in the United States. It is not likely that we shall see, however, a production response equivalent to the price rise, because the depletion of crude oil reflects its geologic occurrence in small, sharply bounded accumulations in the upper part of the earth's crust. The costs of finding and recovering crude oil (and natural gas) follow steep exponential curves, and ultimately energy from alternative sources is bound to be cheaper than that which might be obtained from the last few undiscovered domestic oil pools. High prices will prolong production, but not by much and, should the present producers' cartel collapse and world oil prices fall sharply, the end of the United States production curve will be abrupt. On a world scale, reserves and production capacity continue to increase faster than consumption; in other words, we are still in the waxing phase of the *global* production history of crude oil; current forecasts by oil company officials place the world production peak at about 1990, after which there may be a rather sudden decline in production.

ECONOMIC-LIMIT DEPLETION: COPPER

We have seen that the ores of mercury and silver as well as crude oil are substances for which the economic limits to exploitation tend to coincide with physio-chemical boundaries in nature. The world's major copper mines (as well as many mines of iron and aluminum) are working ore bodies that have economic limits that are not entirely controlled by natural physio-chemical boundaries, and here the question can be posed: will technologic advances, economies of scale, and the utility of the metal continue to increase the depth from which 0.2 percent ore (4 pounds of copper per ton) can be recovered and to decrease the cutoff grade at constant depth, perhaps to, or almost to, the average crustal abundance of copper (0.126 pound per ton)? Consideration of this question brings into clear focus the importance of energy cost in determining the ultimate cutoff grade for any geologic resource.

Fifty-three percent of the world's known copper reserves are in the so-called porphyry copper deposits, of which Toquepala and Cuajone in Peru are typical. The mass of "mineralized" rock in such deposits has the shape of an inverted truncated cone, within which the copper content (at the two Peruvian mines) ranges from 1.32 percent to below 0.45 percent, the present cutoff grade. Not only does the volume of

Table 13.2 Relations of Ore Grade, Copper Content, and Mining and Milling Energy at Cuajone Mine, Peru

Average grade	Tonnage	Copper content (tons)	Energy needed for mining, milling
1.32%	20,000,000	264,000	0.76△*
0.99%	430,000,000	4,257,000	1.01△
0.32%	102,000,000	326,400	3.13△
<0.20%†	1,057,000,000	66,600†	>16,000△

NOTES: *△ = The amount of energy needed per ton of copper when the average grade is 1.00%.

†Mostly barren overburden of post-ore volcanic rocks; copper content and energy needed were calculated at the crustal abundance of copper.

SOURCE: Figures for Average Grade and Tonnage were taken from Hoyle (1974).

ore not increase continually as the grade drops (see table 13.2), but also the copper content drops sharply with decreasing grade. In other words, there are defined lateral geochemical boundaries although they are gradational and, unlike the walls of an ore vein or of a coal bed, do not coincide with a sharp physical discontinuity in the rocks. Downward the diameter of the mineralized cone decreases and the unit cost of mining increases. The energy cost per pound of copper for mining and milling the ore increases inversely to grade and directly with depth. At Cuajone, if the ore cutoff grade were to fall from the present 0.45 percent to 0.20 percent, the total copper recovery would be increased only 7 percent. At Toquepala, the largest mine in Peru, a similar situation exists; lowering the cutoff grade from 0.45 percent to 0.25 percent would have increased the recoverable copper, as of January 1, 1974, by less than 4 percent. Unless mining costs can be lowered, the economic limits of these ore bodies cannot be extended. Since economies of scale appear to have been exploited fully and energy prices are rising, it does not appear likely that the ore reserves will be extended much by technological advances.

Not all copper deposits are like those described above, but enough *are* to throw a gray shadow of doubt over the rosy projections of the Council of Economic Advisers who published a chart in 1970 that indicated that the volume of copper ore increases geometrically as the grade drops arithmetically and that the tonnage of recoverable copper is approximately proportional to price, an implied straight-line inverse relation between grade and quantity of copper available. That such a relation can hold only for a restricted range of values above some lower grade limit is accepted by most students of ore deposits. It appears that this lower grade limit for copper already has been reached, or nearly so. The price of copper in constant dollars has been rising since 1930 and, by 1975, was about 50 percent higher than in 1930, although during that period the price of energy in constant dollars fell considerably. Because there has been no effective cartel, the rising price trend must reflect basic costs of the steadily decreasing grade of ore mined. A reversal of the trend in the price of energy will cause the real costs of producing copper to rise even faster than they have been.

MULTIPLE-SOURCE DEPLETION: IRON, ALUMINUM

The rich "direct shipping" iron ores of the Lake Superior region are almost exhausted (see figure 13.5). These ores, which contributed mightily to the industrialization and economic growth of the United States, were created by a process related to groundwater level and as well as to particular geologic configurations; therefore, they had sharp boundaries that coincided with economic limits and the production history curve illustrates ore-body depletion. In the waning stage of production, beneficiation (physical upgrading) of low-grade material into concentrates that were worth shipping added substantially to the total production, but delayed exhaustion only a few years. Unlike the Comstock Lode, however, the Lake Superior district has been revived through a technical breakthrough that created a very large resource

FIGURE 13.5

The production history of the Lake Superior iron ore district shows overlapping stages of depletion. First the high-grade ore was mined, then low-grade ore, and finally ore that required a new technology.

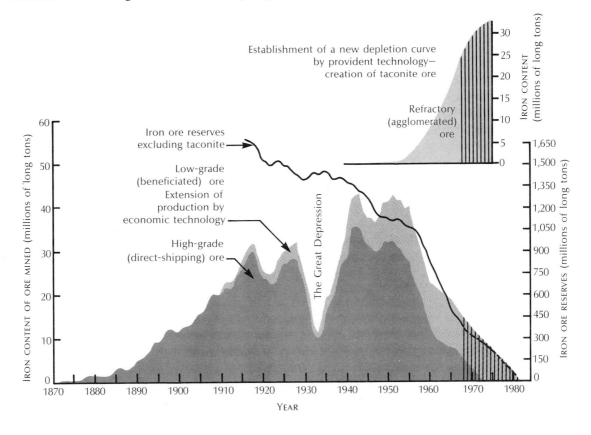

397

out of an iron-bearing rock called taconite, previously worthless. Such provident technology can be effective only where there exists a deposit, already concentrated by natural processes, characterized by a refractory or recalcitrant host or reservoir rock (oil shale and tar sands are other examples). The taconite production-history curve will represent economic-limit depletion, for the boundaries of taconite ores are gradational, but it also will pass through maturity to exhaustion.

Iron and aluminum are abundant elements in the earth's crust. For each there are several kinds of geologic concentrations above the crustal average that represent actual or potential resources, and for each we may expect the depletion history to consist of a series of production-history curves, as availability and cost dictate a steplike descent from high-grade hematite to taconite to iron-rich intrusive bodies and from bauxite to alunite to high-alumina clays. The greater the natural abundance of an element, the greater the probability that, over time, it will prove to have multiple economic sources, and that exhaustion will be long delayed. The greater the possibility of recycling, the more exhaustion can be delayed by reducing the demand for primary resource.

THE IMPORTANCE OF ENERGY COSTS

Although improvement in the technology directly involved in resource exploitation often is granted full credit for reducing the real costs of production, the importance of cheap energy cannot be overlooked. The mining, processing, and transport of mineral resources are energy-intensive activities, in which the energy required, at least through the concentration stage, is inversely proportional to ore grade. The total energy cost per pound of finished copper from 0.20 percent ore is likely to be four times what it is from 1.00 percent ore. The direct energy cost of copper at the present time—even with a fourfold increase in the cost of power plant fuel to some producers during the past two years, is not more than 5 percent of the price of the finished product (at 1.0 percent ore). If prices and costs remained constant and average ore grade were to decline to 0.20 percent, the additional energy required would add only about 15 percent to the cost of copper. Even if energy costs double from present levels while the average grade of mined reserves declines between two- and fivefold, the direct effect on the cost of copper, although substantial, should not be fatal to exploitation. In the short term the key question will not be one of secular rise in energy costs, but one of the steepness of the geochemical gradient for each desired element or compound. If it turns out that there is a sharp reduction in the quantity of metal or material available as the grade of ore declines and we find ourselves truly faced, sooner than we had hoped, with almost barren rock as a source material, the very great increases in the work cost of exploitation (several hundred times for copper, for example) will bring about the cessation of mining for anything but the most common (iron, aluminum, stone, sand, and gravel) and the most energetic (uranium, thorium) of earth substances.

Rises in the work cost of recovering geologic resources are occurring and may be expected to continue as leaner, deeper, and more refractory deposits are exploited in more hostile environments. However, such increases will vary greatly in their economic and depletive effects. There will be dislocations of energy-intensive processing, where practicable, toward regions of lower energy costs. Aluminum, for example, will be reduced in countries possessing sufficient surplus hydroelectric power or cheap fossil energy (Canada, certain African and Persian Gulf nations). The depletion and exhaustion of geologic resources will range from rapid exhaustion far above the crustal-abundance grade of rock for those materials found only or mainly in sharply defined, high-grade deposits to very slow depletion of abundant materials such as iron, aluminum, ceramic materials, and rock aggregate. The actual duration of this process of depletion will be heavily dependent on energy costs. The waning phase of world oil production probably will be identifiable by the year 2000, by which time real scarcity will have replaced artificial scarcity, if the latter has persisted from now to then. The push to develop nuclear power and to supply at least half of industrialized society's energy needs by nuclear power can be expected to intensify. The cost of such power, with environmental, safety, and security costs internalized, will not be nearly as low as that of thermal electricity at present, but conservation could compensate for much of the increased cost. The two largest industrialized nations still have sufficient hydrocarbon reserves for two centuries or more of consumption at present or somewhat higher rates and at costs probably less than twice present prices.

In summary, the effects of increasing costs of energy over the next century or more, although of great competitive importance among companies and nations, are not apt to be severe in terms of shrinking the reserves of nonrenewable resources, whose geologic and geographic parameters will be more important than energy costs in either hastening or prolonging the depletion process.

ESTIMATING UNDISCOVERED RESOURCES

Within the mining industry—including oil and gas mining—the estimation of potential reserves or undiscovered resources traditionally has been based on extrapolation from known or proved reserves based on knowledge derived from productive geologic environments. A hierarchy of reserves reflecting estimated probabilities of valuable material in place and of economic extraction has been established (see figure 4.11). Development loans and delivery contracts commonly are based on proved reserves and some fraction of probable reserves. Exploration expenditures are based on possible and speculative reserves. Reserves in any category are set forth as volumes or weights, with no indication of rates of recovery and depletion.

As the specter of the diminution and ultimate exhaustion of domestic oil and gas resources has taken form, attempts have been made to relate ultimate recovery to time. Most such attempts have followed one of two paths: the fitting of price-weighted demand curves to estimates of ultimate recovery arrived at by geologic analogy commonly biased by technologic forecasts of cumulative recovery efficiency; and projections of exploitation-history records to derive not only an estimate of ultimate recovery but also of the time distribution of future production.

Wide disagreement has developed over the amounts of crude oil and natural gas remaining to be found and used in the United States. The range of explicit and implicit estimates of undiscovered recoverable crude oil, for example, published from 1965 through 1975, adjusted for cumulative production to January 1, 1975, was from 30 to 466 billion barrels (see table 13.3). Even when one replaces the lowest estimate with the latest one by the same author, the largest estimate still exceeds the smallest (seventy-three billion barrels) more than sixfold. Such a spread suggests basic disagreements on appropriate methods of estimation.

In August 1974 a symposium on methods of estimating the volume of undiscovered oil and gas resources was held at Stanford University. As an exploratory venture it exposed differing points of view and featured some vigorous discussion, but resulted in little perceptible advance toward consensus. A number of problems inhibit reconciliation. One is that no model can be verified by a real example on the national scale. Another is a deep division of opinion on the extent to which substantial lowering of exploitation costs (by technological breakthroughs) or substantial inelasticity of demand (which would sustain rising exploitation costs and product prices) would affect supply and thereby the shape and area under the future production curve. In July 1975 a workshop on the same subject was convened by Resources for the Future, Inc., in Washington, D.C. Although most of the participants had not attended the symposium at Stanford, the same difference of opinion was evident. Very little is known about the effect of price on the supplies of oil and gas. The basic disagreement of these two meetings could be resolved only by information that is not available.

The Depletion of Geologic Resources

Table 13.3 Estimates, Explicit or Inferred, of U.S. Crude Oil Recoverable Beyond Known Reserves (1966–1975) (10^9 Barrels)

Author(s)	Year	Method	Oil-in-place initially discoverable	Ultimately recoverable	Recoverable beyond known reserves*
Hendricks and Schweinfurth	1966	Geologic analogy	1,250	500	360
Hubbert	1967	Exploitation history	–	170	30
Elliott and Linden	1968	Exploitation history	–	450	310
Hubbert	1969	Modified exploitation history	–	190[†]	50[†]
Arps, Mortada, and Smith	1970	Exploitation history	–	165[‡]	25[‡]
Moore	1971	Exploitation history	587	353	213
National Petroleum Council	1971	Geologic analogy	727	242[§]	102[§]
Cram	1971	Geologic analogy	720	432[‖]	292[‖]
U.S. Department of the Interior	1972	Geologic analogy	–	549	409
Theobald, Schweinfurth, and Duncan	1972	Geologic analogy	1,895	596[#]	435[#]
Berg, Calhoun, and Whiting	1974	Modified geologic analogy	–	400	260
Energy Policy Report (RFF)	1974	Geologic analogy?	–	682**	466
Mobil Oil Corporation (see Gillette)	1974	Geologic analogy	–	–	88
Hubbert	1974	Modified exploitation history	–	213[††]	73[††]
U.S. Geological Survey	1974	Modified geologic analogy	–	–	200[#]–400[#]
Miller and others	1975	Modified geologic analogy	–	–	50–127
National Academy of Sciences	1975	Not stated	–	–	113[#]

NOTES: *On January 1, 1975; cumulative production (106) and proved reserves (34) not included.
[†]Includes 25 for Alaska.
[‡]Does not include Alaska.
[§]Based on 33.3% cumulative recovery.
[‖]Based on 60% cumulative recovery.
[#]Includes natural gas liquids.
**Sources and method not given; figures are converted from heat-content equivalents of report.
[††]Includes 43 for Alaska.

THE IMPORTANCE OF ESTIMATING UNDISCOVERED RESOURCES

An oil company bases investment decisions on estimates of potential reserves and profit margins. A nation bases its energy policy on estimates of its undiscovered resources and of the relative political and economic costs of supply from foreign and domestic sources. In both cases, an estimate of low potential in one resource may lead to a decision favoring investment in an alternative resource.

At the national level, large estimates (those exceeding past production) of undiscovered recoverable oil and gas may lead to rather different national energy strategies than would low estimates. A large estimate, being interpreted as equivalent to a long *time* of continued high production, may encourage strategies featuring incentives for domestic exploration and production by industry (or, alternatively, direct public investment in exploration and production by a national company), relaxation of environmental controls on exploitative activities, and discouragement of imports. A small estimate—either of the quantity remaining to be discovered or of the time remaining for emplacement of an adequate substitute system—may lead to subsidized development of coal, nuclear power, and shale oil; strong conservation measures; the creation of a strategic economic reserve or stockpile; stringent end-use controls, perhaps to the point of not using domestic oil and gas as fuel and dedicating the remaining reserves to petrochemical and protein production; and continued efforts to secure foreign energy resources at costs less than the costs of domestic resources. The economic and social consequences of basing national energy strategies on high estimates that fail to be translated into production could be dire.

GEOLOGIC-ANALOGY METHODS

There are two basic approaches to estimating undiscovered oil and gas resources: geologic analogy and exploitation-history projections. Geologic analogy has been the basis of estimates of ultimate discovery by the United States Geological Survey, the American Association of Petroleum Geologists, the National Petroleum Council, and most oil companies. In terms of undiscovered recoverable crude oil in the United States, the range of recently published estimates derived by geologic analogy (88 to 409 billion barrels) somewhat exceeds the range of estimates derived by exploitation-history methods (74 to 310 billion barrels).

Simple geologic-analogy methods involve projections of the average oil-in-place content of productive sedimentary masses, the average number of barrels produced per unit rock volume in productive basins, or the finding rate per unit of exploratory footage drilled in productive environments, to potentially productive prisms of sedimentary rock. More complex geologic-analogy methods require the analysis of thirty or more geologic parameters of a virgin basin; probability factors are attached to each parameter by analogy with known relations in productive areas.

The use of production or recovery parameters as the base of analogy locks the resulting estimate into past technological and price history. Accordingly, some groups have preferred to use oil-in-place as the medium of extrapolation, which then allows a range of technological and economic assumptions to be applied to determine the ultimate cumulative recovery factor and thereby the quantity of ultimately recoverable oil and gas.

The National Petroleum Council (1971) published an estimate of 727 billion barrels of oil-in-place initially discoverable in the United States. At that time cumulative recovery efficiency was slightly over 31 percent. If one employed the modest increase in that rate—to 33.3

percent—inferable from another section of the same report, the ultimately recoverable total was 242 billion barrels and the recoverable-beyond-known-reserves figure at the start of 1975 was 102 billion barrels. Using a similar estimate (720 billion barrels) arrived at for ultimately discoverable oil-in-place by a great number of petroleum geologists working under the aegis of the American Association of Petroleum Geologists, Ira Cram in the same year published an estimate of 432 billion barrels ultimately recoverable, which would translate into 292 billion barrels recoverable beyond known reserves at the beginning of 1975. The difference lies solely in the estimate of ultimate cumulative recovery; Cram used 60 percent.

In May 1975 the United States Geological Survey (Miller and others, 1975) released estimates arrived at by geologic analogy but expressed in probabilities. Computer analysis of a great number of increments, each with a subjectively estimated pair of figures, one for 5, the other for 95 percent probability, resulted in the statement of a 95 percent probability that there remained 50 billion barrels of producible crude oil to be discovered and a 5 percent probability that there remained as much as 127 billion barrels.

EXPLOITATION-HISTORY PROJECTIONS

In 1956, M. King Hubbert started to publish a series of projections of the rate of production of crude oil and natural gas in the United States which, by calculation of the area under the completed curves, yield ultimate-recovery figures. In subsequent years (1962, 1967, 1969) Hubbert's estimates, although modified upward by analogy estimates for the Alaska North Slope, remained obtrusively low compared to other estimates. Then in 1970 Arps, Mortada, and Smith published an estimate based on plotting "proved ultimate recovery discovered" as a function of cumulative exploratory footage; their figure of 165 billion barrels of oil for "ultimate resource recoverable" in the conterminous United States was identical to Hubbert's of 1969.

Others (Elliott and Linden, 1968; Moore, 1971) who have used various mathematical models to project finding and production rates in order to calculate ultimate crude-oil recovery have arrived at substantially higher figures than Hubbert's. The reason may lie in Hubbert's control of his projections in tracking the interrelated rates of three variables: cumulative production, proved reserves, and cumulative proved discoveries, from which he derives logistic curves (see figure 13.6) for rates of production, of increase in proved reserves, and of proved discovery. Projection of the production-rate curve to its logistic conclusion not only gives a graphic measure of the total oil

Amount produced $\leqq$ Amount initially present

(A)

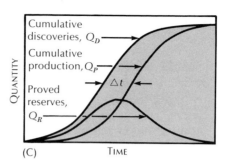

Cumulative discoveries, Q_D

Cumulative production, Q_P

Proved reserves, Q_R

Δt

QUANTITY

(C)　TIME

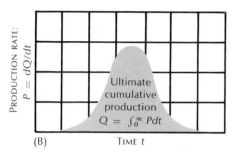

PRODUCTION RATE: $P = dQ/dt$

Ultimate cumulative production $Q = \int_0^\infty P\,dt$

(B)　TIME t

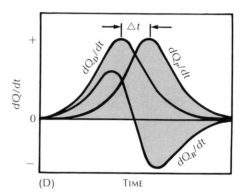

dQ/dt

$+$

Δt

dQ_D/dt

dQ_P/dt

0

$-$

dQ_R/dt

(D)　TIME

FIGURE 13.6

The logical framework for the exploitation history method of depletion forecasting has four links: (A) the amount ultimately produced will be equal to or less than the amount initially present; (B) production will begin at zero and end at zero; (C) there will be a consistent time relation among the three variables of cumulative discoveries, cumulative production, and proved reserves; so that (D) rates of change in these variables may be derived from formulas fitted to the curves of the actual exploitation record (cumulative discoveries, cumulative production, and proved reserves). From the production-rate curve (dQ_p/dt) one can then forecast the depletion of the resource in question.
(From Hubbert, 1956, 1962, 1969.)

that will ever be produced (the area under the curve) but suggests when exhaustion will occur.

Another exploitation-history estimation method involves plotting the finding rate per foot of exploration drilling as a function of cumulative footage of drilling. The ratio is independent of time, as is the method of Arps, Mortada, and Smith. As plotted by Hubbert, it yields a figure for ultimate recovery of 172 billion barrels of crude oil for the conterminous United States.

The exploitation-history method cannot be used for a basin or province in which there has been little or no production. For Alaska, Hubbert has added (1974) 43 billion barrels to his estimate for the conterminous states, an increment derived by geologic analogy from finding rates in exploratory drilling. Some students of the problem (such as Berg, Calhoun, and Whiting, 1974) believe Hubbert should have treated the offshore conterminous United States as he did Alaska.

THE RESOURCE-BASE CONCEPT

Economists of Resources for the Future, Inc. (RFF), and geologists of the United States Geological Survey have promoted the use of the resource-base concept. According to Schurr and Netschert of RFF, the "resource base" includes "the sum total of a mineral raw material present in the earth's crust within a given geographical area (1960, p. 297)." It is, in other words, the crustal content of crude oil, copper, or any other geologic resource. Because estimates, based on considerable sampling, exist of the crustal abundance of all elements and some compounds, this "resource base" can be calculated. From this is subtracted an estimate of the amount of the material that will never be discovered, a second estimate of the amount that will not be recovered if discovered, and the cumulative production to the date of estimation, in order to derive a number corresponding to the remaining recoverable resource.

For crude oil, the resource base and oil-in-place may or may not be synonymous; if oil-in-place is calculated from geological analogy, they are not. The important point is not that there may be a theoretical distinction between the resource base and oil-in-place, but that this approach to resource estimation leads to the practice of regarding essentially all the oil that is believed to be present in the earth's crust as potential reserves.

This method is not the same as calculating speculative reserves, although both methods tend to yield large numbers when compared with the exploitation-history method. The resource-base (or oil-in-place) concept may easily mislead the unwary, as shown by a 1968 publication of the United States Department of the Interior. On the first page of that report's "Summary and Conclusion"—and clearly aimed at the interested layperson—was an estimate of crude oil, natural gas liquids, and natural gas originally in place within the exploitable jurisdiction of the United States, compared with cumulative domestic production of oil and gas through 1967. According to the chart, there were, for example, 2 trillion barrels of oil "originally in place" in the United States and its continental shelf to a water depth of six hundred feet, and of this total only 84 billion barrels, or 4 percent, had been "withdrawn" as of January 1, 1968. The first sentence below the table hammers home the point: "The remaining petroleum resources of the United States are obviously adequate to support consumption for many years into the future." But there is a contradiction in the next sentence: "The real question is whether [these resources] can be located and produced at costs which permit them to compete with other energy sources." This "real question"—whether the so-called resources will ever become true resources—is not obvious to the nonprofessional reader, who might never reach the passage on page twelve of the same report, which reads: "The fact that we have X billion barrels of oil and Y trillion cubic feet of gas in the ground, however, says nothing at all about how much of these quantities will eventually be found and put to use."

Early in 1974, the Energy Policy Project of the Ford Foundation published its first report, which contained a table of United States energy resources prepared by Resources for the Future. It is in resource-base format; when converted to barrels from the heat equivalents given, the figures for petroleum are seen to represent new highs for estimates of oil originally in place ($3,680 \times 10^9$ barrels) and in "recoverable resources" (520×10^9 barrels). Not surprisingly, the report states that "the work done for the project by R.F.F. suggests that energy resources are at least sufficient to meet the year 2000 requirements with major reliance on oil and gas supply (p. 44)."

The statistical basis for rejecting the resource-base model as a guide to an estimation of future recovery of crude oil has been thoroughly presented by Hubbert in his report in 1974 to the Senate Interior Committee, in which he points out the "leverage" represented by changes in estimates of the base or of the ultimate recovery ratio. The recovery ratio, as a manipulable factor of technological forecasting in projections based on estimates of oil-in-place, may unbridle the technological optimist and allow estimates that put extreme demands on the technological cavalry to come riding over the hill in the nick of time to rescue the nation from scarcity.

THE USES OF RESERVE ESTIMATES OR DEPLETION FORECASTS

Perhaps too much attention has been paid to the question of how much crude oil and natural gas the United States has left and not enough to the question of how much time the nation has to replace domestic oil and gas as the main fuels of its economy. If we find more than we expect, we will use it faster. An increase of 50 percent in the remaining undiscovered recoverable oil might mean as few as ten years more before substantial exhaustion. In view of the long lead times, the high capital costs, and the technological uncertainties involved in implementing the domestic alternatives of expanded nuclear power, increased coal use, and the production of shale oil, the nation has no time left—if alternative domestic supplies of energy are to replace the diminishing supplies of crude oil and natural gas. Ample supplies of foreign oil and gas should be available until the year 2000 and perhaps beyond. But at what price? Probably at prices equivalent to the cost of alternatives—if alternatives are truly available. Otherwise, at whatever the producers wish to ask.

The form of presentation of the United States Geological Survey's 1975 estimates of domestic resources of oil and gas is unusually useful. It may be dismaying to those who find some kind of security in a single number, but it should encourage those who know that, in dealing with the future, one needs different numbers for different purposes. The

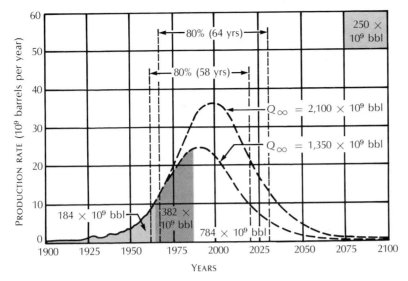

FIGURE 13.7
*By means of the example of the complete cycle of U.S. crude-oil production, Dr. Hubbert here illustrates the relatively small difference in availability that a large difference in the estimate of ultimate resource recovery may have. More than 50 percent increase in ultimate recovery (Q_∞) of U.S. crude oil yields but a few years more during which oil can be a mainstay of the national economy.
(Hubbert, 1969, figure 8.23, p. 196; reproduced by permission of the National Academy of Sciences.)*

voice of prudence suggests that the low estimates (of high probability) should be used in determining national policy and strategies designed to protect the economy and the political independence of the United States against the threat implied by dependence upon imports. The voice of courage argues for testing the improbability of the high estimates. Because there is a low probability of finding much oil off the coast of the eastern states does not mean that we should not look. Only the drill can test subjective probability. The more we find, the more time we will have bought—perhaps only to argue among ourselves, but perhaps to move toward a multicompartmented policy of energy security that would include strategic fuel storage, conservation of energy, government aid to industry in developing a capacity to produce substitute fuels, careful advancement in nuclear power, research in solar and geothermal energy, and negotiations toward a condition of interdependence between one or more oil-producing countries and the United States.

SUPPLEMENTAL READING

Barnett, H. J., and Chandler Morse. 1963. *Scarcity and Growth, the Economics of Natural Resource Availability.* Baltimore: Johns Hopkins University Press.

Presents the case for the continuous creation of resources by economic and technological growth.

Cram, Ira H., ed. 1971. *Future Petroleum Provinces of the United States—Their Geology and Potential,* 2 vols. Memoir 15. Tulsa, Okla.: American Association of Petroleum Geologists.

A massive effort of many geologists to estimate the potentially recoverable oil in the United States.

Brooks, David B., and P. W. Andrews. 1974. "Mineral Resources, Economic Growth, and World Population." *Science* v. 185, p. 13–27.

The notion of running out of mineral supplies is called ridiculous. The authors, who approach the matter from the economist's point of view, relying upon the growth of knowledge to create resources indefinitely, claim that "it is simply not true . . . that average rock will never be mined." A vigorous presentation of the anti-geological argument.

Fisher, John C. 1974. *Energy Crises in Perspective.* New York: John Wiley and Sons.

The perspective is that of a technological optimist who places great faith in very large "official" estimates of potential energy reserves.

Georgescu-Roegen, Nicholas. 1975. "Energy and Economic Myths." *Southern Economic Journal* v. 41, p. 347–381.

A fine review of the concept of entropy in the economic process and explanation of the traditional economist's rejection of it. Acidly cogent, but not encouraging.

Hewett, D. F. 1929. "Cycles in Metal Production." *Transactions of the American Institute of Mining and Metallurgical Engineers* p. 65–93.

Evocative study of the "life cycles" of twenty-eight European mining districts.

Hubbert, M. King. 1974. *U.S. Energy Resources, A Review as of 1972,* part 1. U.S., Senate, Committee on Interior and Insular Affairs, 93rd Con., 2d sess., serial no. 93-40 (92-75).

The definitive presentation of the production history analytical approach to forecasting depletion. Crude oil and natural gas production in the United States have entered their permanent declines.

Kaysen, Carl. 1972. "The Computer that Printed Out W*O*L*F*." *Foreign Affairs* v. 50, p. 660–668.

Another prominent economist who maintains that the finiteness of the earth and even the geology of ore deposits tell us nothing important about the availability of resources.

Lovejoy, Wallace F., and Paul T. Homan. 1965. *Methods of Estimating Reserves of Crude Oil, Natural Gas, and Natural Gas Liquids.* Baltimore: Johns Hopkins University Press.

An informative review of methods, definitions, practices, and problems in estimating reserves.

Lovering, Thomas S. 1953. "Safeguarding our Mineral-Dependent Economy." *Bulletin of the Geological Society of America* v. 64, p. 101–126.

Points out the depletion cost implicit in a government committed to maintaining full employment by subsidizing artificial markets and the production of goods not really needed by the consumer, thereby hastening the exhaustion of nonrenewable resources.

Lovering, Thomas S. 1969. "Mineral Resources from the Land." *In Resources and Man.* Chapter 6. San Francisco: W. H. Freeman and Company.

Presents the case for geological and geochemical limitations on the availability of resources.

McCulloh, T. H. 1973. "Oil and Gas, in United States Mineral Resources." United States Geological Survey, Professional paper 820, p. 477–496.

Contains some interesting facts about the quantitative distribution of crude oil—most is in a few giant fields rather than in many small fields.

McKelvey, Vincent E., 1974. "Approaches to the Mineral Supply Problem." *Technology Review* v. 76, n. 3, p. 12–23.

The director of the United States Geological Survey defends the capability of man's resourcefulness and the market system "to create resources far beyond the possibilities that we recognize now."

Miller, Betty M., and others. 1975. *Geological Estimates of Undiscovered Recoverable Oil and Gas Resources in the United States.* Reston, Va.: United States Geological Survey Circular 725.

Lower than previous estimates from the same organization, these are presented in terms of probability limits.

National Petroleum Council. 1971. *U.S. Energy Outlook, An Initial Appraisal, 1971–1985,* 2 vols. Washington, D.C.: National Petroleum Council.

> The projections are suspect, but the estimates are surprisingly conservative; much useful information, as accurate as can be had.

Nolan, T. B. 1958. "The inexhaustible resource of technology." In Henry Jarrett, ed., *Perspectives on Conservation.* Baltimore: Johns Hopkins University Press.

> The testimony of another geologist who believes that technology will make resources as man needs them.

Schurr, S. H., and B. C. Netschert. 1960. *Energy in the American Economy, 1850-1975.* Baltimore: Johns Hopkins University Press.

> This book, already cited as a source of historical data, contains an early discussion of the resource base concept and its possible utility.

University of Texas. 1968. "Limitations of the Earth: A Compelling Focus for Geology." Proceedings of a symposium held November 16-17, 1967, at the University of Texas at Austin. In *Texas Quarterly* v. 11, n. 2, Summer 1968, p. 5-154; reprinted by Texas Bureau of Economic Geology, Austin.

> Note particularly the papers by Alvin Weinberg, Preston Cloud, Thomas Lovering, and William Pecora, for a range of views from very well-informed people.

William Tenney

Chapter 14 ETHICAL AND MORAL ASPECTS OF ENERGY USE

People should be allowed to do anything they like—provided only that they don't scare the horses in the street.

—Mrs. Patrick Campbell, c. 1896

Morality is a private and costly luxury.

—Henry Adams, 1907

ETHICAL QUESTIONS ABOUT THE USE OF ENERGY

World and national patterns of energy use and flow raise vexing ethical questions, not only about their environmental impact, but also about social equity. One may argue that the fundamental questions involve only the use of power to achieve and defend group objectives and are therefore amoral questions. But since group objectives in the present world almost invariably are stated in terms that imply sharing and invoke human rights, and because shared ethical values are the framework of any viable society and the only basis of humanity (a quality assumed to separate men and women from other animals), the argument of amorality, although perhaps valid in explaining why some things happen, should not prevent us from recognizing that ethical positions and moral issues often have a great deal to do with what happens and why people behave as they do.

The moral aspects of the use of energy tend either to be ignored, as in ordinary economic analysis, or put forward in simplistic form. We of the developed, high energy society, it is said, have a moral obligation to share surplus food, technological knowledge, and ecological insight with our less fortunate brethren in the low energy societies. It is wrong, the chemist contends, to burn beautiful complex molecules that are useful for other purposes but difficult to synthesize. It is unconscionable, the social moralist maintains, to increase profits by selling gasoline or fuel oil at higher prices when shortages appear. Strip mining is to "rape" nature, and oil spills result from the "greed" of corporate managers and major stockholders. American conservationists often refer to a duty to leave a good environment and a stock of resources for future generations; such a duty can only be moral, since future generations will have difficulty in haling present malefactors into court for dissipating the family heritage.

Such positions are sincerely held, can be of great weight in energy decisions, and are often expressed as absolutes, making accommodation with other views difficult. One of the characteristics of present-day high energy society is the lack of a unified moral view or ethic that would furnish a rationale for the allocation of surplus energy; consequently, expediency and group interests rarely are seriously impeded by moral considerations. The wealth ethic of British culture, which gave strength and direction to the Industrial Revolution, no longer unites high energy society, although it is still strong.

DEPLETING THE ENERGY RESOURCES OF OTHERS

Appropriating the energy and power resources of others is one of the main ways of augmenting the energy and power base of a nation or society. Slave traffic through the ages was for this purpose. More recently, food-energy flows from underdeveloped to industrial countries have been a result of political and economic domination. The flow of petroleum from undeveloped to developed nations has raised the question of ethics involved in the practice of advanced nations consuming the finite energy resources of underdeveloped countries.

Between 1945 and 1970, oil from the Middle East, North Africa, Venezuela, and Indonesia was carried in ever-increasing amounts to the high energy countries. Europe would not have made its astonishing recovery from the economic sapping and physical destruction of World War II and Japan could not have exploded into one of the world's major industrial powers without the cheap energy from oil. The postwar prosperity of the industrialized world was based on the *natural subsidy* of cheap oil from underdeveloped but oil-rich nations.

The justification given by the big energy consumers has been that it was necessary to use cheap energy to advance the material welfare of mankind, that they were the ones who could use it most efficiently, and that the resulting social benefits would spread throughout the world, would trickle down to the underdeveloped countries, including those whose petroleum and protein reserves were being profitably exploited. Now that it has become clear that the gap in income and affluence between the high energy and low energy countries has been steadily widening, the argument that benefits would trickle down has lost whatever force it once had. Now one hears the argument that an industrial nation such as the United States has a moral right to use force against the oil-exporting countries in order to avoid mass unemployment and the ultimate starvation of its people. A detailed proposal for the invasion of Saudi Arabia by the United States (Ignotus, 1975), published in *Harper's* magazine, labeled all members of OPEC extortionists, called the Arab members blackmailers, and attempted to justify the proposed invasion and occupation of the Saudi oil fields in terms of restoring jobs to the unemployed in the high-energy nations, of lessening hunger and renewing economic hope in poor countries, of preventing the further development of authoritarian governments in the economically degraded oil-importing countries, and of giving the high-energy world time and capital to develop substitute systems for the supply of energy.

The morality of this argument, which implies a right to take by force what we have grown accustomed to having, regardless of whose heritage or property it is, stands in sharp contrast to the view of the same situation expressed by the Shah of Iran, whose oil revenues are being used for the rapid development of systems of education, transport, and communications in Iran, and to create new industries in that spare land, who believes high prices rather than low will encourage the conservation of ephemeral resources such as oil and gas and will stimulate the search for substitute sources, a search on which all the world's future depends. That the Shah sees the use of resources in a moral light was underlined by his remark that "there are seventy thousand things to be had from oil, including very valuable petrochemicals and pharmaceutical products. We would prefer such things to seeing the noble product of our earth wasted in needless consumption by the affluent industrial societies of the world (1974, p. 34)."

In the world of high consumption, it is often claimed that maximum social benefits are obtained through a system that commonly lowers the value of resources the faster they are being depleted. Add to that a confidence in man's unlimited ability to overcome resource constraints, and we easily reach the proposition that the price of a resource, renewable or not, should be related directly to the cost of exploiting it. But, in the world of limited consumption, where advances in the material conditions of life have been slow—or may not be perceived at all—and where potential advances are seen to be related to the carefully planned expenditure of the geologic capital in nature's vaults, the maximization of social benefits will not be viewed as being obtainable through high rates of production at low prices. Indeed, high production at low prices may be seen as immoral in the producing countries and moral in the consuming countries.

It is not surprising that underdeveloped countries with large energy resources should want to benefit from the consumption of those resources to the greatest extent possible. The Mexican experience after the 1938 expropriation demonstrated that "natives" could manage and develop by themselves an integrated oil and gas industry, and that the social benefits of oil and gas consumption did not need to be shared with foreigners if energy-based industry could be developed within the country. For underdeveloped oil-rich countries today, however, an alternative path is to maximize the return from use of their oil and gas by the developed countries and so to obtain faster industrialization. To do that requires control of production and pricing, which they appear now to have obtained. The long-range goal will be to use their remaining oil and gas to develop domestic industry so as to be able to participate with *other developed* nations in the nuclear and solar energy era.

In North America, a somewhat similar situation is evolving between Canada and the United States. From a basis of self-interest, Canada is restricting exports to the United States of oil and gas and has sharply raised the price of these resources. At the same time, rising opposition to the economic control of Canadian resources by companies in the United States is raising obstacles to further American participation in the development of new Canadian energy resources. Canadians appear to want no part in a "continental energy policy" that would commit them to supply a major part of the growing energy deficit in the United States at the expense of their own future energy supply.

ONE QUESTION LEADS TO MANY

One might pose this question: does a nation rich in resources, as are Saudi Arabia and Canada, have a moral obligation to share those resources with nations that have greater need for them than it does? But one cannot attempt to deal with that question without first considering some related questions:

- Under what conditions, if any, can a country justify the rapid depletion of resources that cannot be replaced, thus mortgaging the welfare of its own future generations?
- Under what conditions, if any, can a country justify the continued development of resource based technologies that incur incalculable but real somatic, genetic, and environmental risks or hazards as does nuclear power, for example?
- Under what conditions, if any, can a nation justify taking for its own use the nonrenewable resources of other nations too weak economically or militarily to use or protect those resources?
- Does a country that has used up both domestic and foreign nonrenewable resources in order to support an increase in its own population, to raise its own material standard of living, and to increase its own military strength, have a moral right to expect other nations to continue to provide it with such resources at the expense of their own potential for economic and cultural improvement?
- Does such a country, especially if it appears to have made no attempt to curb its own appetite for resources or to use resources socially or even technically efficiently, deserve a hearing when it tells the resource-exporting countries that they should bear their share of

the environmental and social costs of the exploitation of their own resources and should act responsibly in restricting their demands for resources and protecting their portion of the environment of the world?

- Does a country that appears to have made no attempt to curb its population growth or to adjust its resource demand to its resource supply deserve to be helped by others who, for whatever reasons, enjoy more favorable supply and demand ratios?

The rapid depletion of nonrenewable resources for the economic benefit of a nation is justified by many as providing the capital required for a society to break out permanently from resource constraints and to be able to develop the technologies and to create the resources that it needs or desires. In this view the critical constraint on a nation's vitality is not the availability of resources but a minimum rate of increase in the gross national product. This credo not only absolves a nation from any guilt that might be associated with the depletion of geochemical concentrations of materials that took thousands or millions of years to form, but actually requires that it do so in leading its citizens toward a material Golden Age. Undergirding this position is an unconquerable faith in provident technology. If we run out of coal, petroleum, and natural gas, we will develop efficient breeder reactors and when the ecological limits of their proliferation are reached, we will have fusion power, geothermal energy, and finally solar energy on which to base our economic and technological well being. Enthusiasts of technology make great use of projections of technological progress, of graphs of GNP against per capita resource consumption, and of charts of agricultural production versus capital input to agriculture. They join with doctrinaire Marxists in rejecting the conclusions of Malthus and neo-Malthusians.

On the other hand, there are some who hold that technological discovery and application are subject to the law of diminishing returns and that, in consequence, there will prove to be some physical limit to humankind's use of resources. If so, a different kind of moral behavior appears called for, one that would curb rather than glorify the appetite for resources.

The second question raised, that of deliberately incurring somatic, genetic, and environmental risks in order to maintain an increasing gross national product and an increasing median standard of living provokes a similar argument of faith in provident technology and the existence of economic man against more mundane or earthbound arguments for reason and caution.

It is often claimed that, in order to bring the disadvantaged of industrial society into the golden economy, it is necessary to run such risks. When these risks become quantifiable, the argument runs, advanced societies will know how to deal with them through technology, and the costs incurred until that time will prove to be more than balanced by accrued benefits of the adopted technology. Doubters point to the importance in western society of the benefit-distribution system in determining what trickles down to the disadvantaged and argue that improvements in this system could eliminate disadvantages within an industrial society without increasing the consumption of resources. Dissenters from the theology of provident technology also draw attention to the possibility of delayed consequences of new technologies that may be catastrophic and uncontrollable, and they ask, what would be the magnitude of human loss had the deformities caused by use of the drug thalidomide not shown up in the first generation, but in the third or fourth generations, an event quite possible in mutagenesis?

Unless we regard as a moral position the contention that to the strong belong the spoils, the justification for strong nations using up the resources of the weak is again the trickle-down philosophy, which holds that the strong will lead all mankind to the land of plentiful energy and the good material life. Doubters point to the increasing gap between the haves and have-nots of this world and to the amply demonstrated reluctance of the haves to share capital with the have-nots or the have-littles as indications that advanced technology cannot be counted upon to stimulate the "trickle-down" mechanism to satisfy the real needs, let alone the rising expectations, of the underdeveloped countries. Some doubters may also probe what they consider the fallacy of linking the *quality of life* causally with the *per capita access to energy and materials.* This link has been the philosophical foundation of western industrial, technological, democratic society. Although philosophers have questioned it and today might vote against it, the silent majority seems to accept it and political decision makers act on that assumed acceptance.

When we address the question of the justification for expecting nations who have surplus resources to continue to supply countries that have depleted their own resources, again we find the cornucopians ranged against the neo-Malthusians or growth-limiters. Here one senses a certain strain in the argument that technological largesse in an abundant world of the future will compensate for present contributions to the ruling theocracy, because this argument frequently is buttressed by somewhat plaintive appeals for support of the big consumer who has protected the little producer from enslavement by communism, fascism, capitalism, or whatever the appropriate demon happens to be. Doubters question the resource addicts' record of amorally used military power, of environmental degradation, of ostentatious consumption, of lack of national moral purpose, and of high rates of crime, drug addiction, and mental illness. Does such a record, they ask, merit continued trust and support, or might it not be the turn of other nations to try to use remaining nonrenewable resources more wisely and in more socially beneficial ways? Perhaps a nation with resources remaining to be allocated would be justified in thinking it could manage those resources not only with greater profit to its people but also with a greater sense of responsibility to the world than has been shown by the past grand masters of resource chess.

Now for the last of this group of related moral questions, the problem of resource aid to the country that does not appear to be willing to restrict its own demand, at least to the extent of controlling its population. This question leads directly to an ethical swamp, for it implicitly subsumes other questions. Should men and women be condemned to lives of poverty and hunger by the circumstances into which they were born? If economic advancement is a result, at least in part, of inequities in the distribution of natural resources and of a kind of social serendipity that allowed the development of a successful acquisitive society only at a certain period of time in a certain area of the world, whence it spread over weaker but perhaps not less good cultures, is it then ethical to demand that the unfortunates in this process make greater sacrifices and show a better ability to make good decisions under stress than the fortunates do in the absence of stress? Is it moral even to try to feed and keep healthy the children in a society in which they and their children seem doomed as adults to despair and misery? There are those neo-Malthusians who contend that the world's resources already are too limited and technology too feeble to bring all the world more than a little way up toward the level of the fortunates (among whom, incidentally, most of the neo-Malthusians are to be found) and that a logical conclusion is to plan for a survival society of the present haves and to leave the have-nots to the fate to which they would consign us all if we "leveled" with them.

These questions suggest the range of moral problems that face a high energy society and may make the free market seem like a panacea. Because the market economy and its pricing mechanism are held to maximize the social return on resource use, one simply eliminates the necessity for political decisions based on ethical or moral codes by letting the market allocate all resources. In this way there is no necessity to distinguish needs, desires, and demands. Social usefulness is equated with ability to consume, and need is measured by the ability to pay. One who persists in believing that the market does not maximize social benefits, especially from nonrenewable resources, is faced with the knotty problems of defining need, of designing or refurbishing a framework for agreement on social goals and strategies, and of discovering effective ways to incorporate social-efficiency factors into the market system—or suggesting a workable replacement system.

SCARCITY, ABUNDANCE, AND RESOURCE MORALITY

When necessary goods are scarce, we reluctantly accept planning for the social allocation and efficient use of those goods. When goods are abundant, we are inclined to reject planning by government in favor of planning by the individual or the interest group. In a growth society, actual or expected abundance tranquillizes our concern for the ethical allocation of goods among the haves and have-nots of a society, or of the world, as well as our concern for the welfare of the yet unborn. It is argued that the expanding economy either does or will offer exceptional opportunities for "all those able to learn and willing to work" either now or in the future. The fact that such is patently not now the case is conveniently ignored. Governmental planning may be viewed with suspicion or even hostility, as a prelude to forced sharing with those who do not deserve what planning might give them. It is held that future generations will be well able to take care of themselves by building on the base we are preparing for them.

In an economy of scarcity (short of general malnutrition), however, a different ethical mosaic may exist. Sharing may be regarded as moral action; planning is accepted as being as necessary in the community and nation as it is in the family; the concept of conservation may change from efficient consumption to the perpetuation of productivity, or supply; and, strangely enough, emphasis on efficient production and material accumulation may be replaced by a concern for the better social use of both work and leisure. The desire for one's children to have something to enjoy can lead to a single-minded drive for family wealth, but it also may lead to an inculcation of family and cultural values and a concern for the preservation of the family, the tribe, and the human community.

In an economy of abundance, there is a range of opinion on the definition of waste. It seems generally agreed that the present energy system in the United States is wasteful, whether in the sense of not being as technically efficient as it could be (with minor decreases in perceived social benefits) or in the sense of not being socially efficient because it uses up nonrenewable resources in garish advertising, in making unneeded products, and unnecessary travel. In an economy of scarcity, it becomes much easier to define waste and consequently to set up a moral code that forbids waste.

Abundance raises the question of whether or not the absolute exhaustion of resources can and will be avoided through man's increasing control of energy. Those who feel that the exhaustion of nonrenewable resources cannot be avoided and that the prospect of the replacement of adequate flows of renewable resources is dubious argue for a morality of scarcity. They tend to see inefficiencies of energy use that are traceable to desires for comfort, convenience, and choice far beyond world levels as reprehensible, because they consume resources for incremental gains in inessential advantages for the few that might be used for gains in essentials for the many, including those who come after us. They would equate prudence instead of prodigality in energy use with moral responsibility for one's fellows and descendants. They may also point out that the faster the present energy resources are used, the faster carbon dioxide will build up in the world's atmosphere and the greater will be the rate of heat formation at the surface of the earth. Because the consequences of either may be catastrophic, it is argued that neither should be accelerated for unnecessary and ephemeral "benefits."

Then there are those who believe that the high energy growth state is the only vehicle that offers hope for the solution of mankind's energy problems, and they advance a morality of abundance. Supporters of this view include many who believe that work is good and that constant increases in production of goods and services are necessary for full employment. The animistic contrast is often made between a "healthy, growing economy" and a "sick, paralyzed economy." It may further be contended that waste is the hallmark of a free society, that freedom is measured by range of choice, and that to supply a range of choice requires waste, by either physical or social criteria. If freedom is a major cultural goal, we must accept waste, as we must accept the philosophy of Mrs. Campbell: that people should be allowed to do as they please, as long as they do not frighten the horses in the street. In other words, as long as their actions do not threaten direct harm to the society. Some economists maintain that the conservation of supply is immoral, although the conservation of losses in the delivery system is not. They argue that the restriction of supply to high energy society (in which, coincidentally, they reside) may prevent the discovery of adequate replacement resources and systems. The resulting collapse of societies and populations would then be the fault of the conservationists who would stretch out the resources who prevailed against the cornucopians.

A moral view of resources has been a strong theme in America's growth. The idea that material and moral progress are linked appealed to the pioneer as well as to those who profited from his efforts. They liked to believe that whatever material success they achieved had been linked inextricably with the development of an American frontier culture, distinguished by the virtues of resourcefulness, fearless liking for productive labor, visionary acuity, cooperative aggression against natural obstacles (including Indians), fierce independence, strength, endurance, and outspoken honesty. A rigid code of frontier morality grew up at the same time that some of the most successful men in America were not following its strictures.

The exploitation of resources, in the frontier morality, is an absolute good, although wrong can be perpetrated if individual rights to the benefits of exploitation are usurped. The rise of the conservation movement in America was not so much a challenge to the dominant resource morality as it was a social movement to spread the benefits of resource exploitation and to preserve a very few extraordinarily scenic areas, without known economic resources, as national or public parks.

THE MORALITY OF THE MARKET IN REGARD TO SCARCITY

Some economists and businessmen, as noted earlier, believe that a market economy allocates resources as well or better than can any government. Let rising prices, they say, reduce demand and direct resources to the uses by which they can most effectively be employed. What this means is that production will be kept efficient at the expense of social equity. Scarce resources will be distributed on the basis of ability to pay rather than on the basis of need. This position appears more tenable in an economy of abundance then in an economy of scarcity. In fact, governments persistently intervene in the market system whenever the uncontrolled pricing mechanism appears to be foreclosing the access of any substantial segment of the population to necessaries; the more affluent the society, the wider the definition of necessaries. One of the characteristics of a democracy may be that it rations scarce goods instead of allowing market allocation and inequitable diversion to those who already control a disproportionate share of the nation's goods.

Government intervenes also to protect public health and the environment against the consequences of market decisions. Pollution was widely regarded (in the United States) as immoral before it became illegal, immoral because it degraded the environment of all for the profit of the few. Now the depletion of nonrenewable resources is becoming a moral issue; where once the development and exploitation of mineral resources was regarded as virtuous, today questions—quite apart from environmental considerations—are being raised about the moral aspects of using up the finite stock of nonrenewable resources. There is a range of moral issues here, but all involve the rights of others than the exploiters to the benefits of such resources. Soviet voices have denounced the United States for wasting the world's small stock of helium; American voices accuse, on moral grounds, the oil industry of a "drain America first" policy. Others question the morality of burning petroleum products in fast, heavy cars instead of conserving them for use in synthetic fabrics and other chemical products of greater or more enduring social value. The moral issues inherent in the depletion of nonrenewable resources will gain contentious voices if not clarity as the social impacts of their scarcity become more troublesome.

The morality of sharing scarce goods superimposes controls on market allocation only within national borders. Scarce necessaries such as protein and petroleum still flow internationally along gradients determined by markets, not need. Attempts at international rationing have not been successful. Pleas to establish a world food bank, mainly of wheat, are countered by the hard market argument that the nations that have a surplus of wheat need to sell it in order to buy oil and therefore cannot afford to commit their grain to a food bank from which the return would be slower and lower.

The tragedy of the commons is that no strong resource ethic is yet common to mankind. Indeed, there are ethical stances and cultural patterns that inhibit the understanding of resource limits, work against population control, and encourage aggressive competition for scarce resources. Wherever one sees perceptive management of a common resource, the ethic that restrains depletive production appears to work because the perceived social benefits are to be shared within a family or other well-knit kinship group. If it were possible to extend the kinship group to include all producers and all consumers, Cottrell maintained that "morality would automatically assure that each would produce according to his ability and would be permitted to consume according to his need (1970, p. 238)." An energy ethic, which has been suggested for the United States, will face the same barriers to widespread adoption that have faced a land ethic or an environmental ethic, the barriers of territorial interests, differing cultural value structures, decision mechanisms geared to consumption rather than conservation, and an uneven distribution of the world's goods. Man usually is a gentleman only when be feels he can afford to be—and then often only within his cultural group. George Gaylord Simpson has remarked that "the acceptability of an ethical system has generally corresponded with wishful thinking in the individual, who firmly believes what he enjoys believing, and with some sort of pragmatic validation in society (1967, p. 269)."

USING ENERGY AGAINST OTHERS

Because energy, from food to fuel, is essential to the existence and well-being of human populations and because it can be used explosively to kill people as well as to destroy their homes and other structures, the greatest moral problems related to its use involve the overt control or destruction of other persons and their life-support systems.

The impact of high energy societies on low energy societies was discussed in chapter 7. One of the curious features of such interactions is the use of moral rationalization within the high energy society to justify the use of force against the low energy society. The forced conversion or slaying of the heathen, the enslavement of those of other colors or religions, the extermination of the American Indian, the ruthless opening of China to the opium trade, and the draining of petroleum and protein from underdeveloped to developed countries all have this in common: a low regard for the lives, pride, and welfare of races and nations deemed inferior to the people with the power, inferior because they are not Christian, or because they have not developed an industrial society, or both. In nineteenth century American literature, the Indian generally was portrayed as a savage, either an ignoble savage preying bloodthirstily on virtuous settlers and therefore to be wiped out without mercy, or a noble savage whose values and way of life, although admirable, had to give way before the greater values of advancing civilization. The morality of the strong in practice often does not harmonize with the welfare of the weak. Victors rarely forswear the use of force.

Warfare, which more and more depends upon the intensive use of energy, is to many the most worrisome moral arena. The use of concentrated packets of energy to rend bodies, shatter homes, sink vessels, and destroy aircraft, to kill an enemy and make his habitat sterile or poisonous has been the cause of some of the world's most hoary but virile rationalizations. That some peoples in some places have erected elaborate moral codes by which to conduct their intraspecific bloodletting suggests that the relation of moral codes to ethical behavior (as defined earlier in this chapter) is apt to be tenuous indeed. Energy and warfare may be related in another way. Sales to other countries of arms made in the United States grew from less than $1 billion in 1970 to $8.6 billion in 1974. One of the principal rationalizations was the need to improve the balance of trade, which was becoming increasingly adverse because of expensive oil imports. Such reasoning will lead to an abundance of lethal materiel in parts of the world where territorial conflict is at best merely dormant.

FARMERS AND MINERS

Farmers and farming have long been regarded as virtuous: miners and mining have seldom been regarded as virtuous. Farmers and miners have had many conflicts. The idea that virtue and vocation are linked is still with us. That the tiller of the soil is more virtuous than other men has been maintained by many, including Thomas Jefferson. That the miner is less virtuous than other men has been held by many, including Lewis Mumford who has claimed that the habitual destruction and devastation of mining brutalizes the miner and that there is "indeed something devilish about the whole business" of mining (1934, p. 72-74). Thoreau, writing in 1863 about gold mining in

California, preached that "God gave the righteous man a certificate entitling him to food and raiment, but the unrighteous man found a facsimile of the same in God's coffers, and appropriated it, and obtained food and raiment like the former (1963, p. 182)."

Listen to Agricola, more than four hundred years ago, recite the farmers' brief against the miner;

> The fields are devastated by mining operations . . . the woods and groves are cut down . . . when the woods and groves are felled, then are exterminated the beasts and birds, very many of which furnish a pleasant and agreeable food for man . . . when the ores are washed, the water which has been used poisons the brooks and streams, and either destroys the fish or drives them away . . . inasmuch as Nature has concealed metals far within the depths of the earth, and because they are not necessary to human life . . . they should not be mined, and seeing that when brought to light they have always proved the cause of very great evils, it follows that mining . . . is harmful and destructive. [1912, p. 8]

Agricola's words and Mumford's environmental determinism call to mind the impassioned arguments of present-day farmers inveighing against strip mining for coal and contemporary environmentalists attributing evil character to the managers of the oil industry.

Environmental politics of the past few years have developed against a background of a new wave of reaction against the excesses of cities and the bad fruits of technology. The leaders of this reaction invest nature with both esthetic and moral qualities. To the nature mystic, miners and oilmen personify the evilness of man as against the goodness of nature. They scar, despoil, desecrate, rape. Mining accords with a development ethic and during the winning of the American West was highly regarded; it conflicts with a preservation ethic and, as the desire to preserve natural portions of our environment grows stronger, mining, including mining for petroleum, tends to be viewed as both morally and ecologically wrong.

The moral distinction between the farmer and the miner, absurd though it is in a society in which we are all miners, is at base an allegory that has very real meaning for modern society: it tells us that we must find a way to farm, husband, and renew our energy resources in place of mining, depleting, and exhausting them.

FOOD AND MORALITY

The United States has been called the "Saudi Arabia of food." Food surpluses have been an American problem until recently when they have turned into an important source of export earnings. If we stopped feeding grain to cattle and hogs, the surplus would become much larger. It is not what the United States can do with its actual or potential food surplus that worries some people, but what it *should* do. The United States can produce an abundance of food for its people; most other countries cannot. Some countries can afford to buy the food that they need; most cannot. Therein lies the moral problem. Should the United States *sell* surplus food to those who can pay for it or *give* surplus food to those who may be undernourished, even starving, but cannot pay?

As is often the case with such questions, there are more than two possible answers. Large as the potential surplus may be in the United States, it still would not raise by very much the daily intake of all the world's underfed people if it were spread among them. Consequently, a choice might need to be made as to which hungry people are given food and which not. It has been suggested that such aid go only to those who are making efforts to control their population and who appear to have a chance of success, since otherwise the additional food would go to increase population and the problem would be magnified instead of solved. There are those who contend that giving away surplus food while buying imported oil is certain to bankrupt our industrial economy and to leave us powerless to help any part of the world by industrial, military, or other forms of aid. The moral questions involving the disposition of surplus American food will become more acute as rapid population growth presses harder on foreign food supplies—and especially if climatic change brings about a substantial decrease in those supplies.

RESPONSIBILITY

Early ethical systems provided ways of adjusting to a world over which man had little control. It was a place of mystery and frequent misery, in which his responsibility lay in the ceremonial propitiation of unseen gods and servile obedience to earthly masters. Later, some ethical systems began to stress the stewardship of man—to care for the earth and his weaker fellows, to improve himself both morally and materially. Responsibility grew as man's power to make choices improved. As George Gaylord Simpson has said (1966, p. 34), the concept of ethics is meaningless unless there exist alternative modes of action and unless man is not only capable of ethical judgment but also free to choose what he judges to be good. Man is aware of his own evolution and that of his environment. He now has the power to alter, in some degree, both organic and social evolution, as well as to control environmental degradation. If man is moral, he must accept responsibility for the results of his choices. Because significant increase in his intelligence ceased perhaps forty thousand years ago, man has desperate need of an ethical system to guide his present choices. There are signs that such a system may be developing in the high energy

Ethical and Moral Aspects of Energy Use

countries, as threats to the stability of human society appear and become credible. Most of these threats involve the use of energy and can be lessened or eliminated if responsible choices are made on the basis of expectable consequences. "Forever" to man means until he changes his mind. Thus survival of his species has little meaning, the entropy death of the universe is unreal, and the depletion of fossil fuels in two hundred years is no threat. But that his children may have a worse life than he—when he perceives that as a threat, he may be ready for the new measure of responsibility that appears to be required for the hard choices that will need to be made.

SUPPLEMENTAL READING

Agricola [Georg Bauer]. 1556. *De Re Metallica.* Translated by Herbert Clark and L. H. Hoover, 1912. London: Mining Magazine.
Book I of this famous work on mining and metallurgy deals with sixteenth century moral views of mining, metals, and agriculture, and contains Agricola's defense of miners as useful and honorable persons.

Cottrell, Fred. 1955. *Energy and Society.* New York: McGraw-Hill.
Chapter 10, a penetrating discussion of the distribution of consumer goods, points out some of the moral problems in the disposition of energy surplus.

Galbraith, John Kenneth. 1958. How Much Should a Country Consume? In *Perspectives on Conservation,* edited by Henry Jarrett, p. 89–99. Baltimore: Johns Hopkins University Press.
In which is raised the ethical question involved in the wasteful consumption of scarce resources entailed by unconstrained pursuit of economic growth.

Leopold, Aldo. 1949. *A Sand County Almanac.* New York: Oxford University Press.
This book closes with a plea for the development of an "ecological conscience" and a land ethic; the argument is both ecologically pragmatic and naturally righteous.

Marsh, George Perkins. 1864. *Man and Nature.* Reprinted Cambridge, Mass.: Harvard University Press, 1965.
The force of this classic in conservation literature derives from its assumption that the welfare of future generations is more important than present desires for wealth and comfort.

Mumford, Lewis. 1934. *Technics and Civilization.* New York: Harcourt, Brace and World.
In one of his superbly biased passages, young Mumford assaults mining, miners, and Agricola as devilish, debased, and lame in refuting the medieval charges against them.

Nash, Roderick. 1967. *Wilderness and the American Mind.* New Haven Conn.: Yale University Press.
An excellent account of the evolution of American thought and political action on natural resources, their utilization, and conservation.

Odum, Howard T. 1971. *Environment, Power, and Society.* New York: John Wiley and Sons.
Chapter 8, on the energetic basis for religion, contains a decalogue of the energy ethic for the survival of man in nature.

Simpson, George Gaylord. 1967. *The Meaning of Evolution.* Rev. ed. New Haven Conn.: Yale University Press.
Chapters 18 and 19 discuss man as a moral animal, his need for an ethical system, the relative nature of ethics, and some different ethical foundations, including one that Simpson favors.

Van Hise, Charles R. 1910. *The Conservation of Natural Resources in the United States.* New York: Macmillan.
Written at a time of national concern about resources running out, this book by a geologist explores the question of how a nation should attempt to conserve a nonrenewable resource; equates conservation with patriotism.

Veblen, Thorstein. 1934. *Essays on Our Changing Order.* New York: Viking.
In the chapter entitled "Christian Morals and the Competitive System" (p. 200–218), Veblen examines the origins and interactions of Christian morality and the principles of pecuniary competition.

Chapter 15 ALTERNATIVE ENERGY FUTURES

I am not charmed with the ideal of life held out by those who think that the normal state of human beings is that of struggling to get on; that the trampling, crowding, elbowing, and treading on each other's heels, which forms the existing type of social life, are the most desirable lot of human kind, or anything but the disagreeable symptoms of one of the phases of industrial progress.

—John Stuart Mill, 1848

ON THE IMPOSSIBILITY OF
SUSTAINING THE GROWTH STATE

One portion of mankind has been in the growth or "progressive" state now for more than a thousand years and the material benefits of growth are apparent, especially to those who have been left out. No wonder that growth in production and growth in population seem to mark the normal, "healthy" state of that society and that those nations that have failed to combine the two are regarded as backward. But the laws of physics do transcend the laws of men, to the effect that perpetual growth in things that occupy space and require energy is not possible. The question for rational discussion, then, is not whether growth of population and production must end, but how it will end, what is most likely to end it, and what subsequent living conditions may be like.

For more than a century there have been those who have argued that a failure of energy supply will undo man. Malthus (1888) held that a limit to food supply would check growth. Jevons (1865) forecast catastrophe for England when her coal gave out. Later authors have pointed to the exhaustion of oil and gas as a limiting factor. But none of these predictions has yet been fulfilled, at least for high energy, industrial society. There are still great coal reserves in some parts of the world, and now we have nuclear power and a real potential for loosening the constraint of fuel supply on the production of electricity by the use of breeder reactors. The limits to growth in the production of food do not appear to have been reached. False alarms and great advances in the discovery and conversion of energy have given rise within present high energy society to complacency about future food and energy supplies. Today, there appear to be many economists who believe that Malthus was wrong and who cite Jevons on the British coal question as the sterling example of a doomsday dud; they also refer disparagingly to gloomy forecasts made in 1910 about forest and mineral resources in the United States and in the 1920s about oil and gas supplies, forecasts that turned out to be very short of the mark.

Why have Malthus and similar prophets been "proved" wrong? Mainly because they did not, indeed could not, assess and forecast the impact on energy supplies of scientific and technological advances. These advances have created energy resources, have been used to find previously unknown deposits of energy materials, and have increased greatly the efficiency of conversion of energy materials into useful heat and work. For almost two hundred years, in the industrializing world, energy supply has not been a problem; in fact, demand usually has been struggling to keep up with supply. Even today, that situation has not changed very much. During the past few years, more crude oil has been added to proved reserves of the world than has been consumed; more uranium ore has been added to proved reserves of the United States than has been consumed in the country; natural gas is being wasted in large amounts in the Middle East; coal reserves, at least in the United States, are ample to supply a greatly increased demand for at least a hundred years; nuclear technology appears about to enlarge present nuclear fuel reserves by fifty times or more. As the cornucopians look at these facts, they see a set of problems related to capital formation, to benefit distribution, and to environmental quality, but not to any failure of energy supply. Insofar as it goes, their analysis is correct. The physical exhaustion, the mining-out, of energy *materials* is not a credible threat to the continuance of high energy growth society over the next hundred years. But the exhaustion of energy *resources* is a credible threat. Similarly, absolute limits to food production in agriculture subsidized by energy lie far above present production levels; however, this fact will not keep world starvation rates from rising.

A resource is defined and limited by energy economics. An energy resource is a natural store of latent energy in material form or a natural energy flux that man knows how to convert to his needs and can *afford* to use. As long as a work profit can be attained an energy resource exists. When the work profit is reduced to zero, the resource disappears—whether or not the droplets of oil and grains of uranium minerals remain in the ground and no matter how much energy flows unused in the lower reaches of the major rivers, in sunbeams, and in hot springs. Two interacting factors determine the physical economy of an energy resource and measure the real *limits* on energy use and potential: the geophysical distribution of the energy material or flow and the conversion efficiencies of the available energy technologies. High concentrations of potential energy at shallow depths in the earth's crust can be resources even though the available technology may be grossly inefficient; coal was one such resource in the early steam age and uranium is at present. Low concentrations of potential energy at considerable depths, or of radiant or heat energy in diffuse flows, may require high efficiencies in appropriate technological systems in order to realize an energy or work profit; natural gas in deep, tight reservoirs and solar radiation are contemporary examples.

From examples of the exponential imperative given in this book, it appears that energy technology is subject to the law of diminishing returns; in other words, each increment of improvement in the efficiency of making energy *available* is purchased at a greater work cost than the preceding increment of improvement. Each increment of improvement in energy utilization produces a lesser gain than did the previous increment of improvement. Consequently, there will come a point where improvements in technology will produce very little work gain; technology will then be unable to overcome the increasing work costs of finding and using geological energy stores that, as they are depleted, become deeper, leaner, and farther away from centers of use. There comes a point where the utilization of natural energy flows, including food production, will no longer be expansible—regardless of technical feasibility. This point or level will be different for different countries, not because the citizens of some are smarter than the citizens of others, but because some can afford to subsidize their food-production systems by use of inanimate energy, whereas others cannot; as the work costs of obtaining inanimate energy rise, subsidies available to agriculture in the former countries will diminish.

Malthus, Jevons, and others were "wrong" because they did not foresee new discoveries of energy materials and increases in efficiency of energy extraction and utilization. Malthus, in addition, did not foresee that population growth rates might decline in a high energy society, nor the potential for an inanimate energy subsidy of agricultural production.

That technological and social changes were not foreseen by the early prophets of doom does not invalidate their reasoning. Malthusian limits have been reached by many societies. In China the food supply limit has been reached several times and has been different at different times. The present steep population growth rate in China does not mean that the food-energy supply limit has been eliminated, but that it is now at a still higher level because of fossil-energy subsidy. The economic usefulness of all energy stocks and flows in the world has finite limits. In all cases, the ultimate limit can be defined the same way: zero energy profit. The largest factors on the energy cost side of the utilization ledger may be those of finding and extracting (natural gas today, petroleum tomorrow), converting (solar power, substitute fuels), or maintaining a safe and healthy environment (coal, nuclear power). Consequently, energy consumption cannot continue to grow indefinitely. As limits are approached, the population can increase only at the expense of a decreased per capita energy consumption. If the limit being neared or reached is that of food energy, population growth will be halted by an increase in the death rate from malnutrition, and there probably will be an actual decline as the fertility of marginal land, pressed into service in a desperate effort to maintain the growing population, is exhausted. If the limit being neared is that of fossil energy, the initial impact will be on living standards, in which severe, even catastrophic declines may be expected, because the size of the population is a function of an energy-consumption rate that can no longer be sustained. Much hand labor will be required again in agriculture to replace the fossil subsidy and to reduce the energy used in food transport and processing. If the limit being neared by a growth society is one of income energy other than food energy, in other words of water or wind power or solar energy, growth will be slowed gradually as it becomes more and more costly to produce marginal goods and services. In all cases growth will be halted ultimately.

It needs to be emphasized that the mechanism that halts the growth state may differ in the three general cases presented. In all three it *may* be self-control based on foresight, in which case a disastrous population decline may be avoided and the virtue of Malthus's "moral restraint" thus be rewarded. If not by self-control, the mechanism in the first case (reaching a food-energy limit) is bound to be starvation and disease abetted by malnutrition, in the second case (the exhaustion of fossil energy) a sharp decline in living level followed by an increase in the death rate, and in the third case (reaching a limit of renewable energy) the straightforward economics of energy will halt growth in the production and consumption of goods and, if self-control is not exercised to keep down the birth rate, the level of living will fall to that required to raise the death rate to equal the birth rate. Whether the growth state is succeeded by an almost steady state, a condition of approximate equilibrium between energy consumption and sustainable energy supply, or by an unsteady state of alternating growth and disastrous decline depends on the wisdom and self-control of the society as well as upon the nature of its energy supply.

Although indefinite growth is not possible and limits to the availability of energy can stop growth, there is no way to forecast when the growth state of the world's industrialized nations may cease and it is by no means certain that an inadequate supply of energy will be the first limiting factor. Social collapse, catastrophic warfare, pestilence, or pollution may stop growth before limits on the supply of energy are reached, but these causes are apt to produce only temporary halts after which man will regroup and build anew from a diminished base. Limits of energy, slower to prevail, perhaps not even catastrophic in effect, will provide more permanent bonds.

Man has "high-graded" his energy resources as he has those of the metals. The most accessible, most easily converted resources have been consumed. But large quantities of crude oil and natural gas that will be inexpensive to produce remain in the ground, are not under the control of the industrialized nations, and have become relatively expensive to them. Until these supplies are exhausted, continued growth in population and production is likely, except that the rates of growth will be higher in the oil-producing countries than in the oil-importing countries. (In 1975 Saudi Arabia banned contraceptives after the World Moslem League ruled that birth control violated Islamic law.) The time will come, probably within fifty years, when growth can no longer be supported anywhere by inexpensive petroleum.

As the world turns to more expensive sources of energy—more expensive because they are deeper and leaner or because they are found in slow or diffuse flows—the natural subsidy of energy will diminish unless improvements in technological efficiency keep pace with the increasing costs of recovery and conversion. The source that offers the best hope is solar energy. Almost a fourth of the energy needs of mankind could be supplied by solar heating and cooling—for which no provident technology is required. If the efficiency of photosynthesis could be improved—and there is no physical law that tells us it cannot—vegetable material for both fuel and food could be grown on a large scale in favorable parts of the world. The efficiency of converting solar radiation to work, very low with present materials and technology, is on that account far below any limits imposed by thermodynamic law. Solar energy is infinitely renewable and absolutely clean. If solar power can be made profitable in terms of energy economics (solar heating and photosynthesis already are), solar energy may serve to redress further the imbalance of material comforts and political power that was created in large degree by the geographic concentration in the northern temperate zone of good coal and fertile land.

Nuclear fusion does not offer the same potential. If it becomes technically feasible, it will produce electrical power only and that power is apt to be so expensive that only the wealthy countries could use it. At present, neither solar energy nor fusion power can offer a sure replacement for the fossil fuels. Only fission power can come close to doing that and it entails environmental costs and social hazards that make its acceptance a true Faustian bargain (Weinberg, 1971). If the growth-state philosophy prevails, the bargain will not be rejected.

ON THE UNDESIRABILITY OF SUSTAINING THE GROWTH STATE

Not only the possibility but also the desirability of sustaining the growth state are beginning to be widely questioned. It is not at all clear that the growth state increases human well-being in any degree commensurate with its social hazards. Especially is it not clear that a

FIGURE 15.1
Graph illustrates three possible futures for present high energy society, the energy sources on which each might depend, and the levels at which those energy sources would support a stable population. In Future I, energy consumption does not reach the stability level because of social and environmental disruptions. In Future II, energy consumption and stability level ultimately coincide. In Future III, energy consumption oscillates above and below the stability level because of the influence of climatic changes on energy supply, mainly food.

doubling of the gross national product will produce a doubling of social benefits and individual happiness. Herbert J. Muller remarked (1970) that the United States has achieved "the highest standard of low living in all history (p. 12)" and that it "is the first nation in history in which more people die of eating too much than eating too little (p. 14)." It may be significant that the argument for the growth state is usually put in terms of the expected defects of the alternative steady state. No one has put the cultural argument against the growth state and for the steady state any better than John Stuart Mill who in 1848 decried the incessant scramble for material position in the growth state, called the opportunity for getting more than one's share through increases in production and accumulation a stimulus of coarse minds, and questioned the assumption that mechanical inventions had lightened the toil of mankind. Mill argued that "the best state for human nature is that in which, while no one is poor, no one desires to be

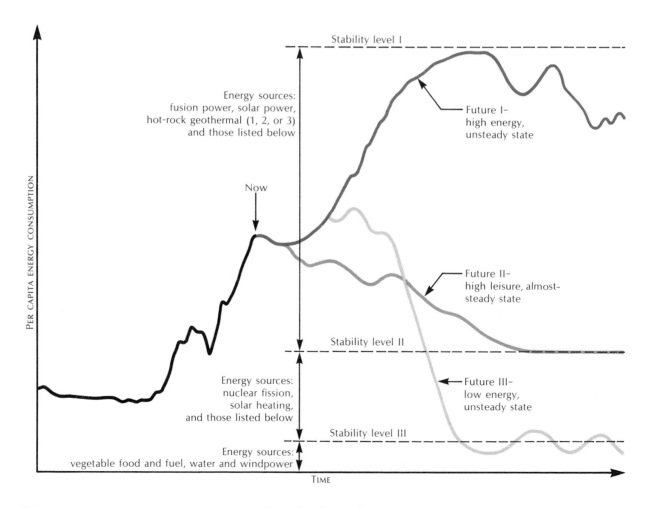

Alternative Energy Futures

richer, nor has any reason to fear being thrust back, by the efforts of others to push themselves forward (1895, v. 2, p. 337)."

In the growth state the structure for reward puts great pressure on industry and government to increase consumption of energy and materials, to build energy converters, to hasten the flight of entropy, and to manage increasingly hazardous wastes in expedient ways. The pressures in a growth state that augment demand force the hasty allocation of resources and poorly examined uses of the environment. They may also produce a higher per capita demand for goods and services that are costly in terms of energy than would the conditions of a steady state, because status in the growth state comes from the accumulation of goods and from power over others, the latter often related to the production of goods, whereas status in the steady state probably would come from the demonstration of special skills and from political power, neither of which need excite the demand for energy.

The growth state has given man comfort and entertainment more than leisure and enjoyment, it has given him mobility and a longer life, but taken from him the sense of social fitness and position; it has given him the insecurity of perpetual change in place of the certainties of traditional culture; it has given him worldwide war and the ability to eliminate his kind; it has given him the illusion of freedom and the prospect of severe restraint. Where do we, in contemporary high energy society, go from here? There seem to be three possibilities (see figure 15.1): continuation of the high energy, unsteady, growth-oriented state; transformation to a high leisure almost-steady state; retrogression to a low energy state. Let us consider each possibility in turn.

CONTINUATION OF THE HIGH ENERGY UNSTEADY STATE

Nothing known to science or engineering precludes the development of an energy system that would produce abundant electrical power at a slowly increasing cost in work, materials, and ecological impact for a very long period of time. On the other hand, there is nothing that guarantees it. The notions that we can accomplish any technological goal "if we put enough money and brains into it" or that "it's just a matter of time" reflect a limited view of the physical economy of nature and the confidence of a bank robber who ascribes his high living entirely to his own skill and daring rather than to his good fortune that the money was there when he broke into the bank.

Energy systems will be needed to replace those that now run on fossil fuels. Heating and cooling for comfort can be accomplished for most of the world through the manipulation of solar heat, aided by minor amounts of vegetable and refuse fuels. To power machines, vehicles, communications networks, industrial furnaces, and electrolytic vats, some combination of fuel and electricity on a large scale will be required. Coal can supply fuels of various kinds, perhaps for several hundred years in North America and the Soviet Union but elsewhere, after oil and gas give out, engines will run by electricity if at all or by hydrogen produced through the use of electricity. Hydrogen will also be needed in great volumes for fertilizer production. Most process heat will be supplied by electricity. The key element in a high energy society of the future is the production of abundant electricity at a net energy profit. The main possible systems appear to be breeder nuclear power, solar power, hot rock geothermal power, fusion power, or some combination of these. Breeder power is feasible technologically, probably feasible from the viewpoint of pecuniary economics, but perhaps not feasible in the terms of social economy. Solar electricity is technologically feasible, but much too expensive with present technology and materials to be used. Controlled fusion has yet to be achieved. We know there is a great deal of heat energy in the earth's crust, but we do not yet know whether more than a tiny fraction of it can be recovered with a profit in energy. There does not seem to be any useful way to rate these dark horses of energy in terms of ultimate success. But let us, for the moment, assume success, the staving off for a very long time of ultimate energy limits. What will be the probable consequences? Will this be the ultimate technologic fix? Probably not. In the first place, the research, development, and emplacement costs of the ultimate energy system or systems will be very large and will be borne by the few nations willing and able to allocate capital to such an enterprise. If they succeed, will they share the fruits of the new power technology with the rest of the world? It seems unlikely, except as it may be deemed expedient to support the rest of the world as consumers of the products and services of the high energy nations.

Another question arises. If power limited only by the costs of maintaining a tolerable environment were to become available, what would be the social and demographic consequences? Would mankind, able to raise food under giant pest-free geodesic domes, then stabilize its numbers and its relations with other elements of the biosphere, and seek to create a world without want, without territories, and without war? Or would mankind continue to grow in numbers, to press on the final ecological constraints (the global heat balance, environmental contamination by radioactivity, a photosynthetic conversion limit, or standing room only), to produce more goods than it could consume (all high energy societies can do this already), and to fight for dominance over others through the control of energy? One can only surmise which road would be followed. But the record yields no cause for confidence in man's ability to attain Teilhard de Chardin's vision of a noosphere (1965, p. 180–184), where man at last master of himself would approach perfection. The one condition mankind has shown no ability to cope with is that of abundance.

Man is a creature of adversity. By skill and cunning developed over many thousands of years, he has fought and scrambled to his present position. We did not get here by passive conformance to natural law; we got here by aggressive adaptation to a constantly changing environment. Our environmental and social behavior is not programmed by genetic codes developed through the evolutionary screening out of the unfit; it comes out of a cultural evolution in which the family has been the key functional element and kinship the social medium. Unless and until the functional concept of kinship extends to the species—and this may require an invasion by little green creatures from outer space—there seems little basis for expecting the creation of a human world in which capital (energy credit) and converters (power plants) would be shared.

Men and women show more endurance, more resourcefulness, and more humanity under conditions of adversity than they do under conditions of affluence. In wartime a people may submerge their differences and pull together; in peacetime, they resume their household squabbles, their racist hoods, and their elbowing for position. In Great Britain and the United States, amid all the shrill cries of conspiracy, the outbursts of irritated inconvenience, and the quiet pain of those truly hurt by rising prices, the energy crisis of 1973 and 1974 generated an undercurrent of relief, an inchoate but deep feeling that here at last was a tangible problem that required sharing, that might pull us back together again, away from that frightening world where we could do all things except agree on what it was worthwhile to do, where our want seemed to increase as our need decreased. A future of "unlimited, cheap power" may seem heavenly to some; to others, it may appear more frightening than a future of limited, expensive power. Unlimited cheap power seems no longer in the cards as man plays out his game. But he may continue to build new power systems that can deliver large amounts of energy at an energy profit. Then what? It is unlikely that any steady state would be attained if energy were abundant. The growth-oriented state will press beyond the capacity of the environment to absorb wastes, of society to consume goods, or of human beings to put up with one another, and high instability is the expectable result. The present emphasis on material and pecuniary measurements of living levels will persist, at the expense of true leisure and social improvement. Recycling will be carried out, not to save energy, but because waste disposal will have become mankind's overriding problem. A great deal of energy will be used in travel and transport, in mechanical entertainment and gadgets, in cosmetic and health care systems. Quality control and performance will be low, because there will always be energy for doing something again and no great incentive for doing it right the first time. Population control may be effected by the pervasive ethic of pleasure, the ineradicability of degenerative disease, and high death rates from suicide and accident.

This is the future of a society that manages to invent and emplace new energy systems that allow to persist the belief that no limits to growth exist, that one need not prepare to cope with scarcity. At the same time no great success in dealing with the problems of satiety is likely to be achieved. The government of such a society can hardly remain republican. Control of waste will require control of production. The conflict between the lessening need for workers in an increasingly automated economy and the necessity to have a job in order to participate fully in the material benefits of production will bring more government control of the conditions of employment and production. Finally, the threat or actuality of revolt by those who are out of work or bored (boredom can be a powerful cause of social unrest and political upheaval) will lead to a planned economy and society, governed in authoritarian fashion. If the growth ethic persists, the high energy parts of the world will be in fierce competition for raw materials, markets, and the dwindling capacity of the oceans and the atmosphere to absorb the noxious wastes of production.

TRANSFORMATION TO A HIGH LEISURE ALMOST-STEADY STATE

The Benedictines, those great teachers by example, introduced technology into the monastery and worked hard at improving it in order to increase the time they could have for "rewarding work," not to achieve a surplus of goods. Those who learned from the Benedictines what system, regularity, and technology could do to increase energy surplus reversed the objective and decided to take their surplus in goods rather than leisure.

If now it happens that present high energy society finds the energy costs of energy use rising and foresees no reversal of these costs, it may be forced into a pattern of conservation that ultimately will produce a high leisure almost steady state at some rate of energy consumption that will be perhaps not too different from that of North America today if the fossil fuels are stretched out by conservation and solar energy in its various forms can be developed to replace them. Such a world might differ strikingly from that of the high energy state.

The steady state of contemporary sociological philosophers is what J. S. Mill called the stationary state (1895, v. 2, p. 334–340). He defined it as the state of society in which there is no increase in capital or wealth and he thought the growth state—which he called progressive—must inevitably culminate in the stationary state. While warning that continued population growth would diminish the quality of life available in the stationary state, Mill optimistically averred that

> A stationary condition of capital and population implies no stationary state of human improvement. There would be as much scope as ever for all kinds of mental culture, and moral and social progress; as much room for improving the Art of Living, and much more likelihood of its being improved, when minds ceased to be engrossed by the art of getting on. [Ibid., p. 339–340]

Today, the concept of a steady state is more apt to be put in the framework of equilibrium—of a balance in energy and materials flows—rather than of capital, and of a constant human population (Daly, 1974, p. 156). Under either the older or newer definition, the steady state has not been attained as long as nonrenewable resources are being consumed. In cosmic terms, the continual degradation of energy, no matter how carefully husbanded, prevents a true steady state from ever being reached. On the other hand, since it is theoretically possible for man to attain an almost-equilibrium with his resource environment, a state of society that could endure for hundreds or even thousands of years, the concept is of real importance.

In the approach to the steady state, emphasis on the efficient use of energy will have brought about great changes in the structure of cities and in transport systems. Cities will be three dimensional as well as smog free. Urban sprawl will be checked. The urban underground will be developed for transport, storage, and manufacturing, and people will again want to live in cities. Present-day cities are flatter than sand dollars; the lateral movement of goods and people takes place mainly on a single level; urban transportation is wasteful of energy and of human living time. The development of several levels underground for different vehicles traveling at different speeds for different purposes, for waste collection and processing systems, for storage of food, fuel, automobiles, and duplicating paper, and for small-scale manufacturing and power production, will make the city a more efficient mechanism, a pleasant place to live and work, and more difficult to sabotage.

A nation's business and industry will be conducted in energy-saving ways. Audiovisual communication will replace business trips and conferences, because the transport of words and images consumes much less energy than does the transport of bodies. Today the transport of energy itself consumes a lot of energy; tomorrow energy will be transported as liquids in pipelines, as electricity in buried cryogenic transmission lines, and possibly as microwave beams. The necessity for speed, that glutton of energy, will be questioned. The fact that it takes forty times as much energy to cross the Atlantic Ocean in an airplane as it does on a ship (Slesser, 1973, p. 328) will have made vacation flights to Europe a thing of the past. But vacations in such a world could be much longer than at present and the speed of travel during a vacation not so important as it is today. There will be many constraints on the use of energy and nonrenewable resources, but these constraints will be implemented more by an ethic of conservation based on national or transnational kinship than by the police power of the state, for the latter can be effective only when the majority allows it to be.

Choice, however, should still be great in human life. There will be a great deal of unemployment by our present definition, for that will be a major social goal. The ingenuity of man will be extended to its utmost in devising myriad ways of using leisure time so that satisfaction and happiness will result instead of boredom and rebellion. The new society will survive only if the old religion idealizing work, production, and growth can be relegated to the attic of past fancies. A nation that has been taught to regard idleness as sinful cannot be counted upon to build a stable society based on the goodness of leisure. If it is found that population growth is inimical to the preservation, or even to the attainment, of a high leisure steady-state society, population will be controlled, probably by a combination of economic and physical disincentives applied by the government and by social disapproval of large families; it seems unlikely that stern measures such as progeny permits or forced sterilization would become necessary or be found practicable.

There is no ideal state. As boredom òn the job and in the kitchen is a bane of present high energy society for which aggressive or escapist behavior is commonly sought to bring temporary relief, so may boredom be the principal ailment of the steady state society. In primitive societies, boredom appears to be avoided because the work the individual does, although using up an average of only about two to four hours a day, is physical and allows rest—and it is only when one cannot rest therapeutically that boredom finds a foothold; because ceremony and ritual are vital elements of primitive life; and because there always exists a social ladder to challenge the ambitious, no matter how stable the population, how unvarying the economy. These important elements of low energy steady state society are greatly diminished in modern high energy growth society. A stable high leisure society may need to rediscover the social benefits of physical exertion, of ceremony and ritual, and of social climbing. It will most certainly require a revolutionary redefinition of wealth and debt. In today's high energy society, wealth is measured by the goods and territory a person has the exclusive and "perpetual" use of and by the debt of all other persons to him as measured by his possession of money. Money consists of tokens that require others to work for the possessor thereof. In tomorrow's high leisure society, it may be that no one has perpetual rights to territory but only transient, renewable-use rights, and that the debt of others to an individual will be measured, not in his or her possession of money, but in his or her elected position of influence and decision. Debt would be social, not monetary, and would derive from gifts of service and goods by the individual to others. In this scheme of things the politician, not the rich campaign donor, would be master. Wealth would consist of social accounts receivable, not of money and possessions.

In the high leisure steady state, agriculture would still be mechanized; animal products would be used, but more sparingly, because they would be priced on the basis of resource-use efficiency. Recycling of metal, paper, and glass would be carried out insofar as energy could be saved thereby or the scarcity of the material required it. The rate of entropy increase would be retarded insofar as practicable. Inefficiency in the social use of energy might come from the desire of people to regulate their activities by the sun's time, thus perpetuating peak-load inefficiencies. The average age of the high leisure population will be substantially greater than in the growth state; generation gaps should disappear and age no longer exclude one from full participation in society.

This appears to be a possible future for an existing high energy nation or society, or for a nation or group of nations that manage to enter high energy society on the basis of their own inexpensive fossil on nuclear fuels. The envisioned society is not without its drawbacks. Occupational mobility would be small, and social mobility, no longer tied closely to occupation, would not be available to many who are now able to move upward in the growth state. Leisure for many might mean the mass viewing of staged activities. Emphasis on the conservation of energy might likewise encourage spectator entertainments in place of sports or diversions requiring a high per capita consumption of energy. Efforts to economize on energy and materials could lead to a lack of choice in building materials, in automobiles, and in ways of getting about, as well as to the centralization of most production and communication facilities. The danger in such centralized sameness is its capacity for facilitating the development of a dual society in which a smaller master group dominates a larger servile group. Certainly the economic pyramid in a country such as the United States will not be frozen in the transition to the almost steady state. The dynamic nature of a society in the growth state is due largely to the widespread perception of opportunities to improve one's material situation. As this opportunity diminishes in the approach to the steady state, there will arise an irresistible demand for more equal shares of the material wealth of the society. It is not reasonable to expect that many will be content with leisure alone, even when leisure has become the socially correct expression of surplus energy. It may prove extremely difficult to accommodate the demand for material equality under conditions of minimized entropy increase. The struggle to do so could result in an upstairs—downstairs society.

Achievement of the almost steady state will require much thought and planning. Thinking not only is hard work but often leads to unpalatable conclusions. Planning carries the threat of future coercion. Neither is apt to be embraced by the political leaders of a nation in the growth state. Attainment of the almost steady state will call for sacrifices of position and wealth by many who now wield power, and it will require mass apostasy from the religion of the growth state. In short, there are formidable, perhaps insurmountable, obstacles in the way of achieving by peaceful means an almost steady state in which material wealth—but not status—would be distributed equitably.

RETROGRESSION TO A LOW ENERGY STATE

There is no reason to expect the transition from a low energy society to a high energy society to be irreversible. If the energy support of a high energy society fails, it must again become a low energy society. The great difficulty is that the low energy phase can be regained only at the expense of degeneration in living standard and life-style.

The sequence of changes is predictable. First, the uses of energy that are not vital to the industrial economy will be restricted and gradually eliminated: pleasure driving, air conditioning, frost-free refrigerators, large lighted advertisements will disappear. At the same time, there will be a sharply increased and unremitted emphasis on energy-use efficiency: car pooling and the use of public transport systems will become general; space heating by electricity, except where alternatives are more wasteful of scarce resources, will cease to exist; buildings will be well-insulated, windows will become smaller or be eliminated; many human activities will be transferred underground, because of the economy with which temperature can be managed underground, protection from storms, and the physical economies inherent in making cities truly three dimensional; bodies will be warmed more by clothing and less by heating the near environment. The service sector of the economy will shrink, except for police and applied research.

The second phase will be marked by the partial dismantling of the industrial plant. Energy-intensive processes, measured in energy consumed per unit of use in the output, will be discontinued, as will the manufacture of unneeded objects and the provision of unneeded services. Mass attendance at entertainment and sporting centers will be discouraged; television will become the complete surrogate. Certain drugs which can be produced and distributed cheaply and whose debilitating effects are negligible, such as caffeine and marijuana, probably will be in widespread use, as will ethyl alcohol, whose effects are less mild. Solar heating and cooling will be widely used; methyl alcohol—from cellulose, algae, and coal—may be the most common fuel.

The third and most drastic phase of retrogression will involve the reduction of the industrial plant to a skin-and-bones system concentrating on the efficient production of fertilizers, pesticides, textiles, cheap construction materials, and cereal foods. Meat animals, except chickens and some pigs, will have disappeared; protein will come from fishmeal and soybeans. Most workers will be back on the farm and the maintenance of high agricultural production will be more important than high productivity per worker. Throughout the entire devolution, birth rates probably will have been dropping, so that by this time the average age of the population will be rather high; because social security in the present sense will be economically impossible, all able-bodied oldsters will be at work, many of them in the fields. There will be no unemployment.

In North America the reversion to a low energy society might take place without a catastrophic decline in population, but in the nations of western Europe, populations already are too high to be supported even by intensive hand cultivation of the available arable land. If the prop of inanimate energy support were removed there, famine would reduce the population drastically. In the United States the amount of arable land is sufficient, under intensive hand cultivation assisted by fertilizers and pesticides, to feed about one billion persons on a vegetarian diet; moreover, remaining supplies of fossil and nuclear fuels could cushion the decline in living standards, the demographic dislocations and the social adjustments required. Democratic government, however, would have a very low probability of survival. Even under a government that is not democratic, social stability will depend upon a shared ethic, although that ethic may be inculcated and enforced by the police power of the state. No matter what the government is called, it will consist, in all likelihood, of a small minority managing the majority. If a world government should emerge under this energy future—and it seems the alternative most likely to result in a world government—that government, too, however parliamentary its façade, probably will be of a small minority that has managed to obtain control of the world's energy resources.

THE HIDDEN ENERGY CRISIS

But these are long-term, wholesale futures. What about the present energy crisis in the United States? If by *crisis* we mean an immediate and dire threat to the welfare of the nation and to the living standard of its people, the United States in the mid-1970s does not have an energy crisis. It has energy problems. It faces shortages and higher prices, not a crisis; shortages of clean fuel to burn under the boilers of electric power plants, shortages and higher prices of fuel oil and natural gas for home, commerce and industry; higher prices for gasoline and jet fuel for passenger transport. For these problems, there are or can be, practicable solutions: alternative fuels, alternative energy sources, alternative technologies, and alternatives of consumption. Some combination of these alternatives will see the nation through the next few decades. How well they serve the nation will depend in large measure on how soon the resolve matures to invest in them, for they will take years and work to implement.

Energy will cost considerably more than it has in the recent past. Despite the great volume of the world's oil reserves and the considerable potential for increasing those reserves by vigorous exploration and technological development, the growing demand for oil and the aspirations of the producing nations should keep the price high until demand finally overtakes discovery and there no longer will be anything artificial about the scarcity or the high cost of the remaining crude oil. Energy from crude oil and natural gas has been very cheap in the United States. In no other industrialized nation have large quantities of such useful energy been available at such low prices. A good deal of whatever competitive advantage the United States enjoyed in world markets during the present century derives from this fact, and that competitive advantage would have been even greater had the United States not taken a substantial share of its natural subsidy in consumer goods and comforts.

The vocal concern about the energy "crisis" suggests that we Americans, as Thoreau wrote (1854, p. 93) more than a hundred years ago, "are determined to be starved before we are hungry." Even with predictable advances in prices between 1970 and 1985—ten times for natural gas at the wellhead or port of entry, three times for gasoline at the pump, three to five times for coal burned in power plants, and five to ten times for uranium as concentrated ore (all in constant dollars)—the total energy fraction in most goods and services is not so high that the prospect of substantial escalation in energy costs has to be in itself a serious threat to the national economy. Much of the increase can be compensated for by economies in the use of energy and materials. Such economies will have an impact on the American life-style and habits of thought. Efficiency and prudence in resource use will be emphasized. Utility and durability of products will be desired. Energy-conserving substitutes will be sought for energy-expensive travel modes and for mechanically powered pleasures. Though life-styles will adapt to new realities, the production of sufficient food, goods, and services to maintain a high level of living in the United States should not be imperiled by energy shortages of the next few decades. Moreover, the principal competitors of the United States in world trade face similar or greater increases in the cost of energy. The Soviet Union, alone among major energy-consuming nations, produces more energy than it consumes but, because the costs of delivering Soviet energy to points of use are high, the energy self-sufficiency of the Soviets appears to give them, in the short term, more political than economic advantage vis-à-vis the other industrialized nations.

There is, however, a real crisis hidden in the present basket of energy problems. Although we began so short a time ago to devour them, the ultimate exhaustion of domestic oil and gas can be foreseen clearly. The faster we use them, the sooner they will be gone. We probably are about midway in using up American crude oil *in quantity,* but more than two-thirds along the path to exhaustion *in time* (because consumption rates are so much higher now than they were while we were consuming the first half). Although we all know that a mining economy is inherently unstable and that exhausted mines are marked by ghost towns, we have always been able either to revive a fading mine by the discovery of a more economical technology or to move on to a newly found deposit when we exhausted the old one. Now, however, we are facing real limits to the discovery of new oil and gas fields, as well as to the ability of technology to extend these resources by more effective search, deeper drilling, improved extraction, and more efficient utilization. The growth of nuclear power presents serious problems of ecological costs and of public health and safety. The replacement of oil and gas by coal and coal products or by shale oil will require large capital commitments, careful attention to the hazards of irreversible ecological damage, and adequate planning for coping with a large decrease in the price of crude oil by the oil-exporting countries.

In the face of these real constraints, there is a continued emphasis on increasing domestic supply rather than on reducing demand. At the same time we deplore dependence upon foreign oil, we are importing an increasing share of our petroleum supply. We seem to be placing our hope for a warm future more and more in provident technological breakthroughs, such as fusion and solar power, in other words, in the technological cavalry's riding over the hill in the nick of time. If we continue draining our finite supplies for inefficient uses, importing more, and failing to implement available alternative means of supply (such as gas from coal)—and the technological cavalry does not show up, the nation could become subservient to its foreign suppliers.

The *hidden crisis* of today is a crisis of attitudes and institutions rather than of energy shortages and higher prices. It involves a shortage of planning and decision capability. We need now to plan for the next hundred years, to move toward an equilibrium society and economy to replace the present unstable growth society and economy. We have not reached the peak of world fossil-fuel production; that will come within the next twenty to thirty years. By that time we may have extended present nuclear-fuel reserves by a hundredfold (by the development of an efficient but not quite safe breeder power plant), but they will still be finite. Only if we have managed to control fusion without the use of lithium or to convert sunlight economically to electricity will we have broken through the availability constraints imposed by nonrenewable energy resources, and neither of these rainbow technologies may prove feasible for large-scale energy supply.

Population dynamics, the galloping consumption typical of the growth state, and the persistence of faith in technology and the economic "laws" of man to prevail over the physical laws of nature combine to create a momentum against which planning for a change in the velocity and direction of society will need to be directed for a long time before significant results can be seen. Now is not too early to start planning for the world of our grandchildren. Their future will not be in their hands, it is in ours. The claim that the "unlimited ingenuity" of man will overcome physical obstacles as they are perceived in the road of progress is a myth that tranquillizes concern, numbs responsibility, and defeats planning.

The localized energy shortages of today are the forerunners of global shortages that will develop in thirty or forty years, when there will be many more people in the world than there are now. In order to be prepared for the global tensions of that time, and to provide some leadership for the survivors, we shall need to change our attitudes and behavior. Our present patterns of thought about and behavior toward resources reflect a mining philosophy that emphasizes increasing supply and consumption as well as faith in new technological bonanzas. If we are to aim for equilibrium between man and the resources available to him, we must abandon efficiency of *production* as a social goal and replace it by efficiency of *consumption*. We need to favor measures of *quality* of achievement over measures of *quantity*. And we must stop confusing momentum with progress and growth with goodness. The medical definition of crisis makes it the time or stage when it becomes a matter of certainty whether the patient is to live or die. The real energy crisis is that kind of crisis, and we have reached the stage in human affairs when a revolution in thinking is required if the patient is to be alive and tolerably healthy a hundred years from now.

ONE WORLD OR SEVERAL?

The present human community can be classified, on the basis of access to, and use of, energy resources, into four "worlds" (see page 257). It seems unlikely that the range of per capita energy consumption throughout the world will be narrowed during the remainder of this century. The underdeveloped countries with scant energy resources of their own and without valuable mineral materials to exchange for energy face a bleak future. The surplus from the world's total natural subsidy in the fossil fuels was beginning to spread to them when such marginal use of the subsidy was abruptly curtailed by steep price increases. Even with such trickled-down subsidies, the overpopulation and the high rates at which their populations were increasing created, for most of these countries, almost insurmountable barriers to industrialization and higher living standards. The nationalization of copper mines in South America and oil fields there and elsewhere reflects a desire to have all the social benefits of a finite wasting asset accrue to the host country. Without such a natural subsidy in its hands, an underdeveloped country has no hope of becoming industrialized but can only struggle to adjust its population to the energy surplus attainable from agriculture and fishing.

The world of OPEC has the opportunity of a national lifetime, even if that opportunity is constrained in all the countries by physical or demographic factors. In the truly arid OPEC countries, population has been small; consequently, the sudden national affluence has been able to create the appearance of a high energy society almost over-night. Populations in Arabia are starting an explosive growth, and the additional human beings will be totally dependent on oil revenues. The Persian Gulf countries are attempting to create industries that can use the cheap indigenous energy (extremely cheap compared with the prices the importing nations must pay for oil and gas from the same wells) and compete successfully in world markets. Their great asset will be cheap energy; their great liability will be lack of fresh water that is required by many industrial processes. Seashore nuclear desalting plants will be built by other, high-energy countries in exchange for oil, and if the world price for oil can be kept high enough, the high cost of process water from the desalting plants will be more than offset by the relative cheapness of energy, and the products of industrialization may compete successfully in a world market. It is not at all impossible that the entire Persian Gulf region might become one of the world's major industrial regions. The maintenance of such a phoenix of the sands beyond the bonanza period of oil and gas avail-ability would require regulation to stretch out the production of hydrocarbons, sustained access to nuclear fuels, intensive irrigated agriculture or hydroponic food production, and successful transition to a base of nuclear and solar energy.

The humid OPEC nations (Indonesia, Venezuela, Nigeria) already have large and growing populations, pressing relentlessly on the food potential of the land. Both Indonesia and Venezuela appear to be nearing the period of declining oil production before the benefits of oil production have been translated into other sustaining enterprises. Oil can be exchanged for food and fertilizer only until the oil runs out. The potential for attaining and maintaining a high level of energy use in these countries is low.

In the communist world, the Soviet Union has ample resources for continuing a steady rise in energy consumption, material level of living, and national power. Her only energy deficit is in food and it is not a critical one, for the U.S.S.R. will be able to purchase food in the world market as long as any country can produce or squeeze out a surplus. China's situation, by contrast, looks precarious but far from hopeless. Efficient management of agriculture and an intensive geological search for new hydrocarbon deposits has produced an improved level of living for the enormous Chinese population, most of which is on the farm, and the possibility of subsidizing the production and importation of fertilizer while the country is starting to industrialize. The improvement in the food supply has not aided government efforts to reduce the population growth rate. Unless population growth in China can be slowed and eventually stopped, or unless a virtually unlimited energy supply is discovered, China faces a fading future. But that is true for the entire world as well.

In the high energy societies of the western world (including Japan), the present problem may be seen as one of sustaining growth but is actually one of preserving present living levels. Countries in this group whose access to energy is controlled by nations that may desire to become successful competitors or even to reverse a historic relation between exploiter and exploited, are not in a favorable position. Only nations that have or can obtain energy from alternative sources at reasonable cost will have much chance of maintaining their material levels of living. As of today, these latter appear to include Canada, the United States, the North Sea nations, and Australia. As long as gold retains a high pecuniary value, South Africa should also survive.

Canada and the United States have great hydrocarbon reserves, proven and potential, in coal, tar sands and oil shale. Large-scale production of so-called substitute fuels from these resources will require heavy investment and fairly long lead times. It appears to be in the national interest of the United States to subsidize the rapid development of substitute fuel systems in order to have an economic alternative to costly or interdicted foreign supplies of petroleum. Until now, however, most of the cost of developing tar-sand and oil-shale technologies has come from private venture capital. Cooperation with the government of Canada, although perhaps not politically feasible, would seem desirable, especially in developing the tar sands of western Canada, which may contain more recoverable oil than all the crude oil that will ever be produced in the United States. Although oil shale in the United States may likewise contain more recoverable energy than the total recoverable domestic crude oil, there will be strong and sustained opposition to the mining of several million tons of oil shale *a day* in the scenic lands of Colorado, Utah, and Wyoming and to the disposal on the surface of the same amount of waste material. Both Canada and the United States have comparatively large potential reserves of uranium and thorium, the natural fuels of a nuclear age. Accordingly, these two countries appear by far the best provisioned for the future of the western high energy countries. The North Sea nations may become self-sufficient in energy by 1985 if discoveries of oil and gas in the North Sea continue at the current pace, but such self-sufficiency is bound to be temporary.

TOWARD AN AFFLUENT SCARCITY

The "people of plenty" (Potter, 1954) now are troubled by the quickening pains of the change from an economy of abundance to an economy of scarcity. Grooves of thought and action engraved by affluence are poor guides to a good life and a stable society under the constraints of scarcity. Cultural adaptation to scarcity is demonstrated by a few societies; in them the adaptation appears to depend upon the absence of a market economy, the centralization of decision making, and the development of an ethic that works for social efficiency of resource use. These, however, are not industrialized societies and they may offer misleading signposts to new paths for contemporary high energy society.

In affluent western society, there are powerful barriers and resistances to change. Entrenched interests linked to decision structures feel threatened by any move to share scarcity. The "would-be" capitalists of the western world, the small share holders in the growth economy, see their hopes of future riches menaced by any proposal to allocate resources according to social utility rather than private profit. The service layers of high energy society, especially those highly paid who enjoy the status of oracles, such as economic advisers to the government, medical specialists in the diseases of affluence, media commentators, professional athletes, counselors of the lustlorn, and pedagogic protectors of the cultural heritage, wince at the suggestion of a reexamination of their social usefulness that may be called for when scarcity replaces affluence. Men and women usually believe what they want to believe, even when the evidence is overwhelmingly against them. A recent poll suggests that most school children in a heartland region of America believe in literal creation not biological evolution. The leaders of the Soviet world, as well as many highly placed persons in the "free" world, proclaim that population growth is not a crucial problem for mankind. Many, if not most, persons in the Moslem world believe that the sun orbits around the earth. Some college graduates in North America appear to believe that sodium chloride produced from the direct evaporation of seawater or from the desalting of vegetable material is fundamentally different from sodium chloride that is mined. Measured by levels of social concern and controversy, there are many people in the United States who believe that the ionizing radiation produced by an X-ray machine is not harmful whereas that produced by a nuclear power plant is. Given such ability and practice in believing, and the childish delight men and women take in conspicuous consumption, in power, and in toys, one can understand the mass resentment, pique, and repudiation that flares up when material shortages occur.

Affluence, freedom, and waste are the three musketeers of the growth society. If affluence succumbs, the others will not be far behind. But one of the important questions is, how will freedom be abridged, by consent or by force? The longer government delays in implementing controls on the use of scarce resources, the greater will be the danger that the remainder will get into the hands of a small number of people and the greater will be the probability that controls when they do come will be harsh and inequitable. But government in an affluent society, no matter what its form, has a great tendency to delay because it itself has a vested interest in growth and continued abundance and because it is afraid of being overthrown by an unconvinced, rebellious constituency. It is far easier to control the allocation of resources in an economy that has long been one of scarcity than to do the same in an economy that has just entered the regime of scarcity.

Perhaps the strongest barrier to adaptation in the highest energy countries is the depth of certain conventional myths: that the market economy allocates even scarce resources more beneficially than can government; that the growth state is necessary for social well-being; that technology is exempt from the law of diminishing returns; that both government and the press inform the citizen so that he can exercise his franchise in a more intelligent fashion; and that, although man may not live by bread alone, he makes decisions as if he did. Although each of these can be refuted from a solid basis of logic and experience, all (especially the second, third, and fourth) have great force in the momentum and inertia of government and business in the affluent society. As John Carver, formerly an Under Secretary of the Interior and member of the Federal Power Commission, said (1973), nothing less than national survival is ultimately at stake in the successful management of our energy problems. At the same time he pointed to some of the constraints on adjustment—national policies to keep prices down to help consumers and to abjure any course of action that might yield a "windfall profit" for an oil company; a social frame of mind bent on identifying scapegoats rather than defining problems; a grass-roots reluctance to centralize controls in Washington; the reluctance of any Congress or any elected executive to assume responsibility for bringing the public face to face with unpleasant realities.

The transition from abundance to scarcity will not be smooth. That scarcity need not mean poverty, and that abundance is not synonymous with affluence, are perceptions not yet widespread. It may transpire that a society of abundance may achieve poverty as well as scarcity because it was not aware that affluence and scarcity can co-exist. There will come about an inadequacy of resources to meet the present consumption levels of high energy society. But Stone Age man was affluent at an energy and resource consumption level less than one-twentieth that of present high energy consumption levels, because his wants could be satisfied in great part by activities that did not require great investments of energy or materials. It is entirely practicable for present affluent society to move in that direction, but it would require a revolution in thought, attitude, and behavior similar to that which Saint Benedict effected. In fact, mankind, by rejecting the production of surplus goods as the goal of technology and substituting productive leisure, would at long last be accepting the Benedictine idea of what technology and inanimate energy should be used for—to increase the time available for rewarding activity.

SELECTED READINGS

Brubaker, Sterling. 1975. *In Command of Tomorrow.* Baltimore: Johns Hopkins University Press.
A clever version of the proposition that the United States must continue to consume a disproportionate share of the world's nonrenewable resources in order "to perfect a technology for escape from resource exhaustion" and "to a assure a tomorrow . . . for all mankind".

Daly, Herman E. 1973. *Toward a Steady-State Economy.* San Francisco: W. H. Freeman and Company. Introduces, in the chapter entitled "The Steady-State Economy: Toward a Biophysical Equilibrium and Moral Growth" (p. 149–174), the concept of depletion quotas, whose price at auction would tend to curb depletion, but whose use would divert windfall rents to the government that sells the quotas.

Georgescu-Roegen, Nicholas. 1971. *The Entropy Law and the Economic Process.* Cambridge, Mass.: Harvard University Press.
This is a challenging book, but chapter X richly repays any effort; it relates entropy to economic value, to industrial development, and to social conflict.

Federal Energy Administration. 1974. *Project Independence.* Washington, D. C.: United States Government Printing Office.
The result of a large and hasty effort, this report attempts to calculate and compare the benefits and costs of four different national energy strategies, each designed to reduce the nation's vulnerability to oil-import disruptions.

Hammond, A. L., W. D. Metz, and T. H. Maugh, II. 1973. *Energy and the Future.* Washington, D. C.: American Association for the Advancement of Science. Well-illustrated, lucid review of the status and prospects of the principal actual and potential sources of energy for high energy society; transmission, conservation, and research are treated, but the social aspects of energy are virtually ignored.

Heilbroner, Robert L. 1972. *The Worldly Philosophers.* 4th ed. New York: Simon and Schuster. Engrossing introduction to the worlds and thoughts of Adam Smith, J. S. Mill, Robert Malthus, and other critics and apologists of the growth state.

Jevons, W. Stanley. 1865. *The Coal Question.* London: Macmillan. A Malthus of energy, Jevons worried about Britain's future once her coal gave out; he did not foresee the degree to which petroleum would replace coal through the internal-combustion engine.

Madden, Carl H. 1972. *Clash of Culture: Management in an Age of Changing Values.* Washington, D. C.: National Planning Association. Contains a readable, persuasive argument for "entropy-offset" as a basis for redefining wealth and for planning the socially efficient use of resources.

Meadows, D. H., D. L. Meadows, J. Randers, and W. W. Behrens, III. 1972. *The Limits to Growth.* New York: Universe. Computer-generated models indicate that, given a continuation of present trends of growth in population, pollution, and depletion, the limits to growth will be reached within the next hundred years.

Portola Institute. 1974. *Energy Primer—Solar, Water, Wind, and Biofuels.* Menlo Park, Calif.: Portola Institute. A book about renewable energy systems that can be built and operated by individuals. Much technical information is given and many publications reviewed. A point made rather too gently in the excellent introduction is that energy conservation and the development of small-scale energy systems can be effective counters to the growing economic and political power of those who control conventional energy resources.

Reid, J. T. 1970. *Physiology of Digestion and Metabolism in the Ruminant.* Newcastle-upon-Tyne: Oriel. An illuminating discussion in the chapter on "The Future Role of Ruminants in Animal Production" (p. 1–22) of the future for animal products in the food of the world; ruminants may still be the main source, but will be fed off land that is not arable; in other words, feedlot beef is on its way out.

Rostow, W. W. 1971. *The Stages of Economic Growth,* 2d ed. Cambridge: Harvard University Press. Adam Smith alive and well after almost two hundred years: productivity is limited in the "traditional society," not by the population:resources ratio, but by "the inaccessibility of modern science, its applications, and its frame of mind."

Watt, Kenneth E. F. 1974. *The Titanic Effect.* Stamford, Conn.: Sinauer Associates. Argues that the abandonment of economic growth as a social goal is necessary to any effective planning for social survival.

Bibliography
A List of Sources Cited in This Volume

Adams, Henry B., 1931 (originally published in 1907). *The Education of Henry Adams.* New York: Random House (Modern Library).

Agricola, Georgius, 1912 (originally published in 1556). *De Re Metallica.* Translated by Herbert Clark Hoover and Lou Henry Hoover. London, England: The Mining Magazine.

American Petroleum Institute, 1973. *Joint Association Survey of the U.S. Oil and Gas Producing Industry.* Washington, D.C.: American Petroleum Institute.

Arps, J. J., M. Mortada, and A. E. Smith, 1970. "Relationship between Proved Reserves and Exploratory Effort." *Journal of Petroleum Technology,* v. 23 (June 1971), p. 671-675.

Associated Universities, Inc., 1972. *Reference Energy Systems and Resource Data for Use in the Assessment of Energy Technologies.* Washington, D.C.: Office of Science and Technology.

Averitt, Paul, 1969. *Coal Resources of the United States—January 1, 1967.* United States Geological Survey Bulletin 1275. Washington, D.C.: United States Government Printing Office.

Averitt, Paul, 1973. "Coal." In Donald A. Brobst and Walden P. Pratt, eds., *United States Mineral Resources,* p. 133-142. United States Geological Survey Prof. Paper 820. Washington, D.C.: United States Government Printing Office.

Ayres, Edward, 1969. *The Economic Impact of Conversion to a Non-polluting Automobile.* Springfield, Va.: National Technical Information Service.

Ayres, Eugene, and C. A. Scarlott, 1952. *Energy Sources: The Wealth of the World.* New York: McGraw-Hill.

Baez, Albert V., 1967. *The New College Physics: A Spiral Approach.* San Francisco: W. H. Freeman and Company.

Bailey, E. H., A. L. Clark, and R. M. Smith, 1973. "Mercury." In Donald A. Brobst and Walden P. Pratt, eds., *United States Mineral Resources,* p. 401-414. United States Geological Survey Professional Paper 820. Washington, D.C.: United States Government Printing Office.

Bell, Daniel, 1973. "Technology, Nature and Society." *American Scholar,* v. 42, p. 385-404.

Berg, Charles A., 1974. "A Technical Basis for Energy Conservation." *Technology Review,* v. 76, n. 4, p. 15-24.

Berg, R. R., J. C. Calhoun, Jr., and R. L. Whiting, 1974. "Prognosis for Expanded U.S. Production of Crude Oil." *Science,* v. 184, p. 331-336.

Bierce, Ambrose, 1911. *The Devil's Dictionary.* New York: Albert and Charles Boni.

Blake, Judith, 1974. "The Changing Status of Women in Developed Countries." *Scientific American,* v. 231, n. 3, p. 136-148.

Boulding, Kenneth, 1966. "The Economics of the Coming Spaceship Earth." In Henry Jarrett, ed., *Environmental Quality in a Growing Economy.* Baltimore, Md.: Johns Hopkins University Press.

Bowen, Richard G., and Edward A. Grotz, 1971. "Geothermal—Earth's Primordial Energy." *Technology Review,* v. 74, n. 1, p. 42-48.

Breland, James E., 1973. *Cost Impact of the Federal Coal Mine Health and Safety Act of 1969.* Preprint 73-F-344. New York: Society of Mining Engineers of the American Institute of Mining, Metallurgical, and Petroleum Engineers.

Brobst, Donald A., and Walden P. Pratt, eds., 1973. *United States Mineral Resources.* United States Geological Survey Professional Paper 820. Washington, D.C.: United States Government Printing Office.

Brown, Harrison, 1954. *The Challenge of Man's Future.* New York: Viking.

Browning, Elizabeth Barrett, 1900 (originally published in 1843). "The Cry of the Children." In Elizabeth Barrett Browning, *The Complete Poetical Works of Mrs. Browning,* p. 156-158. Boston: Houghton Mifflin.

Buck, John Lossing, 1937. *Land Utilization in China.* Chicago: University of Chicago Press.

Byerly, T. C., 1966. "The Relation of Animal Agriculture to World Food Shortages." *Proceedings 15th Annual Meeting Agricultural Research Institute* (Washington, D.C.), p. 36-40.

Byerly, T. C., 1967. "Efficiency of Feed Conversion." *Science,* v. 157, p. 890-895.

Campbell, Ian, 1973. "Ephemeral Towns on the Desert Fringe." *Geographical Magazine,* v. 45, p. 669-675.

Carter, George F., 1975. *Egyptian Gold Seekers and Exploration in the Pacific.* Cambridge, Mass.: Polynesian Epigraphic Society, Occasional Publications, v. 2, n. 27.

Carver, John A., Jr., 1973. *The Lawyer's Concerns with the Energy Crisis.* Unpublished statement before the Section of Natural Resources Law, American Bar Association, Washington, D.C., August 7, 1973.

Childe, V. Gordon, 1951. *Man Makes Himself.* London: Rationalist Press.

Conference Board, 1974. *Energy Consumption in Manufacturing.* Cambridge, Mass.: Ballinger.

Cook, Earl, 1971. "The Flow of Energy in an Industrial Society." *Scientific American,* v. 224, n. 3, p. 134-147.

Cook, Earl, 1974. "Permanence as a Management Goal at Avery Island, Louisiana." In J. G. Nelson and R. C. Scace, eds., *Impact of Technology on Environment: Some Global Examples,* p. 69-86. London, Ont.: University of Western Ontario.

Cook, Earl, 1975. "Ionizing Radiation." In W. W. Murdoch, ed., *Environment,* 2d ed. Sunderland, Mass.: Sinauer Associates.

Cook, S. F., and Woodrow Borah, 1971. *Essays in Population History: Mexico and the Carribean.* Berkeley: University of California Press.

Cottrell, W. Fred, 1955. *Energy and Society.* New York: McGraw-Hill. Reprinted in 1970 by Greenwood Press, Westport, Connecticut.

Council on Environmental Quality, 1973. *Energy and the Environment: Electric Power.* Washington, D.C.: United States Government Printing Office.

Cram, Ira H., 1971. "Summary." In Ira H. Cram, ed., *Future Petroleum Provinces of the United States—Their Geology and Potential,* p. 1-54. Tulsa, Okla.: American Association of Petroleum Geologists, Memoir 15.

Crowe, Beryl, 1969. "The Tragedy of the Commons Revisited." *Science,* v. 166, p. 1103-1107.

Culbertson, William C., and Janet K. Pitman, 1973. "Oil Shale." In Donald A. Brobst and Walden P. Pratt, eds., *United States Mineral Resources,* p. 497-504. United States Geological Survey Prof. Paper 820. Washington, D.C.: United States Government Printing Office.

Daly, Herman E., 1973. "The Steady-State Economy: Toward a Political Economy of Biophysical Equilibrium and Moral Growth." In Herman E. Daly, ed., *Toward a Steady-State Economy,* p. 149-174. San Francisco: W. H. Freeman and Company.

Daly, Herman E., 1974. "Steady-State Economics versus Growthmania: A Critique of the Orthodox Conceptions of Growth, Wants, Scarcity, and Efficiency." *Policy Sciences,* v. 5, p. 149-167.

Darnell, Rezneat M., 1973. *Ecology and Man.* Dubuque, Iowa: William C. Brown Company.

DeCarlo, Joseph A., and Charles E. Shortt, 1970. "Uranium." In *Mineral Facts and Problems,* p. 219-242. United States Bureau of Mines Bulletin 650. Washington, D.C.: United States Government Printing Office.

Dinesen, Isak (Karen Blixen), 1934. "The Dreamers." In Isak Dinesen. *Seven Gothic Tales,* p. 271-355. New York: Harrison Smith and Robert Haas.

Dingle, John H., 1973. "The Ills of Man." *Scientific American,* v. 229, n. 3, p. 76-84.

Duncan, D. C., and V. E. Swanson, 1965. *Organic-rich Shale of the United States and World Land Areas.* Washington, D.C.: United States Geological Survey, Circular 523.

Dunning, R. L., L. C. Geary, and S. A. Trumbower, 1974. *Analysis of Relative Efficiencies of Various Types of Heating Systems.* Unpublished report. East Pittsburgh, Pa.: Westinghouse Electric Corporation.

Durant, Will, 1950. *The Age of Faith.* New York: Simon and Schuster.

Ehrlich, Paul, and Anne Ehrlich, 1972. *Population Resources Environment,* 2d ed. San Francisco: W. H. Freeman and Company.

Elliott, M. A., and H. R. Linden, 1968. "A New Analysis of U.S. Natural Gas Supplies." *Journal of Petroleum Technology,* v. 20 (February 1968), p. 135-141.

Federal Highway Administration, 1974. *1972 Highway Statistics.* Washington, D.C.: United States Government Printing Office.

Federal Power Commission, 1973. *National Gas Survey.* Vol. 5. Washington, D.C.: United States Government Printing Office.

Fell, Barry, 1974. *The Polynesian Discovery of America 231 B.C.* Cambridge, Mass.: Polynesian Epigraphic Society. Occasional Publications, v. 2, n. 21.

Ferguson, Eugene A., 1964. "The Origins of the Steam Engine." *Scientific American,* vol. 210, n. 1, p. 98-107.

Finch, Warren I., and others, 1973. "Uranium." In Donald A. Brobst and Walden P. Pratt, eds., *United States Mineral Resources,* p. 456-467. United States Geological Survey Prof. Paper 820. Washington, D.C.: United States Government Printing Office.

Fisher, John C., 1974. *Energy Crises in Perspective.* New York: John Wiley and Sons.

Food and Agriculture Organization (FAO), 1971. *Production Yearbook 1970* (v. 24). Rome: United Nations.

Forbes, R. J., 1968. *The Conquest of Nature.* New York: Praeger.

Freedman, Ronald, and Bernard Berelson, 1974. "The Human Population." *Scientific American,* v. 231, n. 3, p. 31-39.

Gardner, Frank J., 1974. "Changes Restructuring Worldwide Oil." *Oil and Gas Journal,* v. 72, n. 52 (December 30), p. 105-110.

Garvey, Gerald, 1972. *Energy, Ecology, Economy.* New York: Norton.

Georgescu-Roegen, Nicholas, 1971. *The Entropy Law and the Economic Process.* Cambridge, Mass.: Harvard University Press.

Gillette, Robert, 1974. "Did USGS Gush Too High?" *Science,* v. 185, p. 127-130.

Habbakkuk, H. J., 1962. *American and British Technology in the 19th Century.* Cambridge, England: Cambridge University Press.

Hardin, Garrett, 1968. "The Tragedy of the Commons." *Science,* v. 62, p. 1243-1248.

Heichel, G. H., 1973. *Comparative Efficiency of Energy Use in Crop Production.* Bulletin 739, Connecticut Agricultural Experiment Station (New Haven).

Heichel, G. H., 1974. "Energy Needs and Food Yields." *Technology Review,* v. 76, n. 8, p. 18-25.

Hendricks, T. A., and S. P. Schweinfurth, 1966. *The United States Resource Base of Petroleum.* Unpublished memorandum, September 14, 1966. Washington, D.C.: United States Department of the Interior. Cited on page 494 of Brobst and Pratt (1973).

Heyl, Allen V., and others, 1973. "Silver." In Donald A. Brobst and Walden P. Pratt, eds., *United States Mineral Resources,* p. 581-603. United States Geological Survey Professional Paper 820. Washington, D.C.: United States Government Printing Office.

Hirst, Eric, 1972. *Energy Consumption for Transportation in the U.S.* Oak Ridge, Tenn.: Oak Ridge National Laboratory.

Hirst, Eric, 1973. "Living off the Fuels of the Land." *Natural History,* December 1973. p. 21-22.

Ho, Ping-Ti, 1959. *Studies on the Population of China, 1368-1953.* Cambridge, Mass.: Harvard University Press.

Hoffer, Eric, 1967. *The Temper of Our Time.* New York: Harper & Row.

Hoyle, Daniel Rodríguez, 1974. "Cuajone." *Mensajes,* n. 20. Lima, Peru: Southern Peru Copper Corporation.

Hubbert, M. King, 1956. "Nuclear Energy and the Fossil Fuels." In *Drilling and Production Practice,* p. 7-25. New York: American Petroleum Institute.

Hubbert, M. King, 1962. *Energy Resources.* Washington, D.C.: National Academy of Sciences-National Research Council.

Hubbert, M. King, 1967. "Degree of Advancement of Petroleum Exploration in United States." *American Association of Petroleum Geologists Bulletin,* v. 51, p. 2207-2227.

Hubbert, M. King, 1969. "Energy Resources." In *Resources and Man* (National Academy of Sciences Committee on Resources and Man), p. 157-242. San Francisco: W. H. Freeman and Company.

Hubbert, M. King, 1972. *Survey of World Energy Resources.* Unpublished paper presented to the 26th Annual Conference of the Middle East Institute, Washington, D.C., September 29-30.

Hubbert, M. King, 1974. *U.S. Energy Resources, a Review as of 1972.* United States Congress, Senate Committee on Interior and Insular Affairs, 93rd Congress, 2d Session, Committee Print Serial 93-40. Washington, D.C.: United States Government Printing Office.

Ignotus, Miles (pseudonym), 1975. "Seizing Arab Oil." *Harpers,* v. 250, n. 1498 (March), p. 45–62.

Japan Ministry of Foreign Affairs, 1972. *The Japan of Today.* Tokyo: Ministry of Foreign Affairs.

Jennings, R. D., 1958. *Consumption of Feed by Livestock, 1909-56.* Production Research Report 21, Agricultural Research Service, United States Department of Agriculture. Washington, D.C.: United States Govt. Printing Office.

Jevons, W. Stanley, 1865. *The Coal Question.* London: Macmillan.

Kneese, Allen V., Robert U. Ayres and Ralph C. d'Arge, 1970. *Economics and the Environment.* Washington, D.C.: Resources for the Future, Inc.

Leach, Gerald, 1972. *The Motor Car and Natural Resources.* Paris: Organization for Economic Cooperation and Development.

Leaf, Alexander, 1973. "Getting Old." *Scientific American,* v. 229, n. 3, p. 45-53.

Lee, Tan, and Chi-lung Yao, 1970. "Abundance of Chemical Elements in the Earth's Crust and its Major Tectonic Elements." *International Geology Review,* v. 12, p. 778–786.

Leith, C. K., 1975 (originally published in 1935). "Conservation of Minerals." In Garry D. McKenzie and Russell O. Utgard, eds., *Man and his Physical Environment,* 2d ed., p. 168–171. Minneapolis: Burgess Publishing Co.

Lewis, Archibald R., 1958. *The Northern Seas.* Princeton, N.J.: Princeton University Press.

Linton, D. L., 1965. "The Geography of Energy." *Geography,* v. 50, n. 228, p. 197-228.

Lord, Eliot, 1883. *Comstock Mining and Miners.* Washington, D.C.: United States Geological Survey, Monograph 4.

Madden, Carl H., 1972. *Clash of Culture: Management in an Age of Changing Values.* Washington, D.C.: National Planning Association.

Malthus, T. R., 1888. *An Essay on the Principle of Population,* 9th ed. London: Reeves and Turner.

Marsh, George Perkins, 1965 (originally published in 1864). *Man and Nature,* ed. David Lowenthal. Cambridge, Mass.: Harvard University Press.

Mason, Brian, 1958. *Principles of Geochemistry,* 2d ed. New York: Wiley.

Merrill, A. L., and B. K. Watt, 1973. *Energy Value of Foods.* Agriculture Handbook No. 74, United States Department of Agriculture. Washington, D.C.: United States Government Printing Office.

Mill, John Stuart, 1895 (originally published in 1848). *Principles of Political Economy.* 2 vol. New York: D. Appleton.

Miller, Betty M., and others, 1975. *Geological Estimates of Undiscovered Recoverable Oil and Gas Resources in the United States.* Reston, Va.: United States Geological Survey, Circular 725.

Mitre Corporation, 1972. *Symposium on Energy, Resources and the Environment.* Vol. 1. Washington, D.C.: Mitre Corp.

Moore, C. L., 1971. "Analysis and Projection of Historic Patterns of U.S. Crude Oil and Natural Gas." In Ira H. Cram, ed., *Future Petroleum Provinces of the United States—Their Geology and Potential.* v. 1, p. 50–54. Tulsa, Okla.: American Association of Petroleum Geologists, Memoir 15.

Morrow, Walter E., Jr., 1973. "Solar Energy: Its Time is Near." *Technology Review,* v. 76, n. 2, p. 30–43.

Motor Vehicle Manufacturers Association, 1974. *1973/74 Automobile Facts and Figures.* Detroit, Mich.: Motor Vehicle Manufacturers Association.

Muller, Herbert J., 1970. "Human Values and Modern Technology." *Ingenor 7,* p. 11–21. Ann Arbor, Mich.: University of Michigan College of Engineering.

Mumford, Lewis, 1934. *Technics and Civilization.* New York: Harcourt, Brace and World.

Mumford, Lewis, 1967. *The Myth of the Machine: Technics and Human Development.* New York: Harcourt, Brace and World.

Mutch, James J., 1973. *Transportation Energy Use in the United States: A Statistical History, 1955–1971.* Santa Monica, Ca.: The Rand Corporation.

Nash, A. E. K., Dean E. Mann, and Phil G. Olsen, 1972. *Oil Pollution and the Public Interest.* Berkeley, Ca.: University of California Institute of Governmental Studies.

National Academy of Sciences Committee on Mineral Resources and the Environment, 1975. *Mineral Resources and the Environment.* Washington, D.C.: National Academy of Sciences.

National Petroleum Council, 1971. *U.S. Energy Outlook: An Initial Appraisal 1971-1985.* Washington, D.C.: National Petroleum Council.

Northrup, F. S. C., 1956. "Man's Relation to the Earth in its Bearing on his Aesthetic, Ethical, and Legal Values." In W. L. Thomas, Jr., ed., *Man's Role in Changing the Face of the Earth,* p. 1052–1067. Chicago: U. of Chicago Press.

Ophuls, William, in press. *Ecology and the Politics of Scarcity.* San Francisco: W. H. Freeman and Company.

Phizackerley, P. H., and L. O. Scott, 1967. "Major Tar Sand Deposits of the World." In *Proceedings 7th World Petroleum Congress (Mexico),* v. 3, p. 551-571.

Pimentel, David, and others, 1973. "Food Production and the Energy Crisis." *Science,* v. 182, p. 443-449.

Potter, David M., 1954. *People of Plenty.* Chicago: University of Chicago Press.

Putnam, Palmer C., 1953. *Energy in the Future.* New York: Van Nostrand.

Rappaport, Roy A., 1971. "The Flow of Energy in an Agricultural Society." *Scientific American,* v. 224, n. 3, p. 117-132.

Rice, Richard A., 1972. "System Energy and Future Transportation." *Technology Review,* v. 74, n. 3, p. 31-37.

Rush, Benjamin, 1948 (originally published in 1809). *Autobiography,* ed. G. W. Corner. Princeton, N.J.: Princeton University Press.

Russell, Bertrand, 1935. *In Praise of Idleness.* London: George Allen & Unwin.

Sahlins, Marshall, 1972. *Stone Age Economics.* Chicago: Aldine-Atherton.

Sauer, Carl O., 1956. "The Agency of Man on Earth." In W. L. Thomas, Jr., ed., *Man's Role in Changing the Face of the Earth,* p. 49-69. Chicago: University of Chicago Press.

Schramm, L. W., 1970. "Shale Oil." In *Mineral Facts and Problems,* p. 183-202. United States Bureau of Mines Bulletin 650. Washington, D.C.: United States Government Printing Office.

Sears, Francis W., 1945. *Principles of Physics.* Cambridge, Mass.: Addison-Wesley.

Shah of Iran, 1974. "Why Should We Cut the Price of Oil to U.S.?" *U.S. News and World Report,* May 6, 1974, p. 34-36.

Shortt, Charles E., 1970. "Thorium." In *Mineral Facts and Problems,* p. 203-218. United States Bureau of Mines Bulletin 650. Washington, D.C.: United States Government Printing Office.

Shryock, R. H., 1960. *Medicine and Society in America, 1660-1860.* New York: New York University Press.

Simpson, George Gaylord, 1966. "Naturalistic Ethics and the Social Sciences." *American Psychologist,* v. 21, p. 27-36.

Simpson, George Gaylord, 1967. *The Meaning of Evolution.* New Haven, Conn.: Yale University Press.

Slesser, Malcolm, 1973. "Energy Analysis in Policy Making." *New Scientist,* v. 60, n. 87, p. 328-330.

Smith, Adam, 1850. *The Wealth of Nations,* 4th ed. Edinburgh: Adam and Charles Black.

Smith, Grant H., 1943. *The History of the Comstock Lode, 1850-1920.* Reno, Nevada: University of Nevada Bulletin, Geology and Mining Series No. 37.

Soddy, Frederick, 1933. *Wealth, Virtual Wealth and Debt,* 2d ed. New York: Dutton.

Stanford Research Institute, 1972. *Patterns of Energy Consumption in the United States.* Washington, D.C.: U.S. Government Printing Office.

Steinhart, Carol E., and John S. Steinhart, 1972. *Blowout.* North Scituate, Mass.: Duxbury Press.

Summers, Claude M., 1971. "The Conversion of Energy." *Scientific American,* v. 224, n. 3, p. 148-163.

Tanzer, Michael, 1969. *The Political Economy of International Oil and the Underdeveloped Countries.* Boston, Mass.: Beacon Press.

Taylor, C. C., 1973. *Status of Completion/Production Technology for the Gulf of Alaska and the Atlantic Offshore Petroleum Operations.* Unpublished paper delivered at Resources for the Future, Inc. seminar, Washington, D.C., December 5–6, 1973.

Teilhard de Chardin, 1965. *The Phenomenon of Man.* New York: Harper and Row.

Theobald, P. K., S. P. Schweinfurth, and D. C. Duncan, 1972, *Energy Resources of the United States.* Washington, D.C.: United States Geological Survey, Circular 650.

Thirring, Hans, 1968. *Energy for Man.* New York: Greenwood Press.

Thoreau, Henry David, 1971 (originally published in 1854). *Walden.* Princeton, N.J.: Princeton University Press.

Thoreau, Henry David, 1973 (originally published in 1863). "Life without Principle." in Charles R. Anderson, ed., *Thoreau's Vision,* p. 177-192. Englewood Cliffs, N.J.: Prentice-Hall.

Twain, Mark, 1961 (originally published in 1907). Letter to Mary Benjamin Rogers. In Lewis Leary, ed., *Mark Twain's Letters to Mary.* New York: Columbia University Press.

Ubbelohde, A. R., 1955. *Man and Energy.* New York: Braziller.

United Nations, 1973. *World Energy Supplies 1968-1971.* Department of Economic and Social Affairs Statistical Series J No. 16. New York: United Nations.

United Nations, 1974. *World Energy Supplies 1969-1972.* Department of Economic and Social Affairs Statistical Series J No. 17, New York: United Nations.

United States Bureau of the Census, 1974. *Statistical Abstract of the United States, 1974.* Washington, D.C.: United States Government Printing Office.

United States Bureau of Mines, 1970. *Mineral Facts and Problems.* Bulletin 650. Washington, D.C.: United States Government Printing Office.

United States Bureau of Mines, 1971. *Minerals Yearbook 1969.* Washington, D.C.: United States Government Printing Office.

United States Bureau of Mines, 1974. *United States Energy Use Up Nearly 5 Percent in 1973.* Press Release of March 13, 1974. Washington, D.C.: United States Department of the Interior.

United States Bureau of Mines, 1975. *United States Energy Use Down after Two Decades of Increases.* Press release dated April 3, 1975. Washington, D.C.: United States Department of the Interior.

United States Department of Agriculture, 1956, 1966, 1971, 1972, 1973, 1974. *Agricultural Statistics* (annual). Washington, D.C.: United States Government Printing Office.

United States Department of Agriculture, 1965. *U.S. Food Consumption, Sources of Data and Trends, 1909–63.* Statistical Bulletin 364. Washington, D.C.: United States Government Printing Office.

United States Department of Agriculture, 1970. *National and State Livestock-Feed Relationships.* Economic Research Service Statistical Bulletin 446 (see earlier bulletins in this series for data prior to 1960). Washington, D.C.: United States Government Printing Office.

United States Department of Agriculture, Economic Research Service, 1971. *U.S. Food Consumption Price Expenditures,* Agricultural Economic Report 138. Washington, D.C.: United States Government Printing Office.

United States Department of Agriculture, 1972. *U.S. Food Consumption, Price Expenditures, Supplement for 1972.* Washington, D.C.: United States Government Printing Office.

United States Department of the Interior, 1972. *United States Energy: A Summary Review.* Washington, D.C.: United States Government Printing Office.

United States Department of the Interior, 1973. *United States Energy Fact Sheets 1971.* Washington, D.C.: United States Department of the Interior.

United States Geological Survey, 1974. *USGS Releases Revised U.S. Oil and Gas Resource Estimates.* Press Release, March 26, 1974. Washington, D.C.: United States Department of the Interior. Pages 263–267 in Hubbert (1974).

United States Office of Oil and Gas, 1968. *United States Petroleum Through 1980.* Washington, D.C.: United States Department of the Interior.

Viemeister, Peter E., 1972. *Lightning.* Cambridge, Mass.: MIT Press.

Watt, B. K., and A. L. Merrill, 1963. *Composition of Foods.* Agriculture Handbook No. 8, United States Department of Agriculture. Washington, D.C.: United States Government Printing Office.

Weinberg, Alvin M., 1971. "Can We Live with Fission?" Paper presented to American Association for the Advancement of Science, Philadelphia, December 27, 1971.

White, D. E., 1965. *Geothermal Energy.* Washington, D.C.: United States Geological Survey, Circular 519.

APPENDIX:
A CHRONOLOGY
OF EVENTS RELATED
TO THE USE OF ENERGY

Date		Event
B.C.	**2,000,000**	Stones shaped by man for use as tools
	1,000,000	Fire captured and used by man
	100,000	Food cookery
	80,000	Fire used to harden spears
	25,000	Annealing of stones for chipping
	15,000	Beginning of pottery making by firing clay
	8000	Agriculture in Near East; clay bricks and pots being made; hot working of copper
	7000	Agriculture starts in Mexico
	5300	Mining (turquoise) in Sinai
	4000	Mediterranean seafaring under way; by this time copper had already been reduced from ores by smelting and glazes for pottery had been developed
	3500	Rise of cities in the Tigris-Euphrates delta and Indus and Nile valleys; smelting of copper ore in the Middle East; wheel probably originated about this time; start of Sumerian hydraulic civilization; tin and iron had been reduced from ores by smelting and bronze was being made

Date		Event
B.C.	**3000**	Glass beads appear in Egypt; asphalt being mined by the Sumerians; irrigation in Egypt; sailboats appear on the eastern Mediterranean
	2700	Spoked wheel appears; traction plow already developed
	2680	Great Pyramid of Khufu built, representing a high degree of organization of available power as well as ingenious mechanics of transport and construction
	2600	Construction of Lake Moeris in Egypt, a reservoir created by a dam twenty-seven miles long
	2500	Wine press in use in Egypt; glassmaking flourishes there; walls of Jericho water-proofed with asphalt
	2400	First canal for ships, at Elephantine in Egypt
	2280	Rice irrigation in China; meteoric iron wrought in southwest Asia
	1950	Nile–Red Sea canal built

Date		Event

B.C.

1900 Oxen being used for plowing in Egypt; silver, lead smelting

1550 Lever appears in Egyptian well sweep, also in India

1500 Water clocks being used by Egyptians; plowing in Denmark; silver mining at Laurium (Greece)

1400 Salt mining in the eastern Alps; Armenians discover how to make steel facing on wrought-iron tools

1350 First implements made from smelted iron, in Palestine

1000 Assyrians using battering rams mounted on wheeled fighting towers; steel being made

875 Cavalry introduced into Assyrian army; horses previously used only as draft animals

700 First mention of the pulley; chain of pots used to raise water

690 533-m tunnel driven near Jerusalem

600 Thales, a Greek philosopher, produces static electricity by rubbing amber (*electra* in Greek)

400 Catapult and mechanical crossbow (*ballista*) invented at Syracuse and used against Carthage; oil well completed on an island in Ionian Sea and the oil used in lamps

396 1,800-m tunnel outlet to Lake Albano driven

312 Appian Way constructed, representative of Roman road building

285 The *Pharos* constructed at Alexandria: a lighthouse in which a mirror projected the light of fire for thirty miles

220 Great Wall of China begun

215 Archimedes' catapult used against the Romans

180 Quern or revolving mill invented; turned by slaves or asses

170 Invention of paper in China

150 Force pump appears

50 Screw press introduced

27 Vitruvian waterwheel introduced, the first known instance of the transmission of power through gearing

A.D.

41 5,640-m tunnel driven to drain Lake Velinus

45 Rhine and Meuse connected with a twenty-three–mile canal for Roman ships

180 Heavy plow in use in Rhaetia (Tirol and Graubunden)

250 Gunpowder in use in China, and water mills introduced there

367 Successful sea invasion of Britain by Irish, Saxons, and Picts; by this time Germans are using the heavy, wheeled, mortar-board plow and industry in the Rhineland includes manufacture of cloth, arms, clothing, glassware, jewelry, earthenware, brass, and pewter

400 Irish pirates are established in Wales and Cornwall; a bit later Britons colonize Brittany and Galicia, and Saxon pirates settle in France; the wind-driven prayer mill in use in Central Asia

520 Angles migrate to England

529 St. Benedict founds Monte Cassino; the Benedictine Order, which is to become a powerful force in western Christianity, adopts manual labor as virtuous action

597 St. Augustine, a Benedictine monk, arrives in England, bringing Benedictine life and Christianity to Anglo-Saxon England; in 601 he becomes the first Archbishop of Canterbury

600 Water mills in use in France and Switzerland; the 1,000-km Imperial Canal to connect the Huangho and Yangtze-Kiang started

600 Stirrup invented in China; Arabs develop the windmill; dikes built in Holland

673 Saracen fleet besieging Constantinople is destroyed by Greek fire, invented by an Egyptian architect

712 Paper being made in Samarkand, beginning a long migration through the Moslem world to northwest Europe where it was to become the artifact of communication in the development of an industrial civilization

716 St. Boniface, a Benedictine, goes to Germany from England, founds monasteries at Reichenau (724), Murbach (728), Fulda (744); abbeys and country estates become centers of industry and examples of material progress

730 The stirrup reaches western Europe

732 High water mark of the Moslem Empire; defeat of Saracens between Poitiers and Tours by Charles Martel; the stern post-rudder had been invented, increasing the sailing ability of the northern seafarers

762 Grain watermill known in England

787 Vikings attack Dorchester, start of the Viking domination of the North Atlantic region

793 Paper being made in Bagdad; Vikings attack Northumbria

865–878 Great Danish invasion of Britain

884 Vikings control Scheldt-Meuse region and, at various times, most of Rhine valley

885 700 shiploads of Viking freebooters land at the mouth of the Seine and advance to Paris where they are twice (886, 889) bought off

900 Paper being made in Egypt

911 Vikings unsuccessfully attack Chartres, are granted Normandy as a fief

911–985 England's economy develops rapidly, as had Germany's during the period of the Viking menace. Ireland makes great com-

Date	Event
	mercial progress because Viking commercial traders settling along the coast make Dublin, Limerick, and other centers rich before the year 1000
950	Hand crossbow developed
955	Otto I trounces the Hungarians at Lechfeld, uniting Germany, whose economic life surges
985	Discovery of North America by the Vikings
1000	Successful oil wells drilled by the Burmese
1004	The trebuchet appears in China—a missile-thrower of great force, operated by one hundred men or more
1005	Eilmer, an Anglo-Saxon Benedictine monk, flies a glider for six hundred feet from a tower of Malmesbury Abbey
1013	Wells drilled in China for natural gas, which flows through bamboo pipes to be used perhaps in porcelain manufacture
1043	Canute's Viking maritime empire collapses; it included Anglo-Saxon England, Norway, and Denmark, and represented a last great outburst of aggressive energy by identifiable Vikings; thereafter, the Viking energy appears as Norman energy and much later is melded into the social energy of the British Empire
1066	Norman conquest of England; shoulder-bearing harness for horses, as well as iron horseshoes with nails, being perfected on the continent. Now industry moves from the abbeys and country estates to the towns under the influence of merchant guilds and because of energy economics (water mills and water transport)
1086	The Domesday Book lists 5,624 water mills south of the Trent and the Severn rivers in England
1088	University of Bologna founded, the first of the western universities which, although most were founded by the Church, were to uncloister learning and lead to greater dissemination of knowledge outside the priesthood
1095–1099	The First Crusade; the Crusades, of which there were ten, if one counts the tragic Children's Crusade, reflected a growing energy surplus (as did the universities) in western Europe, not yet channeled into national military establishments
1098	Cistercian order, following the Benedictine Rule, founded as a reaction to the material success of the Benedictines
1100	Paper being made in Morocco
1105	First windmill in Europe (France)
1118	Cannon used by Moors
1130	Foundations of Oxford University being laid as an assembly of learned monks and teachers; Cambridge may have even earlier beginnings
1150	Cistercians introduce use of city garbage and sewer water as fertilizer near Milan; both Benedictines and Cistercians drain swamps and lakes in Germany, France, the Low Countries, and Italy
1150	Paper being made in Spain
1185	Windmill reported in Yorkshire
1191	Paper being made in southern France
1195	Magnetic compass appears in Europe
1202	Coal being mined at Liege, Belgium, and imported into Bruges from England
1226	Coal being mined in northeast England and transported to London for heating
1232	Chinese use rockets against the Tartars, invent hot-air balloons
1258	Rockets appear at Cologne
1267–1268	Roger Bacon's *Opera*, which reflect his interest in natural science and his belief in accurate observation and controlled experiment as the guides to knowledge of the physical world; one of Bacon's interesting conclusions was that air could support a craft in the same manner that water supports boats
1269	Peter of Maricourt writes a treatise on magnetism
1273	Coal smoke in London provokes complaints from the gentry
1276	Paper being made in Italy
1300	Post windmills reach the Low Countries; mechanical clocks probably in use in England and France
1306	Edward I makes it capital offense to burn coal in London
1335	Mechanical clock erected in a tower in Milan; also one in Glastonbury Abbey
1337	Longbow introduced; start of the Hundred Years War between England and France
1345	Division of hours and minutes into sixties (Germany); without linear time, the Industrial Revolution would not have been possible
1346	Battle of Crécy: gunpowder first used in Europe in battle, but longbows won the day
1389	First German paper mill built near Nuremberg
1420	Velocipede invented; likewise the bit and brace
1430	Turret windmill invented
1438	Invention of printing from movable type, by Coster in Holland and Gutenberg in Germany; invention of a wind turbine
1472–1519	Leonardo da Vinci makes the following devices: centrifugal pump, dredge for canal building, breech-loading cannon, rifled firearms, universal joint, rope-and-belt drive, link chains, bevel gears, spiral gears, parachute, standardized mass-production house, and an airscrew or propellor

Date	Event	Date	Event
1474	Wine forced from one cask to another by compressed air, at Nuremberg; first known instance of the use of compressed air to transport materials	*1709*	Abraham Darby manufactures coke from coal, suitable for large-scale use in blast furnace
1477	Watches invented at Nuremberg	*1711*	Sewing machine invented by De Cames
1488	Bartolomeu Dias sails around the Cape of Good Hope, inaugurating an age of discovery powered by wind against sail	*1716*	Wooden railway first covered with iron
		1733	Flying shuttle invented
		1745	Pieter van Musschenbrock invents Leyden jar for storing electrical charges
1503	Explosive mines first used	*1747*	Benjamin Franklin deduces the existence of positive and negative electric charges; two years later, invents the lightning rod
1530	Foot-driven spinning wheel invented		
1539	Death of Vanuccio Biringuccio; his *Pirotechnica* is published in 1540 (Venice), a manual of the methods of using fire in metallurgy	*1767*	Spinning jenny invented by James Hargreaves; Philadelphia streets lighted by whale-oil lamps
1555	Agricola (Georg Bauer) dies; his *De Re Metallica* is published in 1556, a manual of geology, mining, and metallurgy	*1769*	James Watt's main patent on condensing steam engine; three-wheeled steam carriage invented and built by Nicholas Cugnot in Paris
1582	Gregorian calendar revision		
1588	Defeat of the Spanish Armada by the English fleet	*1772*	Robert Boyle: heat is a molecular motion
1600	Timber and fuel wood scarcity in England	*1776*	Reverberatory furnace invented by Cranège brothers; *An Inquiry into the Nature and Causes of the Wealth of Nations,* by Adam Smith, is published, a book to become the philosophical foundation of the growth state and of laissez faire economy
1613	Gunpowder first used in mine blasting		
1619	First use of coke instead of charcoal in a blast furnace		
1624	Experimental submarine travels two miles in test between Westminster and Greenwich; first patent law protecting inventions (England)		
		1778	Modern water closet invented by Bramah
		1782	Double-acting steam engine patented by James Watt
1625	Francis Bacon in his *Essays* states that progress depends on the application of knowledge of the physical world to man's needs and desires	*1783*	First flight by man: ascension at Paris of balloon lifted by heated air
		1784	Glass lamp chimney introduced by Argand
1640	Oil well completed in Italy; kerosine from the oil later used for lighting	*1785*	First steam spinning mill; screw propeller invented by Bramah
1654	Air pump invented by Von Guericke, who also devised the Magdeburg hemispheres, which when the air had been exhausted from within them, could be pulled apart only by two opposing eight-horse teams	*1787*	Iron boat built by Wilkinson; screw-propeller steamboat built by John Fitch and operated on the Delaware River
		1788	Threshing machine invented; paddle-wheel steamboat operated in Scotland
1668	Wallis and Huygens: law of conservation of momentum: $m_1 v_1 = m_2 v_2$	*1790*	First steam-heated factory; House of Commons still heated by charcoal braziers
1680	Differential calculus invented by Leibniz; gas engine using gunpowder designed by Huygens, but never built	*1792*	Coal-gas lighting invented by William Murdock in Cornwall
1682	Law of gravitation announced by Isaac Newton; four years later his *Principia* is published, setting forth the laws of motion as well as gravitation	*1793*	Benjamin Thompson, Count Rumford, shows that work is convertible into heat and vice versa; cotton gin is invented by Eli Whitney
1688	First distillation of gas from coal	*1794*	École Polytechnique, the world's first modern engineering school, founded; the Marquis de Condorcet dies in prison, where he wrote a book of supreme optimism about the progress of the human race, titled *Sketch for a Historical Picture of the Human Mind*
1690	Huygens: wave theory of light and radiant heat		
1694	Oil produced in England by retorting oil shale and cannel coal		
1698	First successful steam water pump built by Thomas Savery		
1705	Thomas Newcomen builds his "atmospheric" steam engine	*1795*	Count Rumford invents the Rumford stove, a close-topped range that economizes on fuel; also devises an economical, wholesome diet for the poor
		1796	Lithography invented; hydraulic press invented by Bramah

Date	Event	Date	Event
1797	Greatly improved lathes invented by Maudslay		Weber; laws of electrolysis announced by Faraday
1798	T. R. Malthus publishes *An Essay on the Principle of Population,* as pessimistic in its conclusions as Condorcet's work was optimistic	*1834*	Practical liquid refrigerating machine built by Perkins
1800	Galvanic cell demonstrated by Volta	*1835*	Electric telegraph invented by Samuel Morse; electric automobile invented by Davenport
1801	Public railroad with horsepower, Wandsworth to Croydon; steamboat *Charlotte Dundas* built by Symington; three-wheeled steam-driven passenger car built by Richard Trevithick, Cornish mining engineer	*1838*	First steam-driven crossings of the Atlantic Ocean, by the *Sirius* (eighteen days) and the *Great Western* (sixteen days), both westbound and both paddle-wheelers; first electric telegraph line installed, England
1802	Richard Trevithick builds first successful steam locomotive	*1840*	Robert Davidson builds electric car that goes four miles per hour
1803	First factory illumination by coal-gas lighting, in James Watt's foundry; streets of Genoa and Parma lighted by kerosine from an oil well at Modena	*1842*	Julius Robert Mayer declares that heat and work are equivalent
1805	Twin-screw propeller developed by Stephens	*1844*	First public telegraph line in the United States completed; carbon-arc lamp invented by Poucault; practical woodpulp paper invented by Keller
1806	Nicholas Appert opens a food-canning factory near Paris and in 1810 publishes *The Art of Preserving*	*1845*	J. R. Mayer formulates the principle of conservation of energy; modern high-speed sewing machine invented by Howe; pneumatic tire invented by Thomson; first crossing of Atlantic by a screw-driven steamship, the *Great Britain*, designed by Isambard Kingdom Brunel, famed British engineer
1807	Fulton's *Clermont* goes up the Hudson; abolition of the slave trade in the British dominions; first patent for a gas-driven automobile		
1813	London Bridge lighted by gas	*1850*	James Young starts to produce "coal oil" (kerosine) from coal
1814	Steam printing press invented by Koenig	*1852*	First successful airship flies over Paris
1815	Commercial oil-shale retorting started in New Brunswick	*1855*	Bessemer steel process invented; 800-hp water turbine installed at Paris
1819	Atlantic first crossed by a steam-powered vessel, the *Savannah*	*1856*	Open-hearth furnace invented by Sir William Siemens; railway sleeping car patented by G. M. Pullman
1820	Incandescent lamp invented by de la Rue; streets of Prague lighted by kerosine	*1857*	First Atlantic cable laid; oil discovered in Romania and Ontario
1821	Lighting by coal-gas successfully introduced into America at Baltimore; houses in Fredonia, New York, heated with natural gas	*1858*	An English lighthouse is illuminated by electric-arc light
		1859	Discovery of petroleum in Pennsylvania by Drake
1824	Sadi Carnot writes *Reflections on the Motive Power of Fire*	*1860*	Etienne Lenoir builds the first practical internal-combustion engine, fueled by illuminating gas
1825	Erie Canal opened; Stockton and Darlington Railway completed, first public steam-powered railway	*1861*	Machine gun invented by Gatling; Russian serfs emancipated; Western Union establishes New York–to–San Francisco telegraph service
1827	Steam automobile invented by Hancock; high-pressure steam boiler (1,400 psi) built by Perkins		
1829	Successful oil well drilled in Kentucky	*1863*	United States Emancipation Proclamation, declaring free all slaves in rebel territory; the first Bessemer steel plant
1830	Steam coaches in use on English roads, but forced off in the following year by restrictive legislation; South Carolina Railroad opens, the first in America		
		1864	Theory of light and electricity published by Clerk-Maxwell
1831	Michael Faraday generates electrical current by magnetic induction, Joseph Henry invents the electric motor, and McCormick invents the reaper	*1867*	Dynamite invented by Nobel; gas engine built by Otto and Langen; two-wheeled bicycle by Michaux; refrigerated railroad car introduced
1833	Slavery abolished in British colonies; magnetic telegraph invented by Gauss and		

Date	Event	Date	Event
1869	Transcontinental railway across the United States opened		first automobile race in the United States, at Chicago, won by a Duryea car
1870	Electric steel furnace invented by Siemens; first commercially practical generator built by Gramme	*1896*	Marconi invents radio-telegraph (wireless), Becquerel discovers radioactivity, and Henry Ford builds his first car
1873	Maxwell's *Treatise on Electricity and Magnetism* published; ammonia-compression refrigerator built by Linde; first workable typewriter produced by Remington; first cable cars in San Francisco	*1897*	First turbine-propelled steamship, designed by Sir Charles Parsons, launched
		1900	Quantum theory proposed by Planck; hydroelectric plant constructed at Niagara Falls
1875	A building in France is illuminated by electricity; standard time is adopted by railroads in the United States	*1901*	Spindletop (Texas) oil gusher blows in
		1902	Radial aircraft engine developed by Manly; Hamburg-Amerika Line adopts fuel oil in place of coal on its liners
1876	The telephone is invented by A. G. Bell; N. A. Otto's four-cycle engine, the first to use initial compression, is built	*1903*	At Kitty Hawk the Wright brothers fly the first man-lifting engine-powered airplane; Pacific cable between San Francisco and China completed; Elwood Haynes builds first rotary-valve gas engine; Great Britain begins to convert its navy to oil
1878	Wanamaker's store in Philadelphia lighted by electricity (carbon-arc lamps)		
1879	Streets of Cleveland lighted by electricity (carbon-arc lamps)		
1880	First successful incandescent electric light	*1905*	Einstein publishes, as a mathematical development of his special theory of relativity, his mass energy equation: $E = mc^2$
1881	First steam-electric central-power stations: Holborn Viaduct, London, and Pearl Street, New York City, designed by Thomas Edison; first hydroelectric power plants: Godalming, England, and Niagara Falls.		
		1911	First transcontinental flight by biplane, across America from New York to Long Beach
1882	Steam turbine invented by de Laval	*1912*	Offshore oil wells drilled in Southern California
1883	High-speed gasoline engine developed by Gottlieb Daimler; elevated electric railroad opened in Chicago	*1913*	Tungsten-filament lamp invented by Coolidge; Henry Ford introduces the assembly line in automobile making
1884	Linotype machine invented by Mergenthaler; first steel frame skyscraper built (Chicago); first large-scale use of natural gas, in Pittsburgh	*1919*	Ernest Rutherford disintegrates nitrogen atoms by alpha bombardment
1885	First electric street railway, in Baltimore; standard time adopted internationally; three-wheeled car propelled by internal combustion engine built by Karl Benz	*1921*	First commercial radio broadcasting station (KDKA) in the United States opened in Pittsburgh
		1924	Diesel-electric locomotives introduced, but use does not become widespread until after World War II
1886	Electrolytic process for producing aluminum discovered by Charles Hall; Welsbach mantle invented, tripling the output of kerosine lamps and gas burners	*1926*	First successful liquid-fuel rocket built and launched by R. H. Goddard
1887	Hertz discovers electromagnetic waves; large-scale use of refrigerator cars begins on American railroads; electric cabs in use in Brighton, England	*1927*	Solo nonstop flight from New York to Paris by Charles Lindbergh; first public television broadcast (England)
		1930	Frank Whittle patents the basic design for the turbojet engine
1892	Duryea "horseless carriage," an internal-combustion automobile, introduced in the United States; first electric passenger elevator	*1932*	Largest hydroelectric plant in the world opened at Dnepropetrovsk; East Texas oil field discovered
1894	Rudolf Diesel's engine built; first commercial European four-wheeled car driven by internal-combustion engine, the Panhard, introduced in France, using the Daimler engine	*1934*	First scheduled transoceanic aviation service for passengers, San Francisco to Manila (the first transatlantic passenger service did not start until 1939)
		1936	Hoover (Boulder) Dam completed
1895	Roentgen discovers X-rays; Paris-Bordeaux automobile race in which the first four finishers were powered by Daimler engines;	*1939*	First flight entirely by rocket power made by Fritz von Opel; first flight of jet-powered airplane
		1941	Grand Coulee Dam completed; wind turbo-generator built in Vermont

Date	Event
1942	First controlled fission of uranium (nuclear chain reaction) produces 200 watts
1943	Uranium-graphite atomic pile produces 1,000 kilowatts, Oak Ridge, Tennessee
1945	Atomic bombs dropped on Hiroshima and Nagasaki
1947	First modern offshore oil well drilled off Louisiana
1952	First regular jet air passenger service, London to Johannesburg; first hydrogen bomb exploded; four thousand persons die in London from smog during air-inversion event
1953	Four hundred die in smog incident in New York City.
1954	Rotary internal-combustion engine developed by Felix Wankel; first nuclear-powered submarine launched
1956	Full-scale nuclear power station begins operation (England)
1957	First man-made orbiting earth satellite, Sputnik I, launched; first commercial nuclear power plant in United States is opened
1958	Commercial passenger flights by jet aircraft are inaugurated
1959	Liquefied natural gas is shipped in cryogenic tanker from Lake Charles, Louisiana, to London
1960	First geothermal power plant in United States is placed in operation
1967	Air Quality Act becomes law in the United States
1968	Supersonic transport plane, the Soviet TU-144, is tested; hydrofoil gunboat is delivered to U. S. Navy
1969	Santa Barbara oil spill occurs; the National Environmental Policy Act is passed by the Congress; man lands on the moon; maiden flight of the Boeing 747 takes place
1970	Production of crude oil in the United States reaches its probable all-time peak
1972	Massive grain sale by the United States to the Soviet Union
1973	Embargo of crude oil shipments to the United States and Netherlands is imposed by Arab members of the Organization of Oil Exporting Countries (OPEC)
1974	The Soviet Union becomes the world's largest producer and refiner of crude oil; an American planetary probe reaches the vicinity of Jupiter, approximately 400 million miles from Earth; the price of crude oil on the world market rises to four times its level two years earlier; per capita real income in the United States falls more than 5 percent.

INDEX

China, 36, 60, 72, 200, 217, 355; energy consumption in, 95, 236, 245, 257; energy reserves in, 233, 234, 237–241, 446; food production and supply in, 8, 154, 168, 227, 231–232, 323, 429–430; historical, 66, 81, 95, 183; population and, 170, 227, 429–430, 446; values of, 201, 345, 346, 422; war in, 195, 196. *See also* Taiwan

chlorophyll, 107

Christianity, 177, 179, 196, 200–201, 220, 346, 379

Cistercians, 179–180, 201n, 270

cities, 8, 178, 186, 207, 213, 437

Cities Service oil company, 254

Civil War (U.S.), 196, 197, 214

Clausius, Rudolf, 33–35

Clean Air Act, 123, 338

climatic variations: energy use and, 311; evolution and, 1, 4, 4–6

clock, mechanical: importance of, 180, 182, 204

coal, 17, 27, 31, 51, 72, 75–84, 94, 281, 334; bituminous, 75, 76, 77; consumption of, 226; depletion of, 118–119; efficiency of, 129, 150, 151; fluid-fuel conversion, 126–128; geography of, 15, 236–237, 298, 307; for heating, 151; historical use of, 9, 346–347; importance of, 1, 18, 63, 81, 149, 173, 301; natural subsidy of, 19, 112, 113; production of, 62, 63, 237, 258, 313, 314, 315; regulation of, 337, 351, 352; reserves of, 259, 260, 262, 263; social efficiency of, 9, 183, 280–82, 287, 346–347; in steam power plant, 29, 36, 129, 150; underground mining of, 78–81, 280–282, 287; wood use compared, 60, 82, 151, 301; in world trade, 249, 288, 339

coal gas, 96, 126

coal oil. *See* kerosine

Coastal States Gas Producing Company, 335

coke, 9, 18, 30–31, 83, 93, 144

coke oven gas, 126

Colombia, 192, 230, 236, 237, 238, 241, 263

Colorado, 91, 306, 307, 309, 310, 311; nuclear power in, 102, 243, 307, 366; oil shale in, 101, 447

Colorado River reservoirs, evaporation from, 284

Columbus, Christopher, 67

combustion, heat of, 33, 34, 107

combustion engine, 23–24, 26–28, 30, 44, 45–46

commercial energy use, 310, 311, 314–315, 322, 328, 330–331

commons, resources of the, 357–363, 421

communications media, 204–205, 215, 365

Communist countries, resource development of, 233, 235, 239, 240, 250, 258. *See also by country*

Compagnie Française des Pétroles (CFP), 254

compressor: in aircraft engines, 44–45; in refrigerators, 43

Comstock Lode, Nevada, 393, 394

Concorde, the, 120

condensers, 36, 39–41, 42, 43

Congo, 231

Connecticut, 309, 310, 311, 312

conservation movement, 337, 359, 420, 436. *See also* environmental impact

conservation of energy, law of, 13–14, 33

Continental oil company, 333

continuous mining machine, 79, 80

conversion efficiency, energy, 11, 152, 153, 229, 315; animals in, 8, 28, 155–159, 228–231, 317, 318, 321–325, 327

cooking, energy for, 172, 244, 322, 330, 331, 334; efficiency of, 150–151, 160, 315

cooling towers, 40–41

Copernicus, 183

copper, 159, 385, 386, 387, 389, 395–399; mining of, 117, 197, 388, 390–391

core samples, in oil exploration, 88

corn, 152, 155, 156, 158, 225; cost of, 316, 317, 318; exportation of, 247–248, 340

corn-equivalent feed unit (CFU), 156–157

Cornwall, 177, 180

corporate energy decisions, 351–353, 365–366, 378

Corpus Christi, Texas, 368–370, 372, 373

Costa Rica, 192

Cottrell, Fred, 22, 154, 174, 176, 204–205, 217, 421

Cram, Ira, 401, 402

Cretaceous age, 77

crime, 219, 221; blackmail, 119–120, 207, 283, 377

crops, 157, 158, 170–71, 198–99, 203, 227, 248; energy returns of, 155, 324–25, 326; exportation of, 247, 340; monoculture danger, 276–77; prices of, 317, 318, 319. *See also by name*

crude oil, 1, 17, 51, 82–100, 173, 258, 360; automobile use of, 244, 329; consumption of, 226, 244, 258; cost of, 119, 120–122, 317–318, 319, 375; depletion of, 118, 119, 391, 393–395; early use of, 95–97; energy from, 62, 63, 277, 313, 314, 315; exploration for, 87–90; geography of, 15, 238–239, 240, 298, 307, 385; geology of, 84–87; production of, 90–92, 143, 238–239, 393–395; refining of, 91–95, 238–240, 250, 326–28; regulation of, 332–337, 347–348, 352; reserves of, 259, 260, 262, 385, 387, 389, 390, 400–407; social efficiency of, 277–78, 287, 412; subsidy from, 19, 412; U.S. use of, 244, 298, 302–303, 393–395; in world trade, 249–57, 286, 329, 332, 336, 339, 412. *See also* oil companies

crude oil, substitute, 96, 101, 127, 128

Crusades, the, 66, 182, 195–96

Cry of the Children (Browning), 220

cryogenic electricity transmission, 361

Cuba, 230, 246, 247, 254

Culbertson, William C., 261, 263

Cyprus, deforestation of, 273

Czechoslovakia, 230, 237, 245, 246, 256

Daimler, Gottlieb, 41

Daly, Herman, 353

Derby, Abraham, 9, 30, 184

Darwin, Charles, 165, 185

death: age of, 206, 209, 210; causes of, 208–211, 281, 287, 329. *See also* population

DeCarlo, Joseph A., 263

Defense, Department of (U.S.), 331–332

Delaware, 309, 310, 311

democracy, 208, 222, 365, 378, 379

Democritian atomism, 32, 66

Denmark, 186, 192, 242, 245, 246, 247; food-energy supply in, 225, 229, 230; historical, 180, 181, 220

depletion allowance, 289–290, 351, 356, 365

Depression, Great, 222, 314, 353, 359

deuterium, 17, 54, 71, 104

Diaz, Bartholomeu, 66, 183

diesel engine, 41–42, 45, 153; fuel for, 42, 93, 329

diet, 205, 219, 263; vegetarian, 154, 158, 171. *See also* kilocalorie consumption

diminishing returns, law of. *See* exponential imperative

Dinesen, Isak, 164

diseases: of civilization, 219; infectious, 207–211, 215, 264, 281

District of Columbia, 309

Dominican Republic, 230

Donkey mill, 29

Drake, Sir Francis, 66, 68, 183

Drake oil well, 96

drugs, role of, 221, 222, 440

Dryopithecus, 2–4

Duncan, D. C., 401

dung, as energy source, 225, 226

Duryea brothers, 216

dyne, defined, 12

dyne-centimeter, equivalents of, 13

East, Far, 228–229, 233, 235, 238–239, 248–250, 258. *See also by country*

East Germany, 149, 230, 236, 237, 245, 246. *See also* Germany

ecological hazards, 207–208, 272–273. *See also* environmental impact

Economic Advisers, Council of, 353, 396

economic development, stages of, 218

economies: in decision making, 353–355; pecuniary, 109, 268–269, 275–276, 294; physical, 107–109, 267–269, 292–294; of scale, 130–131, 207, 278, 377; social, 267–294

Ecuador, 171n, 192, 230, 238, 255, 257

Eddington, Arthur, 14

Edison Electric Institute, 312

efficiency: of energy use, 109, 116–117, 132–163; of exploitation, 390–391; measures of, 134–135; net propulsion, 28, 134, 138–140, 148; social, 161–162, 269, 274, 285–287; in war, 197. *See also* conversion efficiency, energy

Egypt: ancient, 29–30, 61, 62, 66, 115–116, 194, 195; modern, 230, 236, 238, 245, 248, 254, 348

electric furnace, in industry, 144, 145

electric generator, 37, 153

electricity: generating plants, 10, 216; generation, 116, 117, 129, 235–236, 315; generation, efficiency of, 36, 40, 117, 150, 328; generation, energy resources for, 77, 83, 104, 147; transmission of, 143, 351; use, 150–151, 235–236, 287, 313–319; use, government regulation of, 338, 351, 352–353; use, in industrializing societies, 146, 172, 173, 303–304

electric motor, 46, 153

electrolysis, 127, 144

electron volts: equivalents of, 13

Elizabeth I, Queen, 183

Elliott, M. A., 401

El Paso Natural Gas, 334–335

employment, 274–275, 329

endangered species, 52

energetics, defined, 108

energy, 1–2, 6–16, 18–20; conservation of, 13–14, 33, 160–162; consumption, 244–245, 257–258, 307–313; consumption, distribution of, 327–328; consumption, efficiency and, 300–301; consumption, gross national product and, 192; consumption, Industrial Revolution and, 216; conversion of, 14, 107–109, 153; costs of, 19–20, 307–308, 315–319, 321, 398–399, 429; decisions about, 345–379; depletion of, 383–407; in future, 432–441; life-styles and, 205, 222; location of natural stocks of, 15–16; population and, 18, 166, 167; power and, 7, 12; production of, 258, 307–312; reserves of, 258–264; shortage of, 222, 364, 365, 383–407, 435, 441–444; surplus, 185, 189–198, 203, 235, 270–271; in U.S., 297–342; use in agriculture, 155–159; use, efficiency of, 116–117, 300–301, 328; use in food production, 152, 154, 319–326; use in industry, 144–148, 326–328; use, moral aspects of, 411–425; use, residential and commercial, 148–152, 330–331; use, social economy of, 267–294; use in transportation, 328–331; use in U.S., 312–331; in world trade, 245–248, 310–312, 313, 339–340, 351. *See also* conversion efficiency, energy; nonrenewable energy resources; renewable energy resources

Energy Policy Report, 401, 405–406

Energy Research and Development Administration (ERDA), 97, 103, 336, 350–351

engine, defined, 23. *See also by type*

England, 30, 31, 65, 96, 201n, 222, 273; coal in, 81–82, 346–347, 427–428; energy decisions and, 345–347, 352; historical energy sources for, 9, 11, 32; industrialization of, 120, 179, 201, 345–346; piracy and, 67–68, 180, 181; social evolution of, 166, 179–186 *passim;* work force for, 186, 212, 213, 220

ENI (Ente Nazionali Idrocarburi), 254

Enterprise de Recherches et d'Activités Pétroliéres (ERAP), 254–255

entropy, 14–15, 353

environmental impact, 20, 40–41, 198–199, 286, 287, 291–293, 317–318, 345–379 *passim*

Environmental Quality, Council on, 286

epidemics, 208–211, 215

ERAP, 254–255

erg, defined, 12, 13

erosional energy system, 16

Eskimo communities, 59, 61, 175, 194; as hunting societies, 203, 205, 360

Estonia, oil shale in, 241

ethane, in crude oil distillation, 94, 95

Ethiopia, 227, 230, 243, 247

ethyl alcohol, 27, 83, 222, 440

Euphrates River, 8

Europe, 5, 18, 148, 176, 201, 220, 255–256, 441; automobiles in, 140; emigrants from, 302, 306, 355; energy reserves in, 237, 238–239, 241; energy surplus in, 114, 116, 182, 194–196; food-energy supply in, 227, 232, 233; health and, 174, 208; leisure in, 270–271; nonrenewable energy resources in, 15, 79, 93, 124, 126, 237–240, 277; postwar recovery of, 277, 412; renewable energy resources in, 9, 30, 60, 63, 64, 66, 82, 233–234, 243; resource development in, 233–244 *passim*, 258; social evolution of, 174, 177, 179–182; steam engines in, 30, 36; in world market, 92, 248–250, 254–255. *See also by country*

European Economic Community, 256

evaporation, 16, 284

evolution: cultural, 6, 166, 199, 220; genetic, 1–6, 165–166; social, 6, 7, 164–187

exponential imperative, 46, 120–124, 288–289, 291

Exxon, 251–252, 333, 334, 335

family, importance of, 185, 189–190, 200, 203–204, 330

Federal Energy Administration, 350

Federal Energy Office, 338

Federal Power Commission, 82, 312, 318, 337, 338

feed, 28, 58, 154, 245, 261; defined, 25; as energy resource, 54, 56–59, 171. *See also* plant food

Fertile Crescent, 8

fertilizer, 60, 72, 92, 93, 95, 227

Finland, 230, 236, 246, 247

fire: hazard of, 215; man's control of, 1, 6, 7, 17, 165, 166, 167, 169

Fischer Tropsch synthesis, 128

fish, as food, 54, 58, 226, 248, 264

Fisher, John C., 386

fission, 13, 17, 37–39, 101, 431. *See also* nuclear power; thorium; uranium

Flemish guild economy, 181

Florida, 309, 310, 311

food, 11, 15, 17, 18, 24–26, 214–215, 421, 423–424; costs of, 276–277, 292, 321; food-energy supply and, 226–227, 230–232, 247–248, 261, 264–265; kilocalorie requirements, 27, 28, 59, 114, 154, 171; production of, 114–115, 152, 154, 276–277, 292, 319–328; as renewable energy, 51, 54, 55, 56–59; surplus, 115–116, 424. *See also* animals, food; plant food

foot pound, defined, 12

Ford, Gerald, 358

Ford automobile, 138, 213, 216

Ford Foundation, 405–406

foreign oil trade, 245–249, 336

fossil fuels, 9–19 *passim*, 63, 197, 225, 233, 235, 313–319, 445; agriculture subsidy from, 19, 172–173, 319, 429; geography of, 236–237, 307; high energy society and, 204, 217, 359; in power plants, 29, 36, 129, 134, 150, 235; price of, 315–319. *See also by name*

France: automobile use in, 41, 244; deforestation of, 273, 346–347; energy consumption in, 192, 236, 245; food-energy supply in, 227, 230; foreign trade and, 177, 246, 247, 255–256; historical, 9, 64, 116, 178, 180, 195, 196, 210, 220; industrialization of, 177, 180; nuclear power in, 242, 304; resource development in, 233, 234, 237, 238, 239, 241, 242

Francis of Assisi, Saint, 182

Frederick II, 182

free-market economy. *See* market economy

French Revolution, 196

Freon-12, 42, 43

fuel, 214, 290; liquid, 45, 47, 125–128, 153. *See also by type*

fuel cells, 45, 47–48, 59, 153

fuel oil, 36, 93, 96, 149, 364

fuel wood: amount of, 225, 226, 273; coal use compared, 60, 82, 151, 301; early use of, 9, 18, 36; efficiency of, 151; importance of, 15–19, 51, 54, 59–60, 62, 63, 149

fund energy sources. *See* nonrenewable energy resources

furnace, efficiency of: industrial, 144, 145, 160, 328; residential, 151, 153

fusion, 17, 45, 48, 71, 104, 261, 264, 431. *See also* nuclear power

Gabon, 238, 248

Galileo, 130, 183

garbage, municipal, burning of, 61

Garvey, Gerald, 272

gas. *See* gasoline; natural gas

gas laser, efficiency of, 153

gasoline, 42, 93, 364; consumption of, 302, 303, 319, 329; price of, 316, 317, 319

gas turbines, 44–45, 46–47, 153

Gaul, northern, industry in, 177

geared mill, 63–64, 117

Gemini spacecraft, fuel cell for, 47

geologic resources, 383–407

Georgescu-Roegen, Nicholas, 353

Georgia, 309, 310, 311

geothermal energy, 17, 18, 103–104, 110, 226, 235, 243–244; deep reservoir, 70, 103; power plant, 70, 104; reserves of, 260, 261, 263, 264

Germany, 32–33, 149, 186, 244; energy consumption in, 166, 192, 236, 245; food-energy in, 230; in foreign trade, 177, 179–180,

thermodynamics, second law of, 34, 145, 353

Third World, 218

Thirring, Hans, 34

Thomas Aquinas, Saint, 201

Thompson, Benjamin, 33

Thoreau, Henry David, 306, 344, 422–423, 442

thorium, 17, 39, 51, 101, 283; geography of, 102, 142–143, 299; reserves of, 260, 262, 263

tidal energy, 10, 16, 17, 18, 53, 54, 69–70, 233, 261

Tigris River, 8

Tirol, wheeled plow in, 177

TNT, 13, 95

Toltecs, the, asphalt seep use by, 96

tonne, defined, 84

tools, 4, 6, 165

trade: international oil, 250–257; U.S. balance of, 339–340

trains. See locomotives; railroads

transportation: of energy, 124–125, 143–144, 334–335, 336–337, 351; energy use by, 119, 135–141, 308–310, 313, 314, 315, 322, 328–329; kinds of, 215, 328–329

traps, geologic, for oil and gas, 84–86

treadwheel, inclined, 25

Trevithick, Richard, 30, 31

Trinidad, 238

trucks, 322, 328, 329, 330

tungsten, 197, 386

Tunisia, 238, 248

turbine generator, 38, 54

turbines. See gas turbines; steam turbines

turboprop aircraft, 44–45

Turkey, 79, 192, 227, 230, 237, 238, 256

Twain, Mark, 344

Ubbelohde, A. R., 176, 220

Uganda, 247

underdeveloped countries, 202, 206, 412–413, 444–445; energy consumption of, 245, 257–258. See also low energy society

underground mining: of coal, 78–81, 280–282, 287; of uranium, 102

unemployment, 271, 285, 438, 441

Ungava, Eskimos of, 175

Union of Soviet Socialist Republics: energy consumption of, 237, 245, 257, 442–443; energy efficiency of, 148–149; energy production in, 237, 238, 239, 442–443; food-energy supply in, 227, 230, 232, 233; fossil fuels in, 92, 217, 237–241, 254, 256, 348–349; government role in, 355; hydroelectric development in, 5, 233, 234, 284; nuclear power in, 242, 304; population and, 227, 447; reserves in, 237, 240, 434, 446; social economy of energy use in, 273, 284; in world market, 248, 254, 256, 319, 339, 348–349, 420

Union Oil Company, 333, 372

United Arab Emirates, 238, 247, 248

United Arab Republic, 230

United Kingdom, 230, 240, 246, 336; automobiles in, 244; energy consumption in, 192, 236, 245; fossil fuel production in, 237, 238, 239. See also by country

United Nations Food and Agriculture Organization, 226–227, 264

United States, 297–342; agriculture in, 155–159, 172, 205, 227, 441; automobiles in, 134, 138, 139, 141–142, 160, 244, 279; coal in, 77, 79–83, 237, 259, 281, 288; crude oil and, 84, 90–99 passim, 118, 122, 238–240, 251–259 passim, 348, 375, 389; depletion of resources in, 118, 388–390; energy consumption in, 13, 112, 116, 166, 167, 192, 213, 236, 244–245, 300–301, 307–312; energy efficiency in, 109, 117, 135, 148, 154–159, 161–162, 290, 292; energy production in, 53, 90–91, 237–239, 242, 307–312, 393–395; energy subsidy in, 114–115; energy types in, 62, 63, 125, 300–301; food-energy supply in, 8, 28, 58, 154, 227–232 passim, 248, 292, 423–424; foreign trade and, 246, 248, 250–253, 255–256, 288, 348; future for, 428, 435, 439, 441–444, 446–449; geothermal energy in, 103, 243; governmental role in, 303–304, 306, 348; health care in, 206, 209–211; heating in, 148, 149–150, 151; hydroelectric power in, 65, 233, 234, 284; industrialization of, 120, 121, 166, 167, 185, 212–216, 346, 348–379, 384; moral aspects of energy use in, 222, 411–414, 418–422; natural gas in, 72, 91–92, 122, 124, 238, 239, 240; nuclear energy in, 10, 38–39, 102–103, 146, 242; reserves in, 237, 240, 242, 259–263, 402–407; social economy of energy use in, 274–275, 279, 281, 284, 288, 291; solar energy in, 297, 300; transportation of energy and, 91, 124; vegetable refuse energy in, 61; war and, 196, 197, 348; work force in, 186, 205, 220. See also by state

United States Geological Survey, 336, 401–404, 406

uranium, 13, 16–19, 39, 51, 101; cost of, 304; energy of, 13, 17, 18, 118, 120, 314, 315, 388; enrichment of, 38–39, 103, 146; fuel cycle of, 146–147; in nuclear power plant, 37, 38, 39, 103, 150; occurrence of, 16, 101, 102, 226, 299, 307, 386, 387; regulation of, 337, 347; reserves of, 242, 260, 262, 263; social costs of, 283

urbanization, 178, 186, 207, 213, 437; automobiles and, 278–279; energy consumption and, 308–310

Uruguay, 226, 230

Utah: energy consumption in, 309, 310, 311; oil shale in, 101, 447; uranium in, 102, 307

utilities, regulation of, 337–338, 347, 351, 352, 356

vegetable refuse, 17, 18, 54, 55, 60–61, 225, 226, 244; reserves of, 261, 264

vegetarian diet, 154, 158, 171

Venezuela, 192, 230–241 passim, 445–446, 248, 339; in world oil market, 255, 257, 339, 412

Vermont, 310, 311, 312

Versailles waterworks, 29

Vietnam, 236, 237, 247

Vikings, the, 66, 180, 181, 194

Virginia, 212, 307, 309, 310, 311

Vitruvian water mill, 29, 63

volcanic energy system, 16, 53, 243
Volkswagen Beetle, 139
Volta, Upper, 247

Wales, 31, 180, 186
Wall of China, Great, 7
Wankel engine, 153
warfare, 195–198, 272, 422
Washington, 302, 307, 309, 310, 311, 367
water gas, 126, 127
water heating, 56, 104, 172; efficiency of, 134, 149, 150, 151, 315; energy use for, 244, 322, 330, 331, 334
water mill, 9, 29, 54, 63, 180, 182
water power, 8–9, 15–19, 51, 54, 63–65, 225, 226, 285–286, 315; historical use of, 11, 18, 24, 65, 270; reserves of, 261, 263, 315; as transportation, 329
water turbine, 29, 46
waterwheel, 8, 29, 62, 63
Watt, James, 25, 29, 30, 35, 39, 183, 184
watt-hour, equivalents of, 13
weapons, 1, 4, 6, 7, 165
Weber, Max, 179
Wedgwood, Josiah, 185
Welsbach mantle, 96, 302
West (U.S.), 82, 213, 274, 306. *See also by state*
West Germany, 149, 230, 244, 246; energy consumption in, 192, 234, 236, 245; fossil fuel production by, 237, 238, 239. *See also* Germany
Westinghouse study, 149
West Virginia, 87, 307, 309, 310, 311; coal from, 77, 281, 307
whale, 52, 359–360; oil from, 61, 359

wheat, 198, 203, 227, 248, 340
White, D. E., 263
White, Lynn, 179
Whiting, R. L., 401
wildcat wells, 87, 89
windmill, 9, 29, 54, 68, 182
wind power, 8–9, 16–19, 51, 53, 54, 63, 65–69; reserves of, 261, 264
Wisconsin, 309, 310, 311
Wisconsin glacial stage, 5
wood. *See* fuel wood
women: in Japan, 219; labor of, 186, 220, 330–331
work: defined, 12; efficiency of, 28, 134, 138–140, 148; in energy use, 194, 300–301, 313, 322; ethic, 355; heat as, 13–14, 33; natural subsidy and, 117–120
World Moslem League, 431
World Power Conference (1968), 259
World War I, 197
World War II, 44, 197, 219, 222, 314, 412
Wright brothers, 216
Wyoming, 102, 307, 311; coal technology in, 79, 81; energy consumption in, 309, 310, 311; oil shale in, 101, 447

Yemen, 236
Young, James, 82–83
Yugoslavia, 230, 234, 237, 238, 248–249

Zaire, 247
Zambia, 230, 236
Zante: oil well on, 95
zinc, 386, 389, 390